Coping with Climate Change

Ramesha Chandrappa · Sushil Gupta
Umesh Chandra Kulshrestha

Coping with Climate Change

Principles and Asian Context

 Springer

Ramesha Chandrappa
Karnataka State Pollution
Control Board
Biomedical Waste Section
Church Street
560001 Bangalore
India
rameshagowda@rediffmail.com

Sushil Gupta
Risk Management Solutions India
A-7, RMSI, Sector 16
201301 Noida
India
sushilgupta74@yahoo.com

Umesh Chandra Kulshrestha
Jawaharlal Nehru University
School of Environmental Sciences
New Dehli 110067
India
umeshkulshrestha@yahoo.in

ISBN 978-3-642-44745-7   ISBN 978-3-642-19674-4 (eBook)
DOI 10.1007/978-3-642-19674-4
Springer Heidelberg Dordrecht London New York

Cover design: SPi Publisher Services

Printed on acid-free paper

Springer is part of Springer Science+Business Media (www.springer.com)

Preface

During the freedom struggle for India, Mahatma Gandhi was asked whether his country would achieve the standards of living as its colonial power after independence. "It took Britain half the resources of the planet to achieve its prosperity," was his reply. "How many planets will a country like India require?"

After half a century, India and its neighbouring country China considered as world's economic powers. Today, the global thinking is that both these countries will engulf the planet with their economic development. But the situation at present is that all the people in India and China are not holding silver spoons to transfer rice into mouth. In south Asia, the country like India which hosts several religions and which have taught non-violence to the world has the habit of spending towards defence out of reserve revenue collected as taxes from public. Positive aspect of such approach and investment is their possible utilization during war against climate change.

Coping with climate change is a great challenge to all of us living on the planet earth. Climate change is the problem of the century. It has global appearance and effects. No single country alone can win the war against climate change. It needs constructive thinking and collective efforts. In this regard, developed countries have to take lead as they carry responsibility of greater fossil fuel consumption and so the major share in greenhouse gas emissions which is responsible for climate change. According to recent global acceptance, Earth's temperature has increased by about $0.7°C$ since the beginning of industrial revolution. The truth is that only 15% of world's intergenerational population from rich countries has contributed half of the emissions of CO_2 leaving the problem to the current generation to solve.

Owing the richness, developed countries can cope with climate change in easier manner but the shocks of climate change are going to affect poors in devastating manner in developing world. South Asia is one of the most vulnerable regions. Having seen and tasted poverty and climate impacts, we the authors of this book came together to ink our experiences which have not been published before.

We are thankful to Ms Jayalakshmi GK, Project Assistant, Karnataka State Pollution Contorl Board (KSPCB), India, for preparing some of the time consuming illustration with lot of enthusiasm and interest.

We are utmost grateful to Dr. C. Witschel and Ms Schneider Marion from Springer-Verlag GmbH for continuous encouragement right from the beginning till its publication. We acknowledge the help of Mr Gyan Prakash Gupta, Jawaharlal Nehru University for in literature search. We are kind to Mr Satish Garje and Mr Amar Yeshwanth, KSPCB for their help in word processing.

Contents

List of Tables

List of Figures

List of Boxes

Chapter 1
The History of Climate

The term climate change refers to noticeable change in the Earth's regional or global climate system over a longer period of time. The time scale can range from a decade to several million years. From the view point of policy makers, the 'climate change' is usually referred only to the changes in recent climate pointing out rising average temperature of earth known as Global Warming which is caused by human activities. For the variations caused by non-human influence, United Nations Framework Convention on Climate Change (UNFCCC), uses the term 'climate variability'.

The term Climate is used to define average weather and statistics of changes in weather over time identify climate change. Climate change and weather are entangled. Climate was never static. In the past few billion years, our planet has ranged from being a global snowball to warmer climate. Evolution of life has played a significant role in Earth's climate. Over millions of years plants enriched the earth with oxygen, making animal life and a cooler world possible. Changes to the Earth's climate system depends how much energy entering and leaving the system alters Earth's radiative equilibrium. These destabilizing influences are called *climate forcings* which include Sun's brightness, Milankovitch cycles (Box 1.1), large volcanic eruptions, particle pollution (aerosols), deforestation, rise in greenhouse gases (GHG) in atmosphere. A forcing will trigger feedbacks such as melting of snow/glaciers/polar ice which will intensify or weaken the original forcing.

The effects of aerosol on surface temperature and hydrological cycle are of opposite to that of GHG. Aerosols exert a cooling and drying effect whereas GHGs warm the surface and increase rainfall. Aerosols alter the hydrological cycle by repressing evaporation and rainfall. This in turn can provide a feedback on temperatures as the latent heat released in rainfall affects the circulation and the wind patterns around the earth.

Climate changes occur in all sizes. It could be micro-climate change like changes observed in climate next to pond or university campuses in case of lot of trees and no trees scenarios. It could be macro also like changes observed at district or state levels e.g., climate change in coastal India, north India etc. Climate change can also occur as Urban Heat Island (UHI); where in central urban locations will have several degree higher temperature than rural areas in the close proximity of similar elevation.

R. Chandrappa et al., *Coping with Climate Change*,
DOI 10.1007/978-3-642-19674-4_1, © Springer-Verlag Berlin Heidelberg 2011

Box 1.1 Milankovitch cycles and glaciations

The episodic nature of the Earth's glacial and interglacial periods within the present Ice Age (in the last couple of million years) is due to variations in the Earth's eccentricity, axial tilt, and precession. These three dominant cycles, collectively known as the Milankovitch Cycles named after the Serbian astronomer Milutin Milankovitch.

Eccentricity

Earth's orbit around the sun is constantly fluctuating. These oscillations, from more elliptic to less elliptic, are of prime importance to glaciations as it alters distance from the Earth to the Sun.

Axial tilt

Axial tilt is the inclination of the Earth's axis to plane of orbit around the Sun. Oscillations in Earth's axial tilt occurs with a periodicity of 41,000 years. The tilt varies between 21.5 and 24.5°.

Precession

Precession is the Earth's slow tremble as it spins on axis. Due to this tremble a climatically significant alteration take place.

The quantity of energy reaching the top of Earth's atmosphere during day time each second is about 1,370 W/m^2. Because the Earth rotates, we will receive 1,370 W/m^2 for a part of each day and, average radiation across the Earth's entire atmosphere at any particular time is about 340 W/m^2. Roughly 29% of energy from the sun that reaches the top of the atmosphere is reflected back to space (Fig. 1.1). Two-third of this reflectivity is due to clouds and aerosols. Light-colored areas of Earth's surface (like snow, ice and deserts) reflect remaining one-third of the energy from the Sun.

Major volcanic eruptions eject material very high into the atmosphere which is removed through precipitation. It is interesting to note when aerosols originated from a volcanic eruption reach far above the highest cloud, they influence the climate for about a year or two. Thus major volcanic eruptions can cause a drop in mean global surface temperature to the tune of about 0.5°C that can last for months or even years. Similarly, some certain man-made aerosols also appreciably reflect sunlight (Le Treut et al. 2007). Solar illumination varies place to place and time. The annual quantity of incoming solar energy varies considerably from equator to poles.

Earth's climate has varied throughout its history, due to complex relations of the solar, oceanic, terrestrial, atmospheric, and living components on Earth. Over the last one million years, our planet has experienced alternate cycles of warming and cooling of approximately 100,000 years. In each such cycle, global average temperatures have fallen and then risen again by about 5°C. Each cycle has taken Earth into an ice age followed by warming it again. Earth's climate has also been influenced by changes in ocean circulations due to plate tectonic movements. Earth's climate has distorted abruptly at times, sometimes due to shifts in ocean circulation, and events such as massive volcanic eruptions.

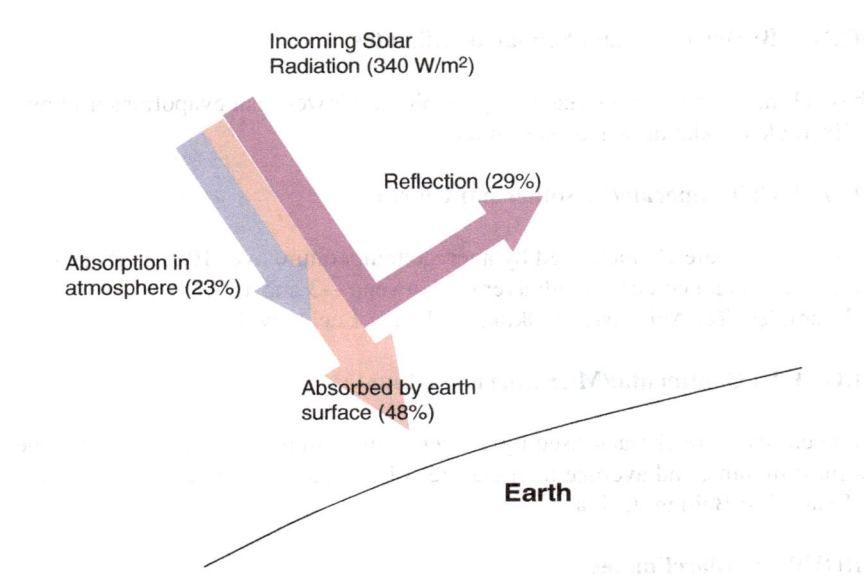

Fig. 1.1 Energy balance at the surface of the earth

Over the past few decades, climate change has gone out of rhythm, bringing international community together to solve the problem. In future, scientists expect climate change to have a rising impact on economy, food, water, raw materials, energy, plant/animal health and biodiversity. Climate model projections imply that negative effects of climate change will significantly overshadow positive ones. It is expected that in future, various countries will be struggling to survive as the ability of a nation to adapt to new conditions, may not be at par with the rate of climate change.

The Köppen climate classification is one of the most widely used climate classification systems which was first published by German climatologist Wladimir Köppen in 1884, with several later modifications by Köppen himself. Later, German climatologist Rudolf Geiger collaborated with Köppen and brought out Köppen–Geiger climate classification system.

Köppen climate classification scheme divides the climates into five main groups and several types and subtypes.

GROUP A: Tropical/Megathermal Climates

These climates are characterized by constantly high temperature at sea level and low elevations with average temperatures of 18°C (64°F) or more during all the months of the year.

Examples: Kuala Lumpur, Malaysia; Singapore; Chittagong, Bangladesh; Mumbai, India.

GROUP B: Dry (Arid and Semiarid) Climates

These climates are characterized by precipitation lower than evapotranspiration.
    Examples: Jodhpur, Rajasthan, India.

GROUP C: Temperate/Mesothermal Climates

These climates are characterised by average temperature over 10°C in their warmest months, and a coldest month average between −3 and 18°C.
    Examples: Tel Aviv, Israel; Okinawa, Japan; Lucknow, India.

GROUP D: Continental/Microthermal Climate

These climates are characterised by average temperature higher than 10°C in their
warmest months, and average temperature colder than −3°C in cooler months.
    Examples: Beijing, China.

GROUP E: Polarclimates

These climates are characterized by average temperatures lesser than 10°C in all
months of the year:
    Examples: Provideniya, Russia.

    Climatic zones of the world can also be broadly categorized as below:

Polar – very cold and dry all year
Temperate – cold winters and mild summers
Arid – dry, hot all year
Tropical – hot and wet all year
Mediterranean – mild winters, dry hot summers; and
Mountains (tundra) – very cold all year

    Interestingly all the above climatic zones can be observed in Asia. The climate is
affected by five factors summarized below:

Latitude

Sun's rays are dispersed over a larger area of land away from the equator due to the
curved surface of the earth.

Altitude

Temperatures decrease with increase in height as air is less dense and cannot hold
heat as easily. The rate at which parcel of hot air looses it temperature as it rises is
called adiabatic lapse rate.

Winds

Winds from hot area like desert will be hot and hence it raises temperatures over which it passes. Similarly winds blown from cold areas will travel with low temperature and hence will bring down the temperature over which it passes.

Distance from the Sea (Continentality)

Land heats and cools faster than the sea and hence coastal areas have a lower temperature range compared to those areas inland.

Aspect

Slopes facing the sun are warmer and hence south facing slopes in the northern hemisphere as well as slopes facing north in the southern hemisphere are warm.

In summary the changing climate is essentially governed by principles of climate science listed below:

- *Sunlight received by the Earth from sun varies from place to place on the land, ocean, and atmosphere*
- *Some of that sunlight is absorbed and other is reflected back to space by the land, clouds, or ice*
- *Sunlight that absorbed by Earth causes warming of the planet*
- *Differential heating of earth due to difference in temperature cases movement of wind*
- *The tilt of Earth's axis and Earth's orbit around the Sun are main reason for changes in the duration of daylight and the quantity of sunlight received at any place on the earth*
- *These changes in quantity of sun light received by Earth cause the seasons and associated temperature changes*
- *Amount of energy received from sun depends on the angle at which sun rays fall on earth. It is maximum at equator and minimum at poles*
- *GHGs absorb heat and enhance the temperature*
- *Gaseous movement is governed by differential concentration, wind speed, wind direction, turbulence in the atmosphere and height of mixing layer*
- *Human activities like combustion of fuels, manufacture of GHGs and accidental/ incidental release of GHGs amplify absorption of heat*
- *Natural phenomenon like decomposition, wild fire, respiration, volcanic emission also contribute to GHGs*
- *Gradual variation in Earth's rotation and orbit around the Sun result in change in the intensity of sunlight received by Earth's polar and equatorial regions. In the last one million years, these changes occurred in 100,000-year cycles resulting in ice ages and the shorter warm periods*

- *Earth's climate is influenced by interactions of the energy from sun, ocean, atmosphere, clouds, ice, land, and life on earth*
- *Bio-geo cycles cause exchange of components essential for life like carbon, water, nitrogen, phosphorous between living and nonliving components of ecosystems and*
- *Aerosols have a complex effect on Earth's energy balance. They can cause both cooling if they can reflect energy from sun back out to space, and warming, by absorbing heat energy in the atmosphere. Reflection and absorption of energy by aerosol depends of characteristics of aerosols*

Sampling and research of ice cores provide key information about past climates. The bubbles trapped provide information of atmospheric chemical composition. With invention of the thermometer in the early 1600s, scientists began to monitor and record the weather. The first deep ice cores from Vostok in Antarctica (Barnola et al. 1987; Jouzel et al. 1987, 1993) exposed a highly related development of temperature changes and atmospheric composition, which was later established over the past 400 kyr (Petit et al. 1999) and now extends to almost one million years. The following paragraphs briefly explain change in climate in the past.

500 Million Years Ago

Katian Age of the Late Ordovician period occurred between 456 and 446 million years ago, reflected the peak of the overall global warming trend of the Ordovician with brief cold spell towards the middle of the Katian (Palaeos 2002). The closing of the northern Iapetus Ocean in the Katian would have affected coastal currents resulting in a sorting out of the endemic marine fauna around the continents (Cocks and Torsvik 2005; Hints et al. 2007).

80 Million Years Ago

The **Cretaceous–Tertiary extinction event**, which occurred about 65.5 million years ago (Ma), was a large-scale mass extinction of living beings in a short period of time, widely known as the **K–T extinction event**. The event is associated with a geological signature known as the K–T boundary, a thin band of sedimentation found in various parts of the world. K is the abbreviation for the Cretaceous Period derived from German name *Kreidezeit*, and T is the abbreviation for the Tertiary Period.

Period about 80 million years ago is recognised as late Cretaceous. A number of the isotopes, paleontology, and paleoclimate and paleo-ocean evidences have indicated that CO_2 concentration in the Cretaceous was approximately four to ten times larger than present concentration (Berner 1991, 1994; Bice and Norris 2002; Yapp and Poths 1996). During the Cretaceous, global yearly mean temperature was about 7–14°C higher than that of the present climate (Huber et al. 2001; Wilson et al. 2002; DeConto 2009).

34 Million Years Ago

34 million years ago hot climate Eocene epoch gave way to the cooler Oligocene epoch, and it has been a cooler, icier planet ever since. The global shift was probably caused by continental drift – the movement of continents, which can rearrange the way air and water flow around the world.

Two Million Years Ago

Earth had roughly 100,000-year cycles of ice age and shorter warm periods that appears to be driven by what's called "orbital forcing". A combination of changes in the shape of Earth's orbit around the sun and the slow tremble of Earth on its spin axis has been cause of *"Milankovitch Cycles"*.

One Million Years ago

Long-term paleo-climatic studies in the Tibetan Plateau demonstrate that both wet and dry periods have occurred in the last millennium (Tan et al. 2008; Yao et al. 2008).

70,000 Years Ago

At this time, Homo sapiens were a sophisticated hunter-gatherer. Eruption of Toba volcano in northern Sumatra about 73,000 years ago resulted in climatic cooling and prolonged deforestation in South Asia (Martin et al. 2009). This eruption might have caused several years of very cool weather.

5,200 Years Ago

The 5.2 ka decrease in Anatolian precipitation, the main source of Tigris-Euphrates stream flow (Cullen and deMenocal 2000), infer that irrigation agriculture in southern Mesopotamia, was reduced during this period (Nutzel 2004).

4,200 Years Ago

This global climate event occurred between 4,500 and 3,500 cal yr BP (Gasse 2000; Weiss 2000; Booth et al. 2005). A severe drought was observed almost everywhere in Eastern Mediterranean region and West Asia. The 4.2 ka event displaced the Mediterranean westerlies and perhaps the Indian monsoon, thereby diminishing the seasonal precipitation required for rain fed cereal agriculture.

21,000 Years Ago

This is also referred as Last Glacial Maximum (LGM). At this period ice sheets covered much of North America and Eurasia. The temperature was 3–5°C cooler than the present.

10,000 Years Ago

The climate continued to warm up from the last ice age and agriculture activities in various parts of the world increased. Many settlements grew faster and transformed to towns and cities. Earth's climate trend continued to gradually warm.

Palaeoclimatic studies indicate that decadal to centennial-scale changes in the regional climate very likely occurred during the past 10,000 yr. However, the mechanisms responsible for these shifts are not well understood (Jansen et al. 2007).

8,200 Years Ago

The period is usually referred as 8.2 kilo year event or 8.2 ka event. Sudden decline in global temperatures that occurred during the period lasted for the next two centuries.

Few Centuries Ago

More than half of the climate-changing volcanic eruptions in past two millennia have happened over the last 700 years. They temporarily cooled Earth's climate.

Fifteen severe droughts have occurred in a region of China over the last 1,000 years (Zhang 2005). The South Asian monsoon, in the drier areas, reversed its trend towards less rainfall. This recent reversal in monsoon rainfall also coincide with an increase in incidental monsoon winds over the western Arabian Sea (Anderson et al. 2002), a change, that can be related to increase in summer heating around the Tibetan Plateau (Bräuning and Mantwill 2004; Morrill et al. 2006).

In the past 1,000 years, three climatic periods had regional variation in the northern hemisphere the Medieval Warm Period, the Little Ice Age, and a renewed period of warming. Assemblages in the Lake Baikal sedimentary record shows that all three periods are expressed in central Asia, in the Lake Baikal region, as evidenced from varying diatom (Anson et al. 2005).

Few Decades Ago

The Industrialisation and urbanisation processes added GHGs to the atmosphere. Huge amount of carbon has been locked up deep in the earth which is unearthed to use as fuel emitting CO_2. Intergovernment Panel for Climate Change (IPCC) was given Nobel Prize for highlighting man made activities during past few decades is the cause of climate change. Eighty-two percent of the glaciers in western China retreated in the last half of the twentieth century (Liu et al. 2006). The glacial area decreased by 4.5% over the last 20 years and by 7% over the last 40 years on the Tibetan Plateau, (CNCCC 2007) indicating an increased retreat rate (Ren et al. 2003). In the last few decades, interseasonal, interannual, and spatial variability in rainfall trends have been seen across Asia. Eastern and central parts of Tibetan Plateau felt increasing trends (Zhao et al. 2004; Xu et al. 2007); the western Tibetan region witnessed a decreasing trend; northern Pakistan has an increasing trend

(Farooq and Khan 2004); Nepal did not had long-term trend in precipitation between 1948 and 1994 (Shrestha et al. 2000; Shrestha 2004). An average of 250 million people a year has been affected over the past decade by natural disasters (Red Cross/Red Crescent Climate 2007). GHGs emissions increased by around 70% during 1970–2004 (TERI 2009).

By 1990s Earth's temperature showed alarming and unnatural warming. The concentration of Carbon dioxide equivalent (CO_2 eq) reached 380 ppm exceeding the natural range of last 650,000 years. Most of the warming during past 50 years is due to human activities (Jansen et al. 2007). As understanding of climate science increased over recent decades, growing evidence of anthropogenic influences on climate change has been found. Many factors continue to influence climate. Principal reasons for human-caused climate change are the amounts of GHGs; changes in small particles (aerosols); and changes in land use. Researchers and policy-makers began calling for controls on GHGs. In the current interglacial period, temperature of Earth reached near highest level which began 12,000 years back. It is important to note that fighting the climate change needs cross generational efforts. This generation seems very serious making such efforts. Hence an effort is made in this generation to make a beginning. Global surface temperatures rose by over $0.7^\circ C$ during the twentieth century Along with the planet's rising temperature, glaciers are melting; the risk of lake-burst floods are rising; water supply of millions of people is affected; rainfall patterns have changed; drying in tropical, subtropical and Mediterranean regions has increased; there is increase in average rainfall and snow in temperate regions; there is rise in heat waves and intense hurricanes (Red Cross/Red Crescent Climate 2007). Atmospheric CO_2 concentrations went up by 100 ppm since their pre-industrial level (TERI 2009).

1.1 Changing the Climate for Development

After humans settled in different region, they started experimenting to evolve new things. This resulted in industrial revolution. During the Industrial Revolution in the eighteenth and nineteenth centuries demand for coal increased. The improvement of the steam engine by James Watt in 1769 resulted in growth in coal use. The history of coal mining and use is inextricably linked with that of the Industrial Revolution, iron and steel production, rail transportation and steamships. Coal was also used to generate gas for gas lights in several cities, which was called 'town gas'. This process of coal gasification saw the augmentation in gas lights across metropolitan areas at the beginning of the nineteenth century (WCI 2009). During last century, various countries and continents competed themselves to develop and make living better. But the period also witnessed world wars destroying themselves and other species in the space ship – *The Earth*.

Development and wars destructed and fragmented biosphere which was core sinks of biogeochemical cycle. The oxides of carbon, water vapour, oxides of nitrogen, and aerosols along with other gases stayed in the atmosphere absorbing and trapping infrared radiation. Subsequent melting of ice and snow resulted in reduction in reflective surfaces on the earth leading to more absorption of heat from sun. This is an effect for albedo effect, depicted pictorially in Figs. 1.2 and 1.3.

Climate change has changed dimensions related health. There is shift in algae, phytoplankton and zooplankton population and distribution. The westerlies and easterlies have undergone changes. Hypolimnic and thermocline temperature has changes affecting flora and fauna of oceans. There is increase in extreme events like typhoons. Some of the semi-arid regions are receiving less rain. While there is decline in lot of species there is increase in ruminant species which is cause for methane generation. Societies have lost resilience. Active layers, Ice-cap, ice-flow, ice-sheet and ice-shelf have changed drastically after industrial revolution. Acclimatization capability species has declined. In spite of introduction of new

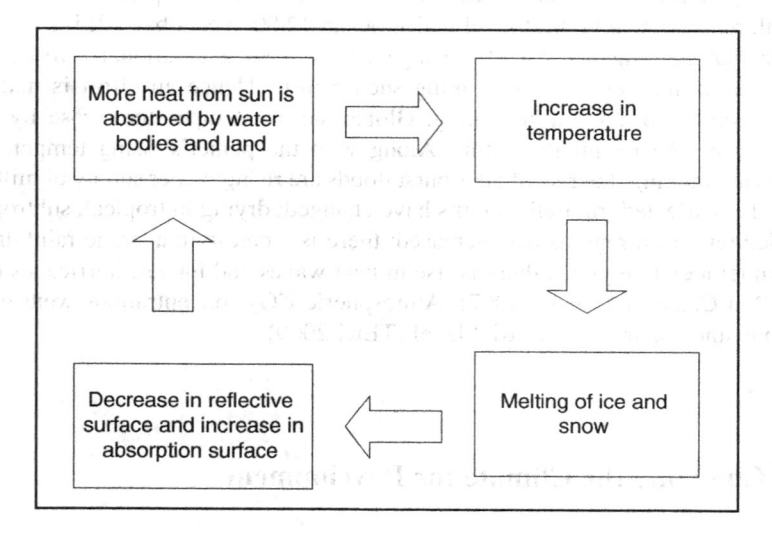

Fig. 1.2 Albedo effect and climate change in snowy area

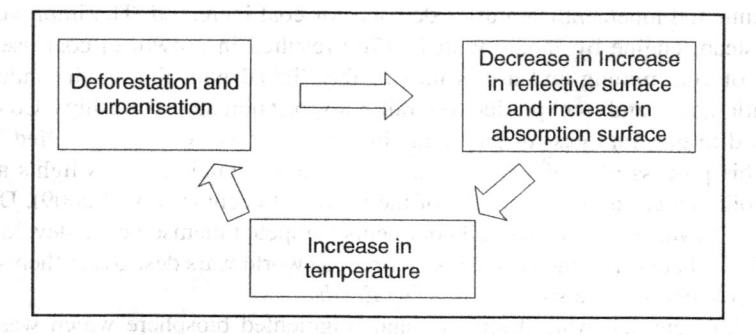

Fig. 1.3 Albedo effect and climate change due to deforestation and urbanization

technologies like fuel cells, wind power, solar energy, hybrid vehicles there is constant change in hydrographic events, hydrologic cycles. On the one hand there is increase in use of halocarbons and on the other hand there is increase in hail storms and urban heat islands. Fire in Peat and peat lands have increased in the recent past contributing to GHGs. There is increase in international agreements and laws. Enforcement mechanisms by penal law have not reduced impact on carbon sinks, sand storm, salinasation, eutrophication, feed backs, food chain and food web. Encouragement to use bio fuel and afforestation is not sufficient to recharge aquifers. Agriculture and aquaculture have been affected drastically in past few decades. Technology has contributed to huge variety of GHGs which can block many wavelengths in absorption spectrum.

Figures 1.4–1.6 shows three major biogeochemical cycles which causes climate change. While Nitrogen cycle contributes to oxides of nitrogen to atmosphere, carbon cycle contributes carbon dioxide and methane to atmosphere. The water cycle contributes to water vapour. Magnitude and direction of Anabatic and katabatic wind movement depicted in Figs. 1.7 and 1.8 are altered in recent years due to destruction of hills for mining and quarrying.

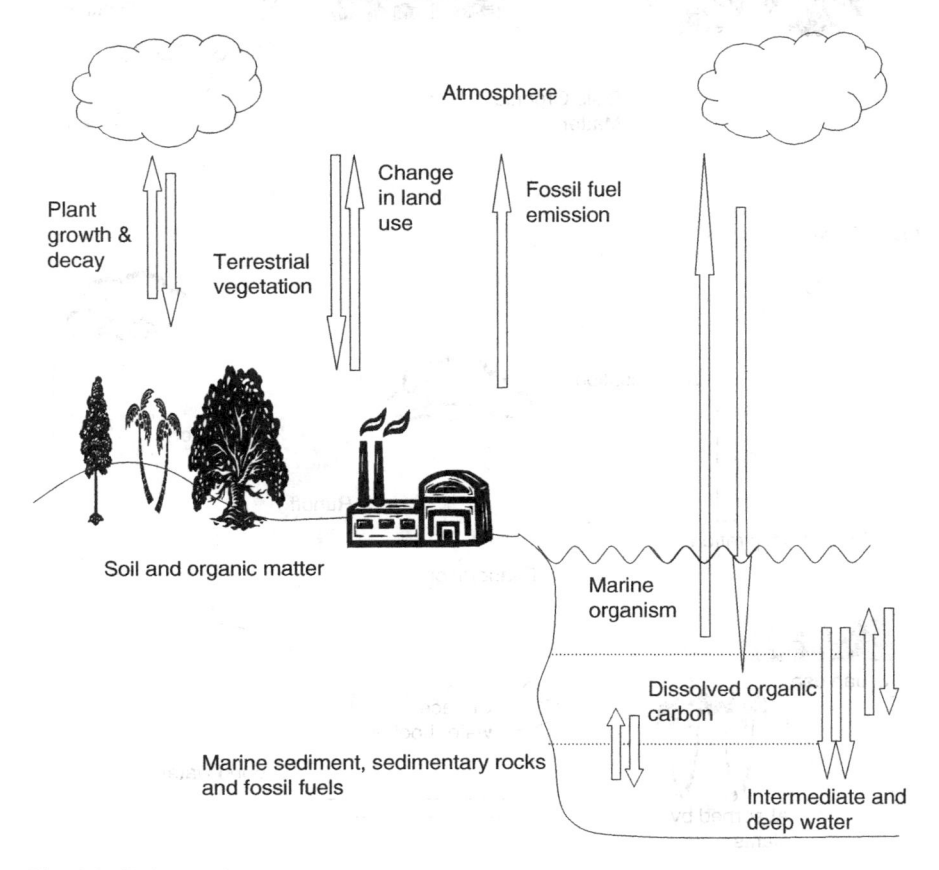

Fig. 1.4 Carbon cycle

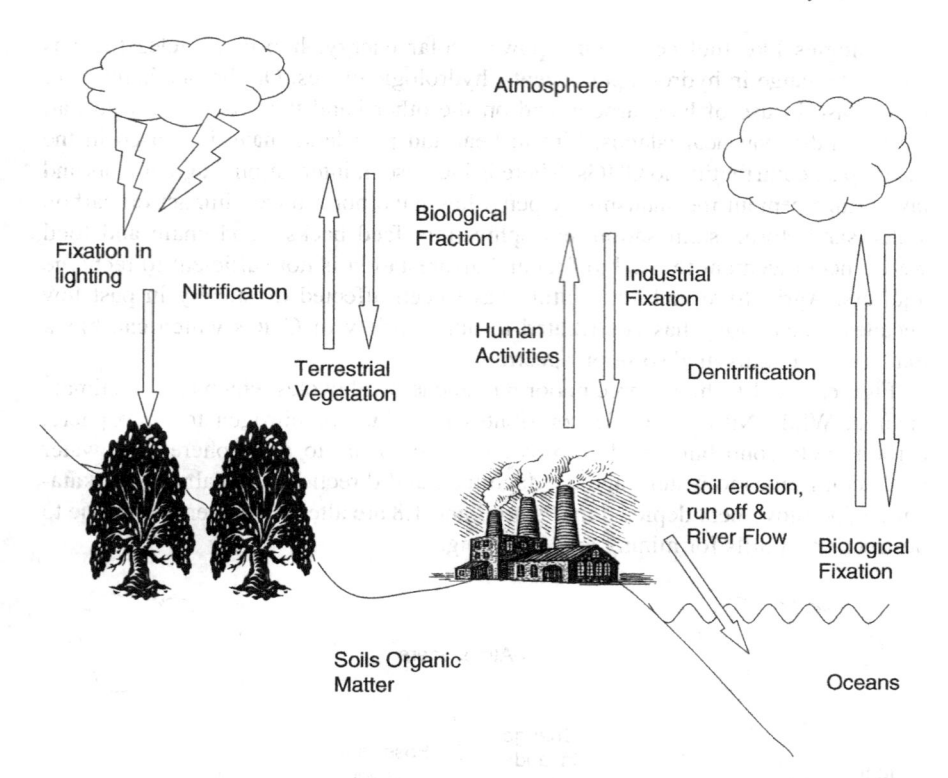

Fig. 1.5 Nitrogen cycle

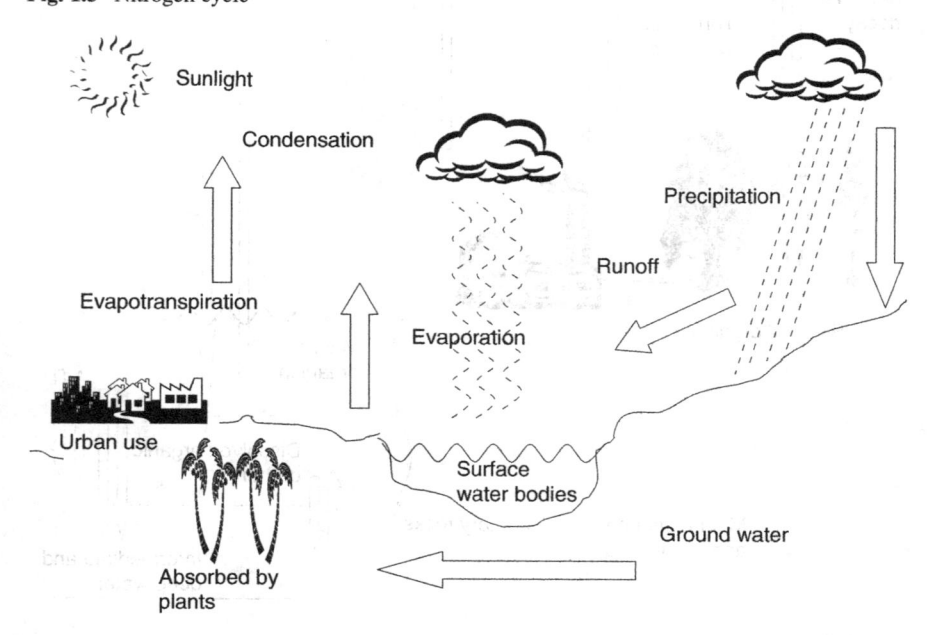

Fig. 1.6 Water cycle

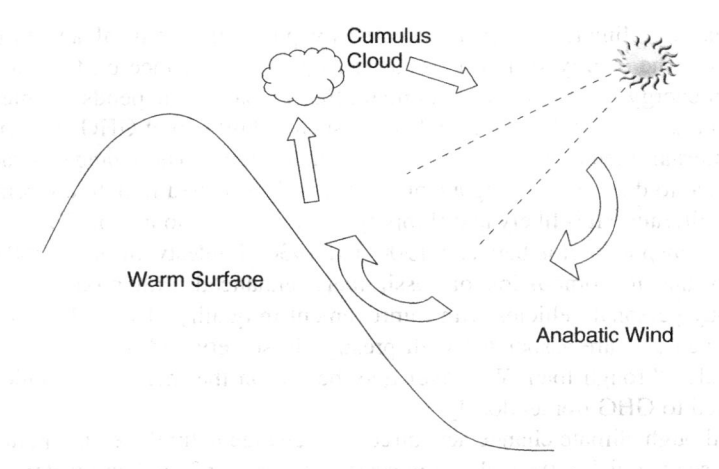

Fig. 1.7 Anabatic winds

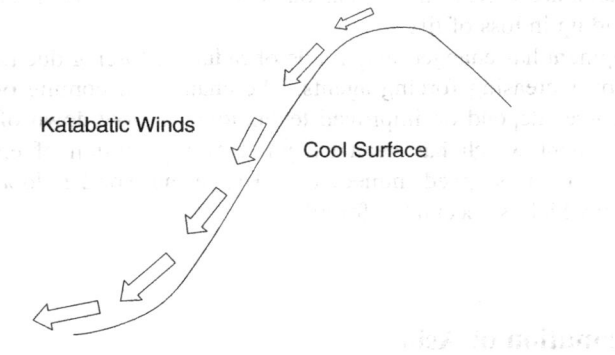

Fig. 1.8 Katabatic winds

The changing environment often towards degradation has questioned the concept and theory of sustainable development. Attempt to overcome poverty and later to become rich has resulted in changing climate, vanishing species, shortage of food, new diseases, and shortage of water resources. Some crisis can be reversed but needs global cooperation. Like diminishing fish population can't be reversed without stopping catching and eating fish. But on the other hand such proposals are not welcome by everyone. Recommendation to slaughter animals immediately after they have gained maximum weight to reduce methane emission is not welcome by animal lovers and certain religion. One of the states in India is already in the process of enacting and enforcing anti cow slaughter bill to save the animals due to respect towards species. Restriction on fuels, travel, vehicles, tourism is not welcome to reduce GHGs and hence usually does not appear in any agenda or recommendations. In spite of agreement that GHGs has to be curbed there is lot of disagreement on how it has to be achieved.

The earth's climate system is affected by physical, chemical and biological interaction. The energy system is regulated by energy balance on the earth. Any change in energy balance changes climate. Energy balance depends on interaction between natural and anthropogenic drivers (such as build up of GHG and aerosols). These external agents are recognised as forcing agents. The change in radiation balance due to different forcing agents collectively is called radiative forcing.

As per literature it is likely that shipping was the largest source of CO_2 emissions from the transport sector between 1900 and 1930 (Fuglestvedt et al. 2008). The emissions due to combustion of fossil fuels, enhanced with modern transport. People buy personal vehicles with improvement in quality of life. But some time personal vehicles are associated with prestige in society and hence many people buy vehicles through loan. Whatever may be reason the increase in vehicles has contributed to GHG tremendously.

Even though climate change acts directly to change natural weather patterns, its effects cascade quickly through many sectors. Scarcity of food and water, damage to ecosystems, adverse effects on human health, disruptions to infrastructure, and drop in tourist are some of the threats that could not only cancel economic benefits but also end up in loss of life.

Development has changed magnitude of radiative forcing due to anthropogenic activities by increasing forcing agents. The chances of coming out of effects of radiative forces depend on improved technology or slow down of economy. But the development which has been triggered giving million of employment and livelihood can't be stopped immediately. Hence the world is looking around for the solution which is acceptable for all.

1.2 Formation of Asia

The major Southeast Asia plates originated during the Proterozoic as parts of Gondwana and detached during Paleozoic time. The detached plate moved northward and the amalgamation of South China plate and Indochina occurred at Early Carboniferous time (e.g., Metcalfe 1998, 2002) or in Silurian-Early Devonian time (e.g., Krobicki and Golonka 2006).

During Triassic time, as a result of the Indonesian orogeny and closure of the Paleotethys Ocean, the Southeast Asian plates joined the Asian continent associated with strong tectonic deformations, metamorphism and magmatic intrusion and extrusion events (Michał et al. 2008).

The onset of the collision of India with Asia occurred near the Paleocene-Eocene boundary (Golonka et al. 2006) and the opening of the Southeast Asia basinal zones occurred due to complex tectonics during Paleogene-Neogene time (Michał et al. 2008).

Ice core studies show that atmospheric CO_2 was in the range of 180–300 ppm in the glacial-interglacial cycles occurred in the last 650 kyr (Petit et al. 1999;

Siegenthaler et al. 2005). The explanation of these CO_2 variations remains unanswered in climate research (Jansen et al 2007).

The current Asia with the present geological setting happened due to series of geological events discussed in following paragraphs.

500 Million Years Ago

During this period in the Cambrian era, great extent of the Arabian landmass was covered by shallow seas that evaporated in the hot climate, resulting in formation of thick salt deposits.

Katian Age of the Late Ordovician period varied between 456 and 446 million years ago. The Katian largely reflected the peak of the overall global warming trend of the Ordovician. However, there are some indications of a brief cold spell towards the middle of the Katian, perhaps coincident with the traditional Caradoc-Ashgill boundary. This is reflected in changes of sedimentation patterns in Baltica (Cocks and Torsvik 2005) and a minor dip, or at least leveling off, in the global diversity profiles of many animal groups. The closing of the northern Iapetus Ocean in the Katian would certainly have affected coastal currents in some fashion, and it clearly resulted in a sorting out of the endemic marine fauna around the three continents (Cocks and Torsvik 2005; Hints et al. 2007). The exact locations of Siberia and Perunica at this time are unclear, but they, too, seem to have been close enough to Baltica for some degree of faunal exchange to take place.

Radiolarian diversity peaked around the mid-Katian. The group seems to have begun its decline well-before the Hirnantian (Webby 2004a). A similar pattern is found in other environmentally-sensitive groups, such as sponges and some brachiopods.

400 Million Years Ago

During this period Afro-Arabian tectonic plate moved near the South Pole where it went through an ice age.

250 Millions Years Ago

Over 250 million years ago, all the earth's land were together forming one single super continent called Pangea which was surrounded by a large ocean. In the major tropical ocean called the sea of Tethys separated the Afro-Arabian continent from the Eurasian landmass. During late Permian to late Cretaceous era that occurred between 250 and 65 million years ago thick layers of limestone and dolomite rocks were deposited. Abundant organic material like algae and other micro-organisms were deposited in the warm tropical seas of the time and was buried deeper leading to oil and gas. Currently United Arab Emirates (UAE)'s oil is found in these rocks.

200 Million Years Ago

This period is also known as Middle Permian Period. During this period, the super continent Pangea began split into different land masses and move apart in various directions. Rivers from both the northern Eurasian land mass called Angara and the southern Indian land mass called Gondwana started depositing large amounts of sediments into Tethys. The Angara and Gondawana moved closer and closer.

130 Million Years Ago

During this period folding and faulting occurred in Arabia.

70 Million Years Ago

This period was referred as Upper Cretaceous period. At this period Angara and Gondawana began to collide with each other resulting folding of shallow seabed resulting in longitudinal ridges and valleys.

65 Million Years Ago

About 66–68 million years ago, basaltic lava erupted from present-day India to produce the Deccan Traps. About 65 million years ago the bed of the Tethys started rising again resulting in elevated high mountain ranges. During 23–65 million years ago sea levels rose and fell several times. The coastal region was at regular intervals covered with warm tropical waters.

50 Million Years Ago

Volcanic centres formed leading to formation of the Maldives during 55–60 million ago. The seafloor in the western portion of the Philippine Sea was developed between 35 and 60 million years ago. Eastern portion of Philippine Sea was formed by back arc spreading from 30 million years ago. About 45 million years ago Indian sub continent collided with the Asian plate resulting in separation of Sri Lanka from the Indian plate.

25 Million Years Ago

This period is refereed as Middle Miocene Period. Geological activity in this period lead to mountain building resulting in formation of the low Shivalik ranges. Mauritius Islands was formed during 18–28 million years ago. Japan was not an independent geographical entity and the strata that formed up to that time were part of the continental edge of Eurasia. Japanese landmasses which were occupied positions parallel to the continental edge and rotated outwards into their present positions between 15 and 19 million years ago (Gina 2003). About 25 million years ago Afro-Arabian plate moved north and was pushed under Eurasia resulting in formation of the Zagros Mountains in Iran. The western part of the sea of Tethys disappeared and the straits of Hormuz closed during the period.

Two Million Years Ago

The UAE coast rose above sea level in during 2–5 million years ago and the Arabian Gulf filled with water again about 4 million years ago.

6,00,000 Years Ago

Periodic mountain building that occurred due to pushing of Indian plate against the Eurasian plates led to rising of the Himalayan ranges further.

8,000 Years Ago

About 8,000 years ago the drop in sea level allowed sand to be blown into Abu Dhabi from Saudi Arabia. Since 6,000 years the Arabian climate has become steadily drier.

1.3 Climate in the Pre-pottery Neolithic

The impact of change in environment to human societies has been a concern of palaeo-ecological and archaeological research. Human migrations, plant and animal domestication, the restructuring of settlement patterns and socio-economic transformations have often been recognized due to the climate change and/or human impact on the landscape and its resources (Redman 1999). In the Middle East, the end of the Early Neolithic period has been linked with climatic worsening characterized by cold conditions bringing about increasing aridity occurring at about 8,200 years cal BP and lasting for < 400 years (Alley et al. 1997; Wiersma and Renssen 2006). The 8.2 ka event is believed to have forced population diffusion and the spread of settlement west wards into Southeast Europe (Weninger et al. 2006).

The Pre-pottery Neolithic A (PPNA, around 9000 BC) succeeds the Natufian culture of the Epipaleolithic. During PPNA, domestication of plants and animals was in its beginning and triggered by younger Dryas.

During, the PPNA and following PPNB, pottery was yet unknown.

The younger Dryas stadial, also referred to as Big Freeze (Berger 1990), was a geologically brief (1,300 ± 70 years) cold climate period between 10,800 and 9,500 BC (Muscheler et al. 2008).

1.4 Climate, the First Technological Revolution and Urbanization

Humans lived as nomads during Paleolithic period. There were no permanent settlements and no food surplus (WCI 2009). The Paleolithic Period is a prehistoric era distinguished by the development of the first stone tools, occurred prior to 10,000 BCE.

With slowing down of nomadic life, people started settling down and thus began urbanization. With urbanization people started using variety of fuel which included wood, agro-waste, coal etc. Coal has a very long history and some historians believe that coal was first used commercially in China. North-eastern China supplied coal for smelting copper and for casting coins around 1000 BC. Coal cinders found amongst Roman ruins in England show that the Romans used energy from coal before AD 400.

The Neolithic or New Stone Age, occurred beginning about 9500 BCE in the Middle East The Neolithic followed the terminal Holocene Epipalaeolithic period, beginning with the rise of farming.

During the Holocene, the Asian continent south of about 45° latitude saw the rise and occasionally abrupt collapse of many agriculture-based societies (Weiss 2000; Weiss and Bradley 2001). Around the 8.2 ka event, farming communities across West Asia were reduced drastically with some habitat-tracking to suitable environments (Weninger et al. 2006). Climate and societal change occurred simultaneously during the late mid Holocene and early late Holocene, roughly 5,500–3,500 yr BP.

In west Asia, at the 5.2 ka event, late Uruk period society in south and north Mesopotamia collapsed (Postgate 1986; Weiss 2003). At the 4.2 ka event, changes among early Bronze Age societies suggest link between the event's precipitation diminutions and collapse of society which dependent upon cereal agriculture (Weiss et al. 1993; Cullen et al. 2000; Staubwasser et al. 2003; Drysdale et al. 2006; Arz et al. 2006).

Mesolithic Period which occurred after last ice age and Neolithic Period 10,000–5,000 years ago witnessed growth of population density and depletion of natural resources. The era was characterized by small widely dispersed semi-permanent settlements and nomads. Village life was more sustainable and common.

Bronze age and Iron Age occurred 5,000 years ago witnessed division of labor which leads to hierarchical power structure. The hierarchical structures developed some form of administrative leadership. New technologies and modes of survival contributed to a food surplus, domestication of plants and animals.

Overgrown villages lead to early forms of human settlement covering few acres and supported a population of about few thousand. Overgrown villages further developed to permanent settlement in dense aggregations.

Population pressures, tools and techniques, organized religion, organized government, transportation technologies became prime factors to ancient urbanization

After the discovery of the thermometer in the early 1600s, efforts began to measure and record the weather (Le Treut et al. 2007).

In a report by Randhawa (1945), progressive desiccation of climate of northern India is described which throws very interesting light how the humid climate in Brij region of Uttar Pradesh changed to desert like climate. This report details that based on several sculptures recovered with the help of Mr F S Growse, Pt Radha Kishan and Dr Fuhrer and others, about 2,000 years ago (ca.500 B.C.–ca.640 A.D.), Brij region was full of wet tropical forests of having evergreen trees of Indo-Malayan affinities such as *Saraca indica, Mesua ferrea and Anthocephalus indicus* which flourish at present in Assam, Bengal, Burma and west coast of India.

1760s-mid-1800s witnessed Industrial Revolution in Western Europe and North America. The era saw improvements in industrial machinery, utilization of the steam engine, use of coal in iron smelting, specialization and division of labor in manufacturing. The era also witnessed decline in mortality, population growth and concentration.

Methodical observations of the weather were being made in nearly all inhabited areas of the world in nineteenth century. Formal international coordination of meteorological observations from ships began in 1853 (Quetelet 1854).

Twentieth century witnessed rapid growth of urban population due to migration and natural increase in poor countries.

It was not possible to perceive warming due to human activity in 1980. Madden and Ramanathan (1980) and Hansen et al. (1981) predicted it would be clear within the next two decades. Wigley and Raper (1990) used a simple energy-balance climate model to show that the observed variation in global-mean surface temperature from 1867 to 1982 could not be explained by natural reason. This discovery was later established by estimation from ocean atmosphere general circulation models (Stouffer et al. 1994).

With increase in population urban agglomeration and combustion of fossil fuel the earth saw rise in temperature followed by rise in sea level and changes in precipitation patterns.

Warming was more over land than over ocean in these recent decades (Hansen et al. 2007; Solomon et al. 2007; Sutton et al. 2007), this in part due to fact that the ocean responds more slowly compared to the land. Land warming was enhanced by a factor of 2–3 in Eurasia; 3–4 in the Arctic and the Antarctic Peninsula; and by about 50% in the United States (Hansen et al. 2010). Warming of the ocean surface was highest over the Arctic Ocean, followed by Indian Oceans; Western Pacific Oceans; and Atlantic Ocean (Hansen et al. 2010).

1.5 Asian Climate After Kyoto Protocol

As per observation made in world's longest monitoring record of carbon dioxide in the atmosphere, is collected at the Mauna Loa observatory in Hawaii, United States the annual average in 1959 was 316 ppm.v (parts per million by volume), it was 377 in 2004, 387 in 2009, and 390 during the first 6 months of 2010 (UNEP 2010).

As on date, throughout the world there is no hard law (law which punishes person for violations) on climate change. At international level United Nations Framework Convention on Climate (UNFCCC) was the major convention which lead to Kyoto protocol which is soft in nature. The Kyoto Protocol does not contain any provision for sanctions but parties rely on the "blame and shame" which is not sufficient in this day and age (UNEP 2010).

The principles behind International Law and Environmental Law are given in Box 1.2. UNFCCC calls for the widest possible cooperation under Principle 21 of the 1972 United Nations Stockholm Declaration. Other paragraphs of the Convention refer to the precautionary principle, sustainable development, national inventories, implementation of mitigation programmes, exchange and transfer of technology, adaptation to climate change and cooperation on science and technology. The principle of "common but differentiated responsibilities" includes all Parties to the Convention.

The Kyoto Protocol (named after city where the protocol was adopted) is a protocol to the UNFCCC. From the beginning of UNFCCC developing countries claimed that the current GHG emission is due to developed countries during industrialisation and post industrialisation development. At United Nations summit in Rio de Janerio in the year 1992, the convention was adopted and signed by 154 countries. The protocol aimed at reducing global warming. The Protocol was initially adopted on 11 December 1997 in Kyoto, Japan and came into force on 16 February 2005. As of September 2009, there were 191 Parties (190 States and one regional economic integration organization) to the Kyoto Protocol.

Kyoto Protocol has provisions that have been questioned by many scholars. The international trade in quotas may not lead to substantial emission cuts as countries that overshoot their emission reduction targets can buy emission rights from nations (UNEP 2010).

Further details of Kyoto protocol are discussed in Sect. 18.3. Some people criticize the protocol as there is possibility for a country to carry forward the surplus of 1 year to another. Another loophole is the concept of joint implementation, in which a country investing in climate friendly projects can claim credits to offset its own emissions.

China, the biggest emitter among all the countries in the world produced about 5.1 tons of carbon emissions per head per year during the post Kyoto period. China is currently a world leader in the production of solar power panels, wind turbines, solar water heaters, energy-efficient domestic appliances and rechargeable batteries. Countries such as China and India which have no reduction commitments in the Kyoto Protocol, are taking major steps towards a world of low carbon dependency (UNEP 2010).

In spite of all the effort from global community to curb GHG, 11 years were among the 12 warmest years on record since 1850 occurred in between 1995 and 2006 (Solomon et al. 2007). Patterns of precipitation change are spatially and seasonally variable compared to variation in temperature. Eastern parts North and South America; northern Europe; and northern and central Asia have become

Box 1.2 Principles of environmental legislation

Legislations that deal with conserving natural resources and pollution control are termed as environmental legislations. This type of legislation has two dimensions:

- International law which are in the form of treaties, conventions, international agreements etc are soft in nature and does not have any punitive action against defaulters and
- National laws which are hard laws with provision for punitive action against defaulters

The principles of International Environment legislations are:

- State sovereignty
- Co-operation
- Preservation and protection of the environment
- Prevention
- Precaution
- The "Polluter Pays Principle"
- Information and assistance in environmental emergencies
- Information and consultation in cross-boundary relations and
- The Rights of Individuals: information, participation and access to justice

The principles of environmental legislations commonly used in national legislations are:

- Sustainable development
- Polluters pay principle and
- Precautionary principle

Principles play an important role in the creation and the development of law in general. All the legislation of judicial procedure in law is based on the principle of the equality of the parties. General principles of law such as the principle of good faith in the execution of obligations imposed by legislation are applicable in all legal systems.

State sovereignty is one of the oldest principles of international law which means State has exclusive jurisdiction on its territory. Article 2 (7) of the UN Charter declares:

Nothing contained in the present Charter shall authorize the United Nations to intervene in matters which are essentially within the domestic jurisdiction of any state or shall require Members to submit such matters to settlement under the present Charter.

Principle 21 of the Stockholm Declaration states that:

States have, in accordance with the Charter of the United Nations and the principles of international law, the sovereign right to exploit their own resources pursuant to their own environmental policies...

significantly wetter, where as it was drier in Sahel, the Mediterranean, southern Africa and parts of southern Asia (Trenberth et al. 2007).

The number of people affected by disasters was one third more in 1996–2005. South Asia suffered massive monsoon rains, heavy flooding, tornadoes and land-slides in 2007 (Red Cross/Red Crescent Climate 2007). July 2010 global surface temperature was more than 5°C warmer than the average July temperature in the 1951–1980 in the eastern Eurasia.

August 05, 2010 witnessed cloud burst in Ladhak of India located in Himalayas. 2010 also witnessed worst flood Pakistan has suffered and second worst flood Delhi has seen.

Unlike Montreal protocol wherein the ozone depleting substances was phased out, Kyoto protocol will not phase out CO_2 and other GHGs. Once emitted, the GHGs will remain in atmosphere for long time. Time lags are important penalty of climate change inertia and even the stringent measures will not affect average changes in temperature till mid-2030 (UNDP 2007).

The GHGs emitted over millions of years do continue to stay and accumulate due to depleting sinks and increase in fossil fuel combustion. Warming will continue until ambient air quality standard and emission standards for CO_2 is stipulated by individual countries and implemented strictly. Further there is no down trend in population growth and with increase in population the demand for food, shelter, clothing, health care, cosmetics, electronic gadgets, transportation also growing. The effectiveness of combating GHGs not only depend on technol-ogy but also on sacrifices of individual countries in terms of aspirations in terms of cars, buildings and luxury achieved by developed countries. The impact of climate change is also mixed with rising prices, diminishing environmental quality, increase in health burden and resource depletion.

The climate change not only depends on GHGs and other radiative forcings but also on the natural reasons like solar activity, volcanic eruption, change in hydro-logic cycles and hydro geochemical cycles. The scientific community is concen-trating more on GHGs and anthropogenic causes of climate change and the world at its limited understanding of the climate science does not have quick-fix solutions to climate change.

References

Alley R. B., Mayewski P. A., Sowers T., Stuiver M., Taylor K. C. And Clark P. U. 1997: Holocene climate instability: a prominent, widespread event 8200 yr ago. Geology 25(6):483–86.

Anson W. Mackay, D.B. Ryves, R.W. Battarbee, R.J. Flower, D. Jewson, P. Rioual, M. Sturm, 2005: 1000 years of climate variability in central Asia: assessing the evidence using Lake Baikal (Russia) diatom assemblages and the application of a diatom-inferred model of snow cover on the lake, Global and Planetary Change 46, 281–297

Anderson D.M., Overpeck J.T, and Gupta A.K, 2002: Increase in the Asian southwest monsoon during the past four centuries. Science, 297(5581), 596–599.

Arz H.W, Lamy F, Pätzold J, 2006: A pronounced dry event recorded around 4.2 kyr in brine sediments from the Northern Red Sea. Quaternary Research 66, 432–441.

Barnola J.M, Raynaud D, Korotkevich Y.S, and Lorius C, 1987: Vostok ice core provides 160,000-year record of atmospheric CO2. Nature, 329, 408–414.

Berger W.H., 1990: "The Younger Dyas cold spell-a quest for causes". Global Planetary Change 3(3):219-237. doi:10.1016/0921-8181(90)90018-8

Berner R A. 1991: A model for atmospheric CO2 over Phanerozoic time. American Journal of Science, 291:339–376.

Berner R A. 1994: GEOCARB II: A revised model of atmospheric CO_2 over Phanerozoic time. American Journal of Science, 294:56–91.

Bice K.L, Norris R.D, 2002: Possible atmospheric CO2 extremes of warm mid-Cretaceous (Late Albian–Turonian). Peleoceanography, 17:1–17.

Bräuning A, and Mantwill B, 2004: Summer temperature and summer monsoon history on the Tibetan plateau during the last 400 years recorded by tree rings. Geophys. Res. Lett., 31(24), L24205, doi:10.1029/2004GL020793.

Booth R.K, Jackson S.T, Forman S.L, Kutzbach J.E, Bettis E.A, Kreig J, Wright D.K, 2005: A severe centennial-scale drought in mid-continental North America 4200 years ago and apparent global linkages. Holocene 15, 321–328.

Cocks L.R.M & Torsvik T.H, 2005: Baltica from the late Precambrian to mid-Palaeozoic times: The gain and loss of a terrane's identity. Earth-Sci. Rev. 72:39-66.

Cullen H.M., deMenocal P.B, 2000: North Atlantic influence on Tigris- Euphrates streamflow. International Journal of Climatology 20, 853–863.

Cullen H.M, deMenocal P.B, Hemming S, Hemming G, Brown F.H, Guilderson T, Sirocko F, 2000: Climate change and the collapse of the Akkadian empire: evidence from the deep sea. Geology 28, 379–382.

CNCCC (China National Report on Climate Change), 2007: Beijing: China National Committee on Climate Change as quoted in Mats Eriksson, Xu Jianchu, Arun Bhakta Shrestha, Ramesh Ananda Vaidya, Santosh Nepal, Klas Sandström, 2009: The Changing Himalayas, Impact of climate change on water resources and livelihoods in the greater Himalayas, International Centre for Integrated Mountain Development (ICIMOD)

DeConto R.M, 2009: Late Cretaceous Climate, Vegetation, and Ocean Interactions: An Earth System Approach to Modeling an Extreme CHEN Junming et al. Earth Science Frontiers, 16(6): 226–23.9

Drysdale R, Zanchetta G, Hellstrom J, Maas R, Fallick A, Pickett M, Cartwright I, Piccini L, 2006: Late Holocene drought responsible for the collapse of Old World civilizations is recorded in an Italian cave flowstone. Geology 34, 101–104.

Farooq A.B, Khan A.H, 2004: Climate change perspective in Pakistan. Proceedings of Capacity Building APN Workshop on Global Climate Change Research, Islamabad, p. 39–46

Fuglestvedt J, Berntsen T, Myhre G, Rypdal K, and Skeie R.B, 2008: Climate forcing from the transport sectors. Proceedings of the National Academy of Sciences of the United States of America, 105, 454–458.

Gina L. Barnes, 2003: Origins of the Japanese Islands: The New "Big Picture" Japan Review, 15:3–50

Gasse, F., 2000. Hydrological changes in the African tropics since the Last Glacial Maximum. Quaternary Science Reviews 19, 189–211.

Golonka J, Krobicki M, Pajak J, Nguyen Van Giang and Zuchiewicz W, 2006: Global plate tectonics and paleogeography of Southeast Asia. Faculty of Geology, Geophysics and Environmental Protection, AGH University

Hansen J, Johnson D, Lacis A, Lebedeff S, Lee P, Rind D and Russel G 1981: Climate impact of increasing atmospheric carbon dioxide. Science, 213, 957–966.

Hansen J, Sato M, Ruedy R, Kharecha P, Lacis A, Miller R.L, Nazarenko L, Lo K, Schmidt G.A, Russell G, 2007: Climate simulations for 1880-2003 with GISS modelE. Clim. Dynam., 29, 661-696, doi:10.1007/s00382-007-0255-8.

Hansen J, Ruedy R, Sato M, and Lo K, 2010: Global surface temperature change, Rev. Geophys., accepted http://data.giss.nasa.gov/gistemp/2010july/ downloaded on 6.9.2010

Hints L, Hints O, Nemliher R & Nõlvak J, 2007: Hulterstad brachiopods and associated faunas in the Vormsi Stage (Upper Ordovician, Katian) of the Lelle core, Central Estonia. Estonian J. Earth Sci. 56:131-142.

Huber B T, Norris R D, MacLeod K G., 2001, Deep-sea paleotemperature record of extreme warmth during the Cretaceous. Geology, 30:123–126.

Jouzel J, Lorius C, Petit R.R, Genthon C, Barvok, N.I, Kotlyakov V.M. Petrov V.M, 1987: Vostok ice core: a continuous isotope temperature record over the last climatic cycle (160,000 years). Nature, 329, 402–408.

Jouzel J, Barkov N.I, Barnola J.M, Bender M, Chappellaz J, Genthon C, Kotlyakov V.M, Lipenkov V, Lorius C, Petit J.R, Raynaud D, Raisbeck G, Ritz C, Sowers T, Stievenard M, Yiou F and Yiou P, 1993: Extending the Vostok ice-core record of palaeoclimate to the penultimate glacial period. Nature, 364, 407–412.

Jansen E, Overpeck J, Briffa K.R, Duplessy J.C, Joos F, Masson-Delmotte V, Olago D, Otto-Bliesner B, Peltier W.R, Rahmstorf S, Ramesh R, Raynaud R, Rind D, Solomina O, Villalba R and Zhang D, 2007: Palaeoclimate. In: Climate Change 2007: The Physical Science Basis. Contribution of Working Group I to the Fourth Assessment Report of the Intergovernmental Panel on Climate Change [Solomon S, Qin D, Manning M, Chen Z, Marquis M, Averyt K.B, Tignor M and Miller H.L (eds.)]. Cambridge University Press, Cambridge, United Kingdom and New York, NY, USA.

Krobicki M and Golonka J, 2006: Caledonian orogeny in Southeast Asia: questions and problems. Geolines, 20, 75–78.

Le Treut H, Somerville R, Cubasch U, Ding Y, Mauritzen C, Mokssit A, Peterson T and Prather M, 2007: Historical Overview of Climate Change. In: Climate Change 2007: The Physical Science Basis. Contribution of Working Group I to the Fourth Assessment Report of the Intergovernmental Panel on Climate Change [Solomon S, Qin D, Manning M, Chen Z, Marquis M, Averyt K.B, Tignor M and Miller H.L. (eds.)]. Cambridge University Press, Cambridge, United Kingdom and New York, NY, USA.

Liu S.Y, Ding Y.J, Li J, Shangguan D.H, Zhang Y, 2006: 'Glaciers in response to recent climate warming in Western China'. Quaternary Sciences, 26(5): 762–771

Madden R.A., and Ramanathan V, 1980: Detecting climate change due to increasing carbon dioxide. Science, 209, 763–768.

Martin A.J. Williams, Stanley H. Ambrose, Sander van der Kaars, Carsten Ruehlemann, Umesh Chattopadhyaya, Jagannath Pal, Parth R. Chauhan, 2009: Environmental impact of the 73 ka Toba super-eruption in South Asia, Palaeogeography, Palaeoclimatology, Palaeoecology 284, 295–314

Metcalfe I, 1998: Paleozoic and Mesozoic geological evolution of the SE Asian region, multidisciplinary constraints and implications for biogeography. In: Hall R., Holloway J.D. (Eds), Biogeography and Geological Evolution of SE Asia. Backhuys Publishers, Amsterdam, pp. 25–41.

Metcalfe I., 2002: Permian tectonic framework and paleogeography of SE Asia. Journal Asian Earth Sciences, 20, 551–566.

Michał Krobicki, Jan Golonka and Khuong The Hung, 2008: Proceedings of the International Symposia on Geoscience Resources and Environments of Asian Terranes (GREAT 2008), 4th IGCP 516, and 5th APSEG; November 24–26, 2008, Bangkok, Thailand

Morrill Carrie, Jonathan T. Overpeck, Julia E. Cole, Kam-biu Liu, Caiming Shen, Lingyu Tang, 2006: Holocene variations in the Asian monsoon inferred from the geochemistry of lake sediments in central Tibet. Quat. Res., 65(2), 232–243.

Muscheler R, Kromer B, Björck S, Svensson A, Friedrich M, Kaiser K.F & Southon J, 2008: Tree rings and ice cores reveal ^{14}C calibration uncertainties during the Younger Dryas. Nature Geoscience 1:263–267, doi:10.1038/ngeo128.

Nutzel W, 2004: Einführung in die Geo-Archäologie des Vorderen Orients. Reichert, Wiesbaden.

Palaeos, 2002: <http://www.palaeos.com/Paleozoic> retrieved on 20 November 2010

Petit J.R, Jouzel J, Raynaud D, Barkov N.I, Barnola J.M, Basile I, Bender M, Chappellaz J, Davis M, Delaygue G, Delmotte M, Kotlyakov V.M, Legrand M, Lipenkov V.Y, Lorius C, PÉpin L, Ritz C, Saltzman E & Stievenard M, 1999: Climate and atmospheric history of the past 420,000 years from the Vostok ice core, Antarctica. Nature, 399, 429–436.

Postgate N, 1986: The transition from Uruk to Early Dynastic: continuities and discontinuities in the record of settlement. In: Finkbeiner U, Rollig W. (Eds.), Đamdat Nasr: Period or Regional style? Reichert, Wiesbaden, pp. 90–106.

Quetelet, A., 1854: Rapport de la Conférence, tenue à Bruxelles, sur l'invitation du gouvernement des Etats-Unis d'Amérique, à l'effet de s'entendre sur un système uniform d'observations météorologiques à la mer. Annuaire de l'Observatoire Royal de Belgique, 21, 155–167.

Randhawa M.S, 1945: Progressive desiccation of northern India; J. Bom. Nat. Hist. Soc. 45 558–565

Red Cross/Red Crescent Climate Centre, 2007: Red Cross/Red Crescent Climate Guide

Redman C. L. 1999: Human impact on ancient environments. University of Arizona Press, Tuscon.

Ren J.W, Qin D.H, Kang S.C, Hou S.G, Pu J,C, Jin Z.F, 2003: 'Glacier variations and climate warming and drying in the central Himalayas'. Chinese Science Bulletin 48(23): 2478–2482

Siegenthaler U, Stocker T.F, Monnin E, Lüthi D, Schwander J, Stauffer B, Raynaud D, Barnola J. M, Fischer H, Masson-Delmotte V, Jouzel J, 2005: Stable carbon cycle-climate relationship during the late Pleistocene. Science, 310(5752), 1313–1317.

Solomon S, Qin D, Manning M, Alley R.B, Berntsen T, Bindoff N.L, Chen Z, Chidthaisong A, Gregory J.M, Hegerl G.C, Heimann M, Hewitson B, Hoskins B.J, Joos F, Jouzel J, Kattsov V, Lohmann U, Matsuno T, Molina M, Nicholls N, Overpeck J, Raga G, Ramaswamy V, Ren J, Rusticucci M, Somerville R, Stocker T.F, Whetton P, Wood R.A and Wratt D, 2007: Technical Summary. In: Climate Change 2007: The Physical Science Basis. Contribution of Working Group I to the Fourth Assessment Report of the Intergovernmental Panel on Climate Change [Solomon S., Qin D, Manning M, Chen Z, Marquis M, Averyt K.B, Tignor M and Miller H.L (eds.)]. Cambridge University Press, Cambridge, United Kingdom and New York, NY, USA.

Shrestha A.B, Wake C,P, Dibb J.E, Mayewski P.A, 2000: 'Precipitation fluctuations in the Nepal Himalaya and its vicinity and relationship with some large-scale climatology parameters'. International Journal of Climatology 20:317–327

Shrestha A.B, 2004: 'Climate change in Nepal and its impact on Himalayan glaciers'. In Hare W. L, Battaglini A, Cramer W, Schaeffer M, Jaeger C (eds) Climate hotspots: Key vulnerable regions, climate change and limits to warming, Proceedings of the European Climate Change Forum Symposium. Potsdam: Potsdam Institute for Climate Impact Research

Staubwasser, M., Sirocko, F., Grootes, P., Segl, M., 2003. Climate change at the 4.2 ka BP termination of the Indus valley civilization and Holocene south Asian monsoon variability. Geophysical Research Letters 30, 1425 doi:10.1029/2002GL016822.

Stouffer R.J, Manabe S, and Vinnikov K.Y, 1994: Model assessment of the role of natural variability in recent global warming. Nature, 367, 634–636.

Sutton R.T, Dong B, and Gregory J.M, 2007. Land/sea warming ratio in response to climate change: IPCC AR4 model results and comparison with observations, Geophys. Res. Lett., 34, L02701.

Tan L, Cai Y, Yi L, An Z, Li, Ai L, 2008: Precipitation variations of Longxi, northeast margin of Tibetan Plateau since AD 960 and their relationship with solar activity. Climate of the Past 4:19–28

TERI (The Energy and Resources Institute) 2009: Simplifying Climate Change, based on the findings of the IPCC Fourth Assessment report, pp 140

Trenberth K.E, Jones P.D, Ambenje P, Bojariu R, Easterling D, Klein Tank A, Parker D, Rahimzadeh F, Renwick J.A, Rusticucci M, Soden B and Zhai P, 2007: Observations: Surface and Atmospheric Climate Change. In: Climate Change 2007: The Physical Science Basis. Contribution of Working Group I to the Fourth Assessment Report of the Intergovernmental Panel on Climate Change [Solomon S., Qin D, Manning M, Chen Z, Marquis M, Averyt K.B,

Tignor M and Miller H.L (eds.)]. Cambridge University Press, Cambridge, United Kingdom and New York, NY, USA.

UNDP, 2007: Human Devloopment Report 2007/2008, Fighting Climate Change: Human solidarity in a divided world, pp 4

UNEP, 2010: Air pollution promoting regional cooperation

WCI (World Coal Institute), 2009: The Coal Resource A Comprehensive Overview of Coal

Wigley T.M.L., and Raper S.C.B, 1990: Natural variability of the climate system and detection of the greenhouse effect. Nature, **344**, 324–327.

Weiss H, Courty M.A, Wetterstrom W, Guichard F, Senior L, Meadow R, Curnow A, 1993: The genesis and collapse of 3rd millennium north mesopotamian civilization. Science 261, 995–1004.

Weiss H, 2000: Beyond the Younger Dryas: Collapse as Adaptation to Abrupt Climate Change in Ancient West Asia and the Eastern Mediterranean. In: Bawden, G., Reycraft, R. (Eds.), Confronting Natural Disaster: Engaging the Past to Understand the Future. University of New Mexico Press, Albuquerque, 2000, pp. 75–98.

Weiss H, Bradley R.S, 2001: Archaeology – What drives societal collapse? Science 291, 609–610.

Weiss H, 2003: Ninevite Periods and Processes. In: Rova E, Weiss H. (Eds.), The Origins of North Mesopotamian Civilization. Subartu IX, Brepols, Turnhout, pp. 593–624.

Weninger B, Alram-Stern E, Bauer E, Clare L, Danzeglocke U, Jöris O, Kubatzki L, Rollefson C, Todorova H, van Andel T, 2006: Climate Forcing due to the 8200 cal BP event observed at Early Neolithic sites in the Eastern Mediterranean. Quaternary Research 66, 401–420.

Wiersma A.P. and Renssen H. 2006: Model-data comparison for the 8.2 ka BP event: confirmation of a forcing mechanism for catastrophic drainage of Laurentide Lakes. Quaternary Science Reviews 25:63–88.

Wilson P.A, Norris R.D, Cooper M.J, 2002: Testing the Cretaceous greenhouse hypothesis using glassy foraminiferal calcite from the core of the Turonian tropics on Demerara Rise. Geology, 30:607–610.

Xu Z, Gong T, Liu C, 2007: 'Detection of decadal trends in precipitation across the Tibetan Plateau'. In Methodology in Hydrology. Proceedings of the Second International Symposium on Methodology in Hydrology held in Nanjing, China, October–November 2005) IAHS Publication 311, pp 271–276. Wallingford: IAHS

Yao T, Duan K, Xu B, Wang N, Guo X, Yang X, 2008: Ice core precipitation record in Central Tibetan plateau since AD 1600. Climate of the Past Discuss. 4:233–248

Yapp C J, Poths H. 1996: Carbon isotopes in continental weathering environments and variations in ancient atmospheric CO2 pressure. Earth Planet. Sci. Lett., 137:71–82.

Zhang D.E, 2005: Severe drought events as revealed in the climate record of China and their temperature situations over the last 1000 years. Acta Meteorol. Sin., **19**(4), 485–491.

Zhao L, Ping C.L, Yang D.Q, Cheng G.D, Ding Y.J, Liu S.Y, 2004: 'Change of climate and seasonally frozen ground over the past 30 years in Qinghai-Tibetan plateau, China'. Global and Planetary Change 43:19–31

Chapter 2
Industrial Revolutions, Climate Change and Asia

Development of civilisation would have not happened without carbon emission in most cases. But carbon emissions are now believed to add to global warming and subsequent climate change events. Scientists believe that the world has already burnt half the fossil fuels necessary to bring about 2°C rise in global temperature.

As the humans began to settle, their energy requirement also increased with wood as main source of energy. By the 1280s, people started using coal for fuel in processes such as limekilns and metalworking which resulted in air pollution having black smoke and oxides of sulphur in its emissions.

Late eighteenth and early nineteenth centuries witnessed major changes in agriculture, manufacturing, production, mining, and transportation. The onset of the industrial revolution marked turning point for climate change. The use of coal gas in street lighting was eventually replaced with the emergence of the modern electric era. With the development of electric power in the nineteenth century, coal's future became closely tied to electricity generation. The first practical coal-fired electric generating station, developed by Thomas Edison, went into operation in New York City in 1882, supplying electricity for household lights.

Oil overtook coal as the largest source of primary energy in the 1960s, with the huge growth in the transportation sector. Coal still plays a vital role in the world's primary energy mix, providing 23.5% of global primary energy and 39% of the world's electricity in 2002.

The industrial revolution had a great effect on the socioeconomic and cultural conditions starting in the United Kingdom, followed by Europe, North America, and eventually the world. The industrial revolution marks a major turning point in human history. Starting in the later part of the eighteenth century, Great Britain's previously manual labour and animal-based economy changed to machine-based manufacturing. It started with the mechanization of the textile industries followed by development of iron-making techniques which lead to increased use of coal. The developments of machine tools in the first two decades of the nineteenth century led to manufacturing of more machines for other industries. The first industrial revolution, which began in the eighteenth century led to Second industrial revolution in nineteenth century, with the development of steam-powered ships and railways. Nineteenth century witnessed internal combustion engine and electrical power generation. With industrial revolution came a series of environmental impacts – air

R. Chandrappa et al., *Coping with Climate Change*,
DOI 10.1007/978-3-642-19674-4_2, © Springer-Verlag Berlin Heidelberg 2011

pollution, water pollution, thermal pollution, noise pollution and degradation of forest and other ecosystems. Increase in carbon dioxide led to global warming due to green house effect.

Industrialized countries have accumulated 'historical' emissions, in the atmosphere since the beginning of the industrial revolution. Very recently, developing countries are only adding to this carbon pool already created in the atmosphere.

With the industrial revolution and subsequent development, climate system was overused because of its natural availability as a resource whose access is open to all free of charge. This resulted in 'free riding', a situation in which development of some individuals (the 'free riders') enjoy the benefits at the cost of others (including other species). Change in climate has significant implications for intra-generational and inter-generational equity, and the application of diverse equity approaches has most important implications for policy recommendations (Halsnæs et al. 2007).

The growing threat of global warming and climate change has alerted attention on the economic growth and environmental pollutants. Intergovernmental Panel on Climate Change (IPCC) estimated that the average global temperature will rise between 1.1 and 6.4°C in the next 100 years (IPCC 2007).

The global trade in goods depends on transportation of freight along complex supply chains (Fred 2009). Many literature published on globalization and the environment focuses on the effect trade treaties and increased global trade on the ecosystem (Boghesi and Vercelli 2003; Chapman 2007; Clapp and Dauvergne 2005; Ehrenfeld 2005; Tisdell 2001). Past studies reveal the economic growth-environmental pollution nexus (Lee and Lee 2009) and the economic growth-energy consumption nexus (Al-Iriani 2006; Apergis and Payne 2009a, b; Chen et al. 2007; Joyeux and Ripple 2007; Lee 2005; Mishra et al. 2009; Mahadevan and Asafu-Adjaye 2007; Mehrara 2007; Lee and Chang 2008; Narayan and Narayan 2008; Narayan and Smyth 2009).

Globally, CO_2 emissions increased from 1971 to 2004 at an annual rate of 2% with the largest regional increases in CO_2 emissions for commercial buildings and residential buildings from developing Asia for 30 and 42% respectively (Levine et al. 2007). Like all other continents, Asia also underwent climate change due to anthropogenic activities.

East China is witnessing abrupt change of summer climate since early 1980s with southward retreat of summer monsoon rainy belt (Qun 2001).

At the beginning of this century, most of the leaders of the world were busy to tackle the problem of climate change. Most of them are eager to solve the problem with exception of a few neutral who have not understood the problem, others are either eager to solve the problem or neutral. Still some of the leaders are looking for funds and advice to overcome the problem. Climate change is one of the major problems faced by present and immediate next generation. If these two generations are not able to solve the problem then it would be too late to solve the change which would have gained inertia. Across the world, millions of people are already forced to cope with the climate change. In the climate change scenario, developing countries suffer more than developed countries.

In spite of constant improvements in energy intensities, global energy use and supply are expected to continue to grow, especially as developing countries follow industrialization (Rogner et al. 2007). Urbanization and growing wealth in developing countries indicate a large increase in demand for energy services in the next few decades, which mean increased carbon emission. How the world reacts to this problem will have direct bearing on the poor and rich at different time, magnitude and type. Poor people are more vulnerable to climate change but all who are vulnerable are not poor. However, eradication of poverty is not the solution of climate change (Kulshrestha 2010). According to recent findings of IPCC, increasing GHG emissions have been correlated with more energy usage especially during past two decades which led to uplifting of human society especially in developing world. If poverty is removed, technology driven population will be increased carrying more per capita CO_2 emissions. It is important to note that poor men society many a times recycles the materials due to poor monetary condition thus emitting limited CO_2.

In many regions, religion, culture and ethnic differences have been source of conflict among the people. But one common thing about all these people is they share same planet and are subject to climate change.

Current patterns of resource depletion, if continued, lead to issues related to the food and energy security of Asian countries. Hence these nations should recognize and accept the links between the expansion of economic activity, resource depletion, and pollution.

2.1 Industrial Revolution and Asia

The industrial revolution was introduced by Europeans into Asia in the last years of the nineteenth and the beginning of the twentieth century which saw the development of industries in India, China, and Japan.

The possible reasons for Asia not catching up with industrial revolution in eighteenth and early nineteenth century could be many countries were under the control of colonial rule. Subsequent struggle of these countries for freedom as well as being busy in participation of wars rather than peace made Asian countries to catch-up industrial activity late. The blessing in disguise to Asia in twentieth century is some of industrial houses were looking forward to outsource polluting activities to avoid stringent environmental legislation and standards in there own countries. Other reasons include cheap labour and search for new market by old companies who have achieved end of growth in their own countries as most of the people in these countries possess materialistic luxury and hence does not create demand for cars, refrigerators, electronic goods etc.,

In the 1960s, about 60% of the Chinese labour force was employed in agriculture, by 1990; the fraction of the labour force employed in agriculture had fallen to about 30% and by 2000 still further. The rapid economic growth is mainly due to growth of the industrial sector in absolute terms, of up to 8% per year during the 1970s. Economic growth and the amplified integration in the world economy of other countries from Asia are contributing to the increase of international marine transport. Globally, large enterprises dominate industrialization, but small- and medium-sized enterprises (SMEs) outnumber large industries in developing countries. In India, SMEs have major shares in the metals, chemicals, food industries (GOI 2005). China has 39.8 million SMEs accounting for 99% of the country's enterprises (APEC 2002). One disadvantage with SMEs is lower adherence to environmental laws. Many of SMEs are unorganized and will not register with government bodies to evade trade restrictions, bureaucracy, taxes, bribes and paperwork. Production of Natural Gas in the Middle East and Asia Pacific increased in the year 2009 due to growth in Iran, Qatar, India and China (BP 2010).

China has become the world's largest energy-related GHG emitter surpassing the United States. In 2007, China's energy sector accounted about 6.1 billion tons of CO_2 or about 21% of entire global energy-related CO_2 emissions (WEO 2009). China is now the world's principal producer of cement, glass, steel, and ammonia (CSIS 2010). If Asian countries continue to rapidly industrialize, transport demand will grow with extreme rapidity over the next several decades.

It is very likely that combined radiative forcing from CO_2, CH_4 and N_2O concentration augmentation, have been at least five times quicker over the period from 1960 to 1999 than over any other 40-year period for the duration of the past two millennia prior to the industrial era (Jansen et al. 2007).

Economic and population growth in Asia over the last 30 years has been extraordinary. While conventional economic indicators have been growing constantly, indicators of resource and environmental qualities are decreasing, the reason often referred to out dated technologies, numerous unorganised small scale units and intuitional failure (see Box 2.1). Extremely urbanised populations and poorly planned municipal development; 2.5 times augment in the use of passenger cars over last two decades; haze pollution from forest fires from Indonesia and other countries has resulted in urban air pollution in Asia.

Box 2.1 Pollution control and prevention: institution failure

The international community often refers institution failure as key reason for failure of pollution prevention but usually do not elaborate the reasons for failure. Given below the some gist why pollution control institutions usually fail.

- **Increase in legislation and policy**: After ratification of international environmental legislation the countries usually bring out new national legislation or policy. But institutions do not grow in proportion to responsibility imposed on them due to enacting new legislation or announcement of new policy.
- **Lower manpower**: institutions often fail to recruit man power with increase in legislation and some institutions take decade between recruitments. The number of officer to country's population some time varies from one to ten officers for every million people. The officer to

(continued)

geographical area varies from one to ten officers for every 1,000 km^2. Similarly the ratio of number of enforcing and monitoring staff to coastal stretch, river length, number of water bodies, and number of industries in developing country is far less than developed countries.

- **Non-implementation of laws in holidays and at nights**: Unlike police and fire and emergency services, pollution control institutions usually work only in week days and time in office hours. The departments usually do not work in public holiday and festivals.
- **Use of pollution control monitoring officer to other duties**: The officers and staff of pollution control authorities are used for other duties like election duties and other committees not related to pollution control and prevention.
- **Corruption**: Developing countries have comparatively more corruption and pollution control agencies are not an exception.
- **Leadership**: Most of the institutions responsible for pollution control and prevention are political nominees and quality of institution depends on quality of leadership.
- **Slow decision and communication**: Institutional procedures differ in countries to country and many country still use old paper based records and communication instead of electronic versions.
- **Technology shyness**: Many developing countries do not use remote sensing technology and continuous monitoring equipment for monitoring environmental degradation. Manual procedures are prone to errors, inconsistency and can be easily manipulated.

The term institutional failure is usually not elaborated in many literatures. Failure as understood is not delivering functions that is expected from government agencies which should act as 'trust' and the officers of these agency should act as trustees to protect natural resources. But it is often seen these goals are not achieved due to lack of transparency. Coping with climate change can't be achieved if the national action plans does not include measures to curb institution failure. Economic growth in coming days is considered to be dependent on ensuring an efficient use of natural resources, and at same time striving to reduce environmental damages (Reddy and Goldemberg 1990; Byrne and Shen 1996).

In an urge to catch up with Europeans and Americans in terms of development, Asian countries are liberating the economy and regions with in the countries are competing for industrial growth. Many states in India often conduct *Global Investors Meet* to lure them to invest in the state. Such meeting often promise uninterrupted power supply, fast track clearances from all government department, adequate land and water.

In 2009, global oil use declined by 1.2 million barrels per day (b/d), or 1.7%, however global refining capacity in the period grew by 2.2%, or 2 million b/d with the Asia-Pacific region accounting for more than 80% of the global growth, mainly due to increases in India (BP 2010).

The software Industry boom was also blessing in disguise for India which could provide manpower at six to ten time cheaper than the manpower cost in USA. The fluency in English and ability to work hard with little added incentive has been the success story of software Industry in India. The development centres which work round the clock often demand a work of more than 12 hours from their employees. As a result the working population were able to generate an income 10–20 times that of their parents. This has increased purchasing power of youngster in growing population of cities like Bangalore, Pune, Delhi, Noida, Gurgoan, Mumbai and Hyderabad. The software giants have also started opening development centre in less populated cities like Mysore, Indore and Thiruvanathapuram to cut down costs.

From 1900 to 2000, world primary energy increased more than tenfold, where as world population rose only fourfold from 1.6 to 6.1 billion (Sims et al. 2007). The highest growth rate in the last decade was in Asia. A large number of the world's energy-intensive industries are now located in developing countries with China being world's largest producer of steel (IISI 2005), aluminum and cement (USGS 2005).

As a result of urbanisation, there is tremendous pressure on environment due to increase in vehicles and housing activities. The industrialisation has also resulted in formation of multiple nuclear families by splitting of joint family. Such development has created demand for personal vehicles and house hold articles. Growing pressure to deliver goods and service has also reduced time to cook and hence has increased fast food industry and restaurants and hotels in many cities. This means there is more activity and combustion of fossil fuel which has ultimately contributed to environment degradation. There is also problem of low quality of constructions due to mushrooming apartments. Increased construction activities diminished availability of quality sand around the cities. Construction companies in such area are mixing soil with high fine particles to make concrete and mortar which results in low bonding. The long term impact could only be assessed during disasters when apartments fall like castle of playing cards.

Other practices during construction like use of freshwater for buildings, and washing fertile soil to separate course material for construction will have impact in future due to depletion of water resource and loss of fertility in soil. There is tremendous pressure on the hills due to mining and quarrying activity for supplying construction material.

The environmental degradation is not serious issue for the governments and often ready to compromise with certain extent for development. It is also noteworthy that such acceptance has paid high dividends in terms of economic growth in some area. On the other hand opposition from local community to some of the mega projects have resulted in stagnant economic activities.

Three decades of industrial development in Asia, has translated into resource depletion and environmental impacts. Rather than reducing environmental problems, rapid economic growth has taken its cost on the environment. Hence Asian nations should direct efforts towards making and implementation of policies that are based on long-term accommodation of population and economic growth ambition to the limitations of our biophysical world (Pradeep et al. 2001).

Of most concern in these anthropogenic factors is the increase in CO_2 levels due to emissions from fossil fuel combustion, followed by aerosols (particulate matter in the atmosphere).

Projected climate change-related exposure are likely to affect health of people through rise in malnutrition; increased deaths, disease and injury; rise in diarrhoeal disease; the increased frequency of cardio-respiratory diseases; change in spatial distribution of vectors (IPCC 2007). This adds to already existing disease burden due to nosocomial infections, zoonosis, poor sanitation and deteriorated environment. Further, in present days changing food habits and increase in alcohol consumption will boost the unhealthy citizens in Asia.

Increased industrialisation in Asia does not in any way imply that all the people are benefitted by the change. Figure 2.1 shows relationship between rich, poor and resources. Irrespective of the countries rich people have been benefitted due to economic growth where as poor people have suffered with pollution and other negative consequences of development. Poor is pushed and exposed to climate change. Walking more than 10 km/day each way to farms, schools and clinics is usual in parts of Asia (Kahn Ribeiro et al. 2007). While some people are enjoying the benefits of development others are being affected by negative impact. The growth in China and India as depicted in statistics is not really bringing to benefit to those who are unable to cope with changing world. Around 500 million people in India are living without access to electricity, in a country which is considered as world's global economic threat.

Development does not mean always luxury and negligence towards environment. Some Asian cities with strong governments, actively and effectively pursuing strategies to slow motorization by providing high quality public transport

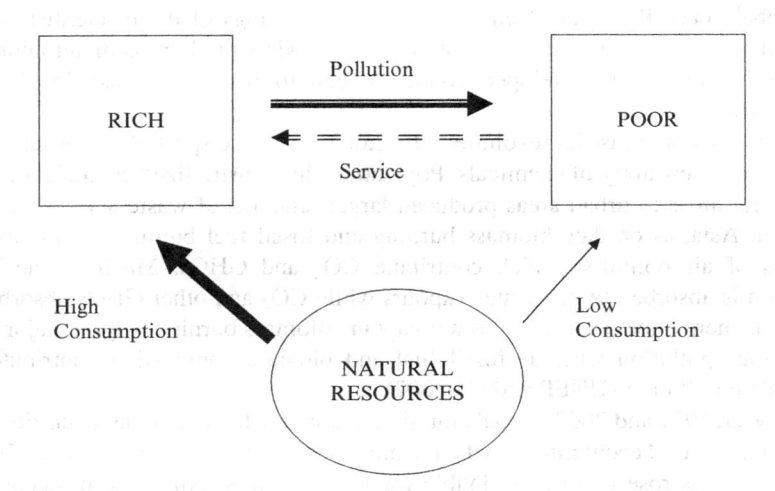

Fig. 2.1 Relationship between rich, poor and resources

(Cullinane 2002; Willoughby 2001; Cameron et al. 2004; Sperling and Salon 2002). But there are also people in the same country who exit one air-conditioned building to enter another air conditioned building after travelling in air conditioned car. There is a huge gap with in the society. The economic growth in many areas due to is taking away land of poor by compulsory acquisitions while the lands of influential people are not touched for acquisition. The poor in such situation have been pushed to further poverty and exposed to climate impact more severely than before.

2.2   Air Pollution and CO_2

Pollution is the introduction of contaminants into environment that causes harm or discomfort to the ecosystem, human health and property.

According to Air (Prevention and Control of Pollution) Act 1981 of India,

> *"Air pollutant" means any solid, liquid or gaseous substance present in the atmosphere in such concentration as may be or tend to be injurious to human beings or other living creatures or plants or property or environment;*
> *"Air pollution" means the presence in the atmosphere of any air.*

According to Environment Protection Act 1990 of United Kingdom,

> *"Pollution of the Environment" means pollution of the environment due to the release (into any environmental medium) from any process of substances which are capable of causing harm to man or any other living organisms supported by the environment.*

It is about hundred years ago that scientists raised a concern regarding burning of fossil fuels during the nineteenth century which they warned would change climate. Most outstanding among these scientists was Svante Arrhenius of Sweden who won the Nobel peace Prize for Chemistry in 1903. Warnings of these scientists were ignored by society in an ambition to enjoy luxury. Mass production of automobiles and its ownership in developed countries lead to further increase in CO_2 in atmosphere.

With dawn of industrial revolution, the globe witnessed spurt of a new activities along with a new array of chemicals. Populations in industrialized countries moved from rural areas to urban areas produced larger amounts of waste and CO_2 emissions. In Asia, as on date biomass burning and fossil fuel burning are the major sources of air pollution which contribute CO_2 and GHGs. Much of the heat radiation is absorbed by the water vapours while CO_2 and other GHGs absorb the radiative energy not absorbed by water vapour. Biomass burning plays a major role in gaseous pollution whereas fossil fuel and biomass combustion contribute to particulate pollution (UNEP and C4 2002).

Between 1971 and 2007, global emission doubled with developing countries, led by Asia, increased economic activity at a much faster rate. Between 1990 and 2007, CO_2 emissions rose more than double for Asia due to striking rate of economic development particularly within China and India (IEA 2007). Natural gas production in the Middle East and Asia Pacific was increased during 2009, driven by growth in Iran, Qatar, India and China (BP 2010). New coal-fired power stations are being put in operation in China each week making implementation of the ambitions of the UNFCCC and the Kyoto Protocol challenge to merge desired economic development with environmental protection (UNEP 2010). Asia has also some unique problem with respect to population distribution. The South Asian Region is one of the most densely populated in the world. It possesses 3% of the world's land mass with 20% of world population. Population in cities of Asia and the Pacific region is above WHO guidelines. It was as high as 98% for India and 99% for China (UNEP and C4 2002). Such population density is easily vulnerable to air pollution and climate change.

Human activities result in emissions of four principal GHGs: carbon dioxide (CO_2), methane (CH_4), nitrous oxide (N_2O) and the halocarbons (a group of gases containing fluorine, chlorine and bromine). These gases accumulate in the atmosphere, causing concentrations to increase with time.

- Carbon dioxide concentrations are increased due to fossil fuel use, deforestation and decay of organic matter
- Methane concentrations are increased as a result of agriculture, natural gas distribution, landfills, and anaerobic decomposition in wetlands
- Nitrous oxide is emitted by chemical/fertilizer manufacture and fossil fuel burning. Natural processes like, volcanic activity, forest fire, reaction in atmosphere and oceans also release N_2O
- Halocarbon gas concentrations are increased primarily due to human activities. Main halocarbons include the chlorofluorocarbons (e.g., CFC-11 and CFC-12), which were used extensively as refrigeration agents and in other industrial processes
- Ozone is continually produced and destroyed in the atmosphere by series of chemical reactions. In the troposphere, human activities have increased ozone by release of gases such as carbon monoxide, hydrocarbons and nitrogen oxide that chemically react to produce ozone
- Water vapour is generated through chemical destruction of CH_4 in the stratosphere, producing a small amount of water vapour and
- Aerosols are generated by both anthropogenic activity (such as surface mining, fossil fuel combustion and industrial processes) and natural sources (mineral dust released from the surface, sea salt aerosols, biogenic emissions from the land and oceans; and dust aerosols produced by volcanic eruptions)

Apart from GHGs (like CO_2, CH_4, O_3), particles and cirrus clouds are also making significant contributions to the climate impact. Chemically active gases in the atmosphere distribute themselves with large spatial and temporal variations. Air Pollutants can change responses of ecosystems to specific climatic change impacts (Andrzej et al. 2007). The climate-chemistry is therefore characterized by regional differences. Regions such as Southeast Asia is considered as future key region due to large increase in energy consumption and pollution emission (Akimoto 2003; Kupiainen and Klimont 2007; Amann et al. 2008; Klimont et al. 2009; Isaksen et al.

2009). Similarly, ship and air traffic are important sectors because of increases in emissions in recent years (Eyring et al. 2005a, b, 2007; Dalsøren et al. 2009).

Carbonaceous aerosols in South Asia has in recent times received great attention due to its rising emissions (Dickerson et al. 2002; Reddy and Venkataraman 2002a, b; Kulshrestha et al. 2005). In less than 20 years, South Asia has become one of the most polluted areas in the world (WHO/UNEP 1992). Oil consumption in the Asia-Pacific region was more than 80% of the global growth, due to increases of consumption in India (+19.5%, or 580,000 b/d) and China (+10.5%, or 820,000 b/d) (BP 2010).

With its many low-tech industrial activities and poorly controlled combustion, high quantities of air pollutants are emitted into the air. While regulations have resulted in lower pollutant emissions in the westernized nations, rapid industrialization has generated high emissions in South Asia (Kato and Akimoto 1992; Akimoto and Narita 1994; Arndt et al. 1997; Garg et al. 2001). Increasing population and industrial activity will generate even higher emissions in the near future. This is going to affect not only the air quality, but also global climate (Venkataraman et al. 1999; Reddy and Venkataraman 1999, 2000).

While GHGs rise the temperature, non-dark particulate matters reflect the energy from atmosphere and contributes to cooling effect. But dark particles do absorb heat energy and contribute to increase in atmospheric temperature. Effects of the haze include cooling of the land surface; increase in the frequency and strength of the thermal inversion; change of the winter time rainfall patterns; and reduction in the average tropical evaporation as well as precipitation.

The most visible impact of air pollution in Asian countries is the haze, a brownish layer of pollutants and particles that pervades many regions in Asia (UNEP and C4 2002). Atmospheric brown cloud (ABC) created world over during last decade is driving global changes in the Earth's atmosphere. More visible impacts are:

- Changes of rainfall patterns with the Asian monsoon leading to weakening Indian monsoon
- Displacement of the thermal equator southwards via cooling of the air over East Asia and
- Retreat of the Hindu Kush-Himalayan glaciers and snow packs and deposition of black carbon decreasing the reflection and exacerbating the retreat

Aerosol pollution due to dust storms can modify cloud properties to reduce or prevent rain in the polluted region. Aerosol with black carbon can reduce formation of clouds. The decrease in precipitation can cause drier soil, which in turn rises into the air providing a potential feedback loop to further decrease rainfall. Anthropogenic changes of land use exposing the topsoil can start desertification feedback process.

Per capita basis, GDP, energy consumption and carbon emissions in developing Asia are much lower compared with the developed countries of the west. But over the year industries discharged foul, toxic, solid, liquid, and gaseous wastes disturbing ecosystems. The industrial revolution also sparked the first significant wave of industrial pollution litigation in history suing industrial polluters under the common law of nuisance. Common law allowed individuals as well as governments to take legal action against nuisances. But no body took action on any body for emitting CO_2. While other gases like oxides of sulphur and nitrogen found its place in discharge standards, CO_2 still does not find its place in either emission standards or ambient air quality standards.

The reason is evident in the fact that the first visible impact due to air pollution was by particulate matter which soiled the cloth and contributed to cough. The next culprits were toxic and acidic gases like oxide of sulphur which also irritated industrial workers as well as people in the neighbourhood. But CO_2 was considered as constitute of air and no body thought about its ill effect until recently when the snow started melting and temperature increased dramatically.

In spite of wide data published, there is still a large gap with respect to accurate data with regard to non-CO_2 GHGs, black carbon, and CO_2 from sources, such as deforestation, decay of biomass and peat fires; military activities and wild animals. Numerous data gaps still exists as shown in Fig. 2.2 where practices like application of degradable wastewater on land result in higher or lower emission than theoretically predicted as accurate emission factors is yet to be evolved.

Box 2.2 Why bother about CO_2

All gases have a unique pattern for absorption of energy, absorbing some wavelengths of energy and letting out others. Even though water vapour absorbs numerous wavelengths of infrared energy, it is transparent to others. Carbon dioxide and other GHGs absorb energy that water vapour does not. This partially closes the "window" through which heat radiated by the earth's surface would otherwise escape to space. The Earth's climate system depends on how much energy enters or leaves the Earth. Imbalance in Earth's radiative equilibrium will lead to rise or fall of temperatures.

In spite of natural phenomena like absorption by oceans and vegetation, the globe is accumulating CO_2 in the atmosphere. The concentration changed from 280 ppm (during pre pre-industrial revolution) to 379 ppm in 2005 (TERI 2009). The uptake of CO_2 in land variable is because of land use change, pollution, fire, and other factors affecting soil and biomass.

CO_2 is greatest contribution to rising global temperatures followed by black carbon. Carbon dioxide is responsible for 40% of the earth's warming and black carbon is responsible for 12% (Pachauri 2010).

It will be unfair not to mention the concern about GHGs inventory. The calculation of GHGs does always make series of assumption and use data available in records. But in reality numbers in records deviate from existing scenario. The GHGs emission due to fuel consumption and explosions in military activity is usually not included in GHGs calculation. Accepting the limitations of theoretical calculation,

Fig. 2.2 Application of degradable wastewater on land – unconventional practices not considered in GHGs calculation result in higher or lower emission than theoretically predicted

over the 1900–2005 periods, the US was the world's largest cumulative emitter of energy-related CO_2 emissions, and accounted for 30% of total cumulative emissions (IEA 2007). The second largest emitter was the EU, at 23%; the third largest was China, at 8%; fourth was Japan, at 4%; fifth was India, at 2%. The rest of the world accounted for 33% of global, cumulative, energy-related CO_2 emissions (IEA 2007).

Improper discharge of waste and experimental discharge of waste usually contributes different emissions than what is calculated. Another setback of theoretical calculation is fuel adulteration which is very common in many Asian countries. Usually, fuel adulteration is not considered during the calculation.

Coal burned in cities for power generation, industrial applications, domestic and commercial boilers and stoves may pollute remote area too. Chinese cities experience very high airborne particle. Measurements made at background sites in the countryside outside of Beijing show particulate concentrations more than half as high as in the city indicating large-scale regional air pollution problem of large geographic extent covering North China and not restricted to Beijing alone. However, in India atmospheric soil dust rich in calcium carbonate gives some relief from acidic effect of sulphur dioxide and nitrogen dioxide. Kulshrestha et al. (2003a, b) demonstrated that in the atmosphere rich with soil dust, calcium carbonate forms calcium sulphate which is not harmful to ecosystem. In addition, Kulshrestha (2009) proposes atmospheric soil dust in India as a natural geo-engineering tool to combat climate change in the region.

The top five producers of coal are China, the USA, India, Australia and South Africa. Much of global coal production is used in the country in which it was produced; only around 18% of hard coal production is sold in international coal market.

Over 4,030 Mt of coal is currently produced and used throughout the world which is more than 38% increase over the past 20 years (WCI 2009). Between 1971 and 2007, global emission doubled with developing countries, led by Asia, increased at a much faster rate. Between 1990 and 2007, CO_2 emissions rose more than double for Asia due to striking rate of economic development particularly within China and India (IEA 2007). Coal production and use has grown fastest in Asia, while Europe has actually seen a decline in production. New coal-fired power station is put into operation in China each week making implementation of the ambitions of the UNFCCC and the Kyoto Protocol challenge to merge desired economic development with environmental protection (UNEP 2010).

Coal has been the major source of CO_2. Global coal production is likely to reach 7 billion tonnes in 2030 – with China accounting for nearly half the increase over this period. Steam coal production is anticipated to have reached around 5.2 billion tonnes; coking coal 624 million tonnes; and brown coal 1.2 billion tonnes (WCI 2009).

Vehicles are very less in small towns/cities of Asia. But it does not mean that pollution will not reach these small towns/cities. Due to dispersion the concentration of air pollution in small urban dwellings and rural settings are also increasing.

Irrespective of damage to climate coal trading is still a lucrative business with lot of "*investors blessing*". Initial Public Offer to sell 10% stake (offer for sale) by Coal India Limited (CIL), was over subscribed in 2010.

The company made a public offer of 631, 636, 440 equity shares aggregating up to Rs.151,994.40 million. CIL is the major coal producing company (at the lowest cost) in the world and also has the highest coal reserves with dominant market share in India (82%) which is currently facing acute coal shortage. CIL has 471 mines in 21 major coalfields across eight states in India. Out of the 471 mines, 163 are open cast mines and 273 are underground mines. Its current coal production capacity of CIL is about 450 MTPA and beneficiation (washing) capacity is 22 MTPA.

CIL is planning to increase its production capacity to 640 MT and beneficiation capacity to 300 MT by 2017. This means there will "*some more carbon in the line*" to enter atmosphere. Oil, coal, automobile is still preferred choice of investment throughout the world. These businesses also have "*good history*" with a track of continuous net profit and dividends to investors. So carbon emissions are likely increase in coming days with added damage to environment and climate in particular.

Fig. 2.3 There is no shortage of energy in the world as long as people can afford it. Unnecessary energy waste can be avoided to mitigate carbon emission

Asian roads which are almost empty 20 years back has now filled with variety of vehicles of all sizes and shapes releasing GHGs. Similarly there is increase of CO_2 across all sectors resulting in CO_2 concentration levels many fold and hence it is imperative that near-term investment decisions should be taken keeping long-term emissions consequences.

Further there is no clear cut boundary between energy use and energy waste. As shown in Fig. 2.3 there is no shortage of energy in the world as long as people can afford it. Such unnecessary energy waste can be avoided to mitigate carbon emission.

References

Andrzej Bytnerowicz, Kenji Omasa, Elena Paoletti, 2007: Integrated effects of air pollution and climate change on forests: A northern hemisphere perspective, Environmental Pollution 147, 438–445

Al-Iriani M. 2006: Energy-GDP relationship revisited: an example from GCC countries using panel causality. Energy Policy;34:3342–50.

APEC, 2002: Profiles in SMEs and SME Issues, 1990-2000. Asia-Pacific Economic Cooperation, Singapore, World Scientific Publishing.

Apergis N, Payne J. 2009a: Energy consumption and economic growth in Central America: evidence from a panel cointegration and error correction model. Energy Econ ; 31:211–6.

Apergis N, Payne J. 2009b: Energy consumption and economic growth: evidence from the Commonwealth of Independent States. Energy Econ; 31:641–7.

Akimoto H, 2003: Global air quality and pollution. Science 302 (5651), 1716–1719. doi:10.1126/science.1092666.

Akimoto H, Narita H, 1994: Distribution of SO_2, NO_x and CO_2 emissions from fuel combustion and industrial activities in Asia with $1° \times 1°$ resolution. Atmospheric Environment 28, 213–215.

Amann M, Bertok I, Cofala J, Heyes C, Klimont Z, Rafaj P, Scho" pp W, Wagner F, 2008: National emission Ceilings for 2020 based on the 2008 Climate & Energy Package. NEC Report 6. Final report to the European Commission. Available at: www.iiasa.ac.at/rains/reports/NEC6-final110708.pdf.

Arndt R.L, Carmichael G.R., Streets D.G, Bhatti N, 1997: Sulphur dioxide emissions and sectoral contribution to sulphur deposition in Asia. Atmospheric Environment 31, 1553–1582.

Boghesi S, Vercelli A, 2003: Sustainable globalisation. Ecological Economics 44, 77–89.

BP (British Petroleum), 2010: BP Statistical Review of World Energy, BP Oil Company Ltd., London

Byrne J, Shen B, 1996: The challenge of sustainability. Balancing China's energy, economic and environmental goals. Energy Policy 24 (5), 1–8 Reprint.

Cameron I, Lyons T.J, and Kenworthy J.R, 2004: Trends in Vehicle Kilometres of Travel in World Cities, 1960-1990: Underlying Drives and Policy Responses. Transport Policy, 11, pp. 287–298.

Chapman L, 2007: Transport and climate change: a review. Journal of Transport Geography 15, 354–367.

Chen ST, Kuo HI, Chen CC, 2007: The relationship between GDP and electricity consumption in 10 Asian countries. Energy Policy;35:2611–21.

Clapp J, Dauvergne P, 2005: Paths to a Green World: The Political Economy of the Global Environment. The MIT Press, Cambridge, Massachusetts.

CSIS (Centre for Strategic International Studies), 2010: Asia's Response to Climate Change and Natural Disasters. Implecations for an Evolving Regional Architecture.

Cullinane C, 2002: The Relationship between Car Ownership and Public Transport Provision: A Case Study of Hong Kong. Transport Policy, 9, pp. 29–39.

Dickerson R.R, Andreae M.O, Campos T, Mayol-Bracero O.L, Neusuess C, Streets, D.G, 2002: Analysis of black carbon and carbon monoxide observed over the Indian Ocean: implications for emissions and photochemistry. Journal of Geophysical Research 107 (D19), 8017. doi:10.1029/2001JD000501.

Dalsøren S.B, Eide M.S, Endresen Ø, Mjelde A, Gravir G, Isaksen I.S.A, 2009: Update on emissions and environmental impacts from the international fleet of ships: the contribution from major ship types and ports. Atmospheric Chemistry and Physics 9, 2171–2194.

Ehrenfeld D, 2005: The environmental limits to globalization. Conservation Biology 19.2, 318–326.

Eyring V, Harris N.R.P, Rex M, Shepherd T.G, Fahey D.W, Amanatidis G.T, Austin, J, Chipperfield M.P, Dameris M, Forster P.M, De F, Gettelman A, Graf H.F, Nagashima, T, Newman P.A, Pawson S, Prather M.J, Pyle J.A, Salawitch R.J, Santer B.D, Waugh D.W, 2005a: A strategy for process-oriented validation of coupled chemistry-climate models. Bulletin of the American Meteorological Society 86, 1117–1133.

Eyring V, Köhler H.W, van Aardenne J, Lauer A, 2005b: Emissions from international shipping: 1. The last 50 years. Journal of Geophysical Research 110, D17305. doi:10.1029/2004JD005619.

Eyring V, Stevenson D.S, Lauer A, Dentener F.J, Butler T, Collins W.J, Ellingsen K, Gauss M, Hauglustaine D.A, Isaksen I.S.A, Lawrence M.G, Richter A, Rodriguez J.M, Sanderson M, Strahan S.E, Sudo K, Szopa S, van Noije T.P.C, Wild O, 2007: Multi-model simulations of the impact of international shipping on atmospheric chemistry and climate in 2000 and 2030. Atmospheric Chemistry and Physics 7, 757–780.

Fred Curtis, 2009: Peak globalization: Climate change, oil depletion and global trade, Ecological Economics 69, 427–434

Garg A, Shukla P.R, Bhattacharya S, Dadhwal V.K, 2001: Sub-region (district) and sector level SO2 and NOx emissions for India: assessment of inventories and mitigation. Atmospheric Environment 35, 703–713.

GOI, 2005: Annual Report, 2004-2005 of the Ministry of Environment and Forest. Government of India, New Delhi.

Halsnæs K, Shukla P, Ahuja D, Akumu G, Beale R, Edmonds J, Gollier C, Grübler A, Ha Duong M, Markandya A, McFarland M, Nikitina E, Sugiyama T, Villavicencio A, Zou J, 2007: Framing issues. In Climate Change 2007: Mitigation. Contribution of Working Group III to the Fourth Assessment Report of the Intergovernmental Panel on Climate Change [Metz B, Davidson O.R, Bosch P.R, Dave R, Meyer L.A (eds)], Cambridge University Press, Cambridge, United Kingdom and New York, NY, USA

IISI, 2005: World steel in figures, 2005: International Iron and Steel Institute (IISI), Brussels.

Intergovernmental Panel on Climate Change (IPCC), 2007a: Climate change Synthesis report 2007.

IEA, 2007: World Energy Outlook 2007 Edition- China and India Insights. International Energy Agency (IEA), Head of Communication and Information Office, 9 rue de la Fédération, 75739 Paris Cedex 15, France. pp. 600. ISBN 9789264027 (International Energy Agency)

Isaksen, I.S.A, Dalsøren S.B, Li L, Wang W.C, 2009: Introduction to special section on 'East Asia Climate and Environment'. Tellus 61 (4), 583–589.doi:10.1111/j.1600-0889.2009.00432.x.

IPCC, 2007b: Summary for Policymakers. In: Climate Change 2007: Impacts, Adaptation and Vulnerability. Contribution of Working Group II to the Fourth Assessment Report of the Intergovernmental Panel on Climate Change, M.L. Parry, O.F. Canziani, J.P. Palutikof, P.J. van der Linden and C.E. Hanson, Eds., Cambridge University Press, Cambridge, UK, 7–22.

Jansen E, Overpeck J, Briffa K.R, Duplessy J.C, Joos F, Masson-Delmotte V, Olago D, Otto-Bliesner B, Peltier W.R, Rahmstorf S, Ramesh R, Raynaud D, Rind D, Solomina O, Villalba R and Zhang D, 2007: Palaeoclimate. In: Climate Change 2007: The Physical Science Basis. Contribution of Working Group I to the Fourth Assessment Report of the

Intergovernmental Panel on Climate Change[Solomon S, Qin D, Manning M, Chen Z, Marquis M, Averyt K.B, Tignor M and Miller H.L(eds.)]. Cambridge University Press, Cambridge, United Kingdom and New York, NY, USA.

Joyeux R, Ripple R.D, 2007: Household energy consumption versus income and relative standard of living: a panel approach. Energy Policy 2007;35:50–60.

Kahn Ribeiro S, Kobayashi S, Beuthe M, Gasca J, Greene D, Lee D. S, Muromachi Y, Newton P.J, Plotkin S, Sperling D, Wit R, Zhou P.J, 2007: Transport and its infrastructure. In Climate Change 2007: Mitigation. Contribution of Working Group III to the Fourth Assessment Report of the Intergovernmental Panel on Climate Change [Metz B, Davidson O.R, Bosch P.R, Dave R, Meyer L.A(eds)], Cambridge University Press, Cambridge, United Kingdom and New York, NY, USA.

Kato N, Akimoto H, 1992: Anthropogenic emissions of SO2 and NOx in Asia: emission inventories. Atmospheric Environment 26A, 2997–3017.

Kupiainen K, Klimont Z, 2007: Primary emissions of fine carbonaceous particles in Europe. Atmospheric Environment 41/10, 2156–2170. doi:10.1016/j.atmosenv.2006.10.066.

Klimont Z, Cofala J, Xing J, Wei, Wei, Zhang C, Wang S, Kejun J, Bhandari P, Mathura R, Purohit P, Rafaj P, Chambers A, Amann M, Hao J, 2009: Projections of SO2, NOx, and carbonaceous aerosols emissions in Asia. Tellus B.doi:10.1111/j.1600-0889.2009.00428.x.

Kulshrestha 2010: <http://www.icsu-visioning.org/open-forum/open-forum-input/comment-page-1/> retrieved on 21 November 2010

Kulshrestha U.C, Kulshrestha M.J, Sekar R, Sastry G.S.R and Vairamani M. 2003a Chemical characteristics of rain water at an urban site of south-central India. Atmospheric Environment, 37 (21), 3019–3026.

Kulshrestha M.J, Kulshrestha U.C, Parashar D.C and Vairamani M. Estimation of SO_4 contribution by dry deposition of SO_2 onto the dust particles in India. 2003b. Atmospheric Environment, 37 (22), 3057–3063.

Kulshrestha U.C, Sreedhar B and Kulshrestha M.J 2005: Carbon as major constituent of atmospheric aerosols at urban sites in India. Proceedings of Asian Aerosol Conference held at Mumbai during Dec 13–16, 2005.

Kulshrestha U.C 2009: Atmospheric dust in India- A natural geo-engineering tool to combat climate change. ENVIS Newsletter SES JNU ISSN-0974-1364, vol 14(3), pp 2–5.

Levine M, Ürge-Vorsatz D, Blok K, Geng L, Harvey D, Lang S, Levermore G, Mongameli Mehlwana A, Mirasgedis S, Novikova A, Rilling J, Yoshino H, 2007: Residential and commercial buildings. In Climate Change 2007: Mitigation. Contribution of Working Group III to the Fourth Assessment Report of the Intergovernmental Panel on Climate Change [Metz B, Davidson O.R, Bosch P.R, Dave R, Meyer L.A(eds)], Cambridge University Press, Cambridge, United Kingdom and New York, NY, USA.

Lee C.C, Chang C.P, 2008, Energy consumption and economic growth in Asian economies: a more comprehensive analysis using panel data. Resour Energy Econ; 30:50–65.

Lee C.C, 2005: Energy consumption and GDP in developing countries: a cointegrated panel analysis. Energy Econ 2005;27:415–27.

Lee C.C, Lee J.D, 2009: Income and CO2 emissions: evidence from panel unit root and cointegration tests. Energy Policy;37:413–23.

Mahadevan R, Asafu-Adjaye J, 2007: Energy consumption, economic growth and prices: a reassessment using panel VECM for developed and developing countries. Energy Policy;35:2481–90.

Mehrara M. 2007: Energy consumption and economic growth: the case of oil exporting countries. Energy Policy;35:2939–45

Mishra V, Smyth R, Sharma S, 2009: The energy-GDP nexus: evidence from a panel of Pacific Island countries. Resour Energy Econ;31:210–20.

Narayan P.K, Narayan S, 2008: Does environmental quality influence health expenditures? Empirical evidence from a panel of selected OECD countries. Ecol Econ;65:367–74.

Narayan P.K, Smyth R, 2009: Multivariate granger causality between electricity consumption, exports and GDP: evidence from a panel of Middle Eastern countries. Energy Policy;37:229–36.

Pachauri R.K, 2010: Climate Change and the Forthcoming Energy Revolution, Manorama Year Book 2010, 198–201

Pradeep J. Tharakan, Timm Kroeger, Charles A.S. Hall, 2001: Twenty five years of industrial development: a study of resource use rates and macro-efficiency indicators for five Asian countries, Environmental Science & Policy 4, 319–332

Qun Xu, 2001: Abrupt change of the mid-summer climate in central east China by the influence of atmospheric pollution, Atmospheric Environment 35, 5029–5040

Reddy A.K.N., Goldemberg J, 1990: Energy for the Developing World. Readings from Scientific American. Energy for Plant Earth. Freeman, New York.

Reddy M.S, Venkataraman C, 1999: Direct radiative forcing from anthropogenic carbonaceous aerosols over India. Current Science 76, 1,005–1,011.

Reddy M.S, Venkataraman C, 2000: Atmospheric and radiative effects of anthropogenic aerosol constituents from India. Atmospheric Environment 34, 4511–4522.

Reddy M.S, Venkataraman C, 2002a: Inventory of aerosol and sulphur dioxide emissions from India: (I) fossil fuel combustion. Atmospheric Environment 36, 677–697.

Reddy M.S, Venkataraman C, 2002b: Inventory of aerosol and sulphur dioxide emissions from India: (II) biomass combustion. Atmospheric Environment 36, 699–712.

Rogner H.H, Zhou D, Bradley R. Crabbé R, Edenhofer O, Hare.B (Australia), Kuijpers L, Yamaguchi M, 2007: Introduction. In Climate Change 2007: Mitigation. Contribution of Working Group III to the Fourth Assessment Report of the intergovernmental Panel on Climate Change [Metz B, Davidson O.R, Bosch P.R, Dave R, Meyer L.A(eds)], Cambridge University Press, Cambridge, United Kingdom and New York, NY, USA.

Sims R.E.H, Schock R.N, Adegbululgbe A, Fenhann J, Konstantinaviciute I, Moomaw W, Nimir H.B, Schlamadinger B, Torres-Martínez J, Turner C, Uchiyama Y, Vuori S.J.V, Wamukonya N, Zhang X, 2007: Energy supply. In Climate Change 2007: Mitigation. Contribution of Working Group III to the Fourth Assessment Report of the Intergovernmental Panel on Climate Change [Metz B, Davidson O.R, Bosch P.R, Dave R, Meyer L.A(eds)], Cambridge University Press, Cambridge, United Kingdom and New York, NY, USA.

Sperling D. and Salon D, 2002: Transportation in developing countries: An overview of greenhouse gas reduction strategies. Pew Center on Global Climate Change, Arlington, 40 pp.

TERI (The Energy and Resources Institute), 2009: Simplifying Climate Change, based on the findings of the IPCC Fourth Assessment report, pp140

Tisdell C, 2001: Commentary: globalisation and sustainability: environmental Kuznets curve and the WTO. Ecological Economics 39, 185–196.

UNEP and C4, 2002: The Asian Brown Cloud: Climate and Other Environmental Impacts UNEP, Nairobi

UNEP, 2010: Air pollution promoting regional cooperation

USGS, 2005: Minerals Yearbook 2004. US Geological Survey Reston, VA, USA. <http://minerals.usgs.gov/minerals/pubs/myb.html>, accessed on 31/05/07.

Venkataraman C, Chandramouli B, Patwardhan A, 1999: Anthropogenic sulphate aerosol from India: estimates of burden and direct radiative forcing. Atmospheric Environment 33, 3225–3235.

WEO (World Energy Outlook), 2009: Annex A: Tables for Reference Scenario Projections, 647.

Willoughby C, 2001: Singapore's Motorisation Policies: 1960-2000. Transport Policy, 8, pp. 125–139.

WHO/UNEP, 1992: Urban Air Pollution in Mega Cities of the World. Blackwell, Oxford.

Chapter 3
Changing Environment: Where Will This Lead to?

The term "environment" is new in many languages. In French, its origin goes back to twelfth century verb "environner". In other languages new words were created during 1960s to express the concept: "Umwelt" in German, "Paryavaran" in Hindi, "Parisara" in Kannada, "Milieu" in Dutch, "Medio ambiente" in Spanish, "Meio ambiente" in Portuguese, "Al.biah" in Arabic, "kankyo" in Japanese, etc. Approximately 40 years ago, a large part of the world at the same time discovered a new phenomenon which posed a major challenge to modern society.

General meaning of the term "environment" is one's surrounding. It can describe a limited area very adjacent to a micro organism or the entire planet, including the outer space which surrounds it. The term "biosphere" used in particular by UNESCO (1988). International and national legislations include various definitions of the environment.

Council of the European Economic Community adopted legal text on 27 June 1967 which defines environment as:

Water, air and land and their interrelationship as well as relationships between them and any living organism.

Article 48A of the Indian Constitution related to environmental protection states that the word "environment" means "the aggregate of all the external conditions and influences affecting life and development of organs of human beings, animals and plants"(Jariwala 1980).

Convention on Civil Liability for Damage Resulting from Activities Dangerous to the Environment, drafted in the framework of the Council of Europe, adopted in Lugano on 21 June, 1993 defines environment as:

For the purpose of this Convention... 'Environment' includes:

- *natural resources both abiotic and biotic, such as air, water, soil, fauna and flora and the interaction between the same factors;*
- *property which forms part of the cultural heritage; and*
- *the characteristic aspects of the landscape.*

Section 1 of the British Environmental Protection Act 1990, defines environment as:

R. Chandrappa et al., *Coping with Climate Change*,
DOI 10.1007/978-3-642-19674-4_3, © Springer-Verlag Berlin Heidelberg 2011

The environment consists of all, or any, of the following media, namely the air, water and land, and the medium of air includes the air within buildings and the air within other natural or man-made structures above or below ground.

Chapter I of Indian Environment Protection Act define Environment as:

"environment" includes water, air and land and the inter-relationship which exists among and between water, air and land, and human beings, other living creatures, plants, micro-organism and property.

United Nations Declaration on the Human Environment, adopted in Stockholm in June 1972, that states:

Man is both creature and moulder of his environment, which gives him physical sustenance and affords him the opportunity for intellectual, moral, social and spiritual growth.

Care for nature appeared early in different civilizations and beliefs. The World Charter for Nature, adopted and solemnly declared by the United Nations General Assembly, on 28 October 1982, states:

Mankind is a part of nature and life depends on the uninterrupted functioning of natural systems which ensure the supply of energy and nutrients.

The preamble of the Declaration on Environment and Development, adopted on 14 June 1992 at Rio de Janeiro mentions "the integral and interdependent nature of the Earth, our home". According to its Principle 4:

In order to achieve sustainable development, environmental protection shall constitute an integral part of the development process and cannot be considered in isolation from it.

World Summit on Sustainable Development convened in Johannesburg, South Africa, from 26 August to 4 September 2002, adopted a Declaration confirming their will to:

assume a collective responsibility to advance and strengthen the interdependent and mutually reinforcing pillars of sustainable development – economic development, social development and environmental protection – at local, national, regional and global levels.

As the human prepares to move from uncomfortable living to comfortable living he forgets the damage caused by him to environment. On the other hand environment has its own way of reacting to stress enforced on it by any agent whether it is natural or anthropogenic. Rightly mention that climate change can change the effects of air pollutants on ecosystems, and vice versa. Table 3.1 gives some examples of stress on environment and counter stress of environment. Accumulated stresses may lead to loss of ecosystem and services they bring in (CCCD 2009).

The environment is changing and so as the climate. The climate information can be used for many purposes by policy makers. The guiding principles for such use include:

- Reduce vulnerabilities or enhance the resilience of population and ecosystems affected by climate change
- Mitigate climate change
- Help community to migrate to safer place

Table 3.1 Examples of stress on environment and anti-stress from environment

Stress due to anthropogenic activity	Anti-stress from environment
Change in water course	Desertification, change in ground water level
Change in weight on earth due to building dam, displacing mineral ores	Earth tremors
Deforestation	Human-animal clashes
Over reproduction	Shortage of food
Pollution	Disease
Over use of medicine	Drop in natural resistance to diseases and side effect to health
Over grazing forest	Eco imbalance leading to disruption of food chain and loss of food
Killing of wild life (including aquatic life)	Eco imbalance leading to disruption of food chain and loss of food
Too much anthropogenic activity disrupting natural harmony and unnatural distribution of energy	Energy imbalance, climate change
Damage to natural setting of soil and rocks	Increase in probability and incidence of land slides
Damage to ozone layer	Skin cancer to human beings

- Forecast probable conflicts
- Prepare for possible disasters
- Identify research needs and fund them
- Create infrastructure to cope with predicted climate change
- Increase institutional strength

The impacts due to climate change in recent decades are:

- Melting of ice sheets and glaciers
- Thermal expansion of seawater as the oceans warm
- Sea level to rise
- Seawater intrusion contaminate leading to coastal fresh water sources and submergence of coastal facilities
- Reduction in depth of freshwater lenses in coastal area and islands
- Increase in climate induced disasters
- Change in precipitation and subsequent run off leading to changes in water resources (both ground water and surface water)
- Increase in global heat extremes occurrence/intensity
- Increase in frequency/distribution of droughts
- Acidification of oceans leading to coral bleaching and threatening the survival of shell-building marine species and the food web of which they are a part
- Increased in climate induced diseases
- Change in yielding pattern of crops
- Climate induced extinction and expiration of some species
- Changed phenology of some species
- Change in migration patterns/periods of migratory birds

- Changes in aquatic and marine ecosystem leading to drop in fish and other species
- Changes in coastal geomorphology
- Change in coastal wetlands
- Increase in energy demand for cooling and climate adoption
- Increase in consumption of cream, beer, cool drinks, cold cream, etc.
- Change in tourism destination/pattern in some countries and
- Drop in fish catch and subsequent impact on fish based economy and livelihood

Understanding how climate change will affect the earth is a key issue worldwide. Questions addressing the penalty of climate change are now high on the list of main concern for international agencies. Development often leads to degradation due to change in land use, pollution and resource depletion which has direct impact on climate due to GHG emission, albedo effect and other climate phenomena. The climate change leads to disaster which destroys infrastructure demanding redevelopment. Figure 3.1 shows relation between climate change, disaster, development and Environment degradation.

The climate induced impacts are not uniform and never will be uniform all over the world and so as in Asia. The Least Developed Countries (LDCs) and Small Island Developing States (SIDS) are the most vulnerable to climate impacts as severe floods and droughts, will affect millions of people in LDCs and SIDS to

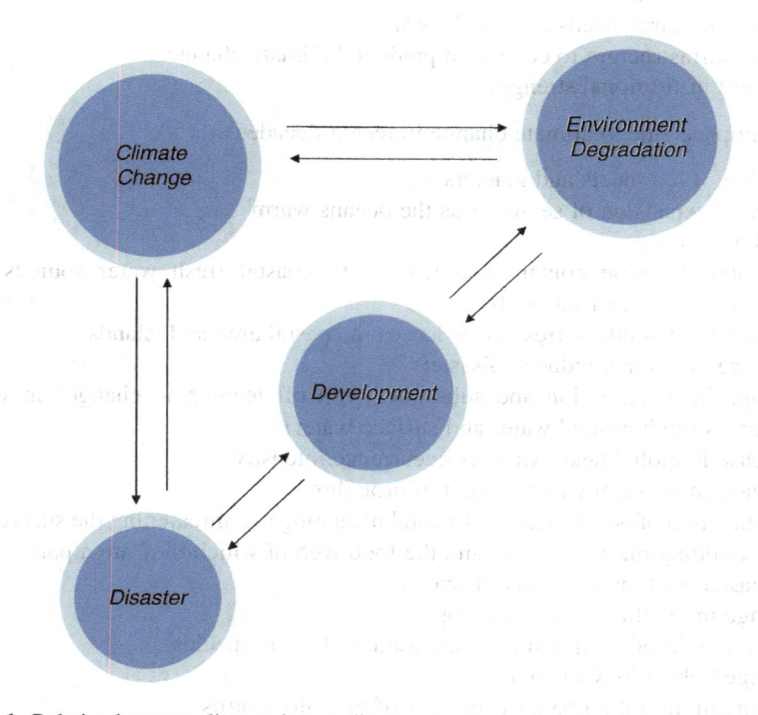

Fig. 3.1 Relation between climate change, disaster, development and environment degradation

poverty, hunger and disease. The rising sea level, along with, threatens the very survival of many SIDS. LDCs and SIDS emit relatively meagre amount of GHGs compared to developed countries and other developing states but LDCs and SIDS will suffer the most. Over 70% of the population in the LDCs live in rural areas and depend on agriculture therefore are more affected than in other countries (UN-OHRLLS 2009).

LDCs are a group of 49 countries: Afghanistan, Angola, Bangladesh, Benin, Bhutan, Burkina Faso, Burundi, Cambodia, Central African Republic, Chad, Comoros, Dem. Republic of Congo, Djibouti, Equatorial Guinea, Eritrea, Ethiopia, Gambia, Guinea, Guinea-Bissau, Haiti, Kiribati, Lao People's Democratic, Lesotho, Liberia, Madagascar, Malawi, Maldives, Mali, Mauritania, Mozambique, Myanmar, Niger, Republic Nepal, Rwanda, Samoa, Sao Tome and Principe, Senegal, Sierra Leone, Solomon Islands, Somalia, Sudan, Timor-Leste, Togo, Tuvalu, Uganda, United Republic of Tanzania, Vanuatu, Yemen, Zambia.

About 860 million people in LDCs and SIDS will be affected by climate impact, and many of them will become environmental refugees (UN-OHRLLS 2009). Hence recognizing the vulnerability of LDCs to climate change and their low adaptation capacity, special LDC Fund has been established in the Marrakech Accords for the purpose of assisting LDCs to adapt to climate change.

There are 51 Small Island Developing States: Antigua and Barbuda, American Samoa, Anguilla, Aruba, Bahamas, Barbados, Belize, British Virgin Islands, Cape Verde, Commonwealth of Northern Marianas, Comoros, Cook Islands, Cuba, Dominica, Dominican Republic, Federated States of Micronesia, Fiji, French Polynesia, Grenada, Guam, Guinea-Bissau, Guyana, Haiti, Jamaica, Kiribati, Nauru, St. Kitts and Nevis, Maldives, Marshall Islands, Mauritius, Montserrat, Netherlands Antilles, New Calendonia, Niue, Palau, Papua New Guinea, Puerto Rico, Samoa, Sao Tome and Principe, Seychelles, Singapore, Solomon Islands, St. Lucia, St. Vincent and the Grenadines, Suriname, Timor-Leste, Tonga, Tuvalu, Trinidad and Tobago, U.S. Virgin Islands, Vanuatu.

Eleven countries among SIDS are LDC. They are Comoros, Kiribati, Sao Tome and Principe, Tuvalu, Guinea-Bissau, Maldives, Solomon Islands, Vanuatu, Haiti, Samoa, and Timor-Leste.

Although SIDS vary in many aspects, they have some common characteristics which make them vulnerable to climatic change (Maul 1996; Leatherman 1997; UN-OHRLLS 2009): (1) limited physical size; (2) limited natural resources; (3) high susceptibility to natural hazards; (4) relatively thin water lenses; (5) small economies and high sensitivity to external market shocks; (6) high population densities and in some countries high population growth rates; and (7) poorly developed infrastructure.

It is predicted that surface runoff will decline in the arid and semi-arid zones of Asia. The average annual runoff in some basins could reduce by 27% by 2050 (IPCC 2001). On the other hand one meter sea level rise can result in extinction of many Asian species, such as Indian otters, Bengal tigers, estuarine crocodiles and mud crabs (UN-OHRLLS 2009). Hence adaptation strategy in these areas shall include the diversification of livelihood activities, institutional architecture, adjustments in farming practice, income generation projects, and the shift towards non-farm livelihood incomes (Bryceson 2004).

Landlocked developing countries (LLDC) are developing countries that are landlocked they are Botswana, Burkina Faso, Burundi, Central African Republic, Chad, Ethiopia, Lesotho, Malawi, Mali, Niger, Rwanda, Swaziland, Uganda, Zambia, Zimbabwe, Afghanistan, Armenia, Azerbaijan, Bhutan, Kazakhstan, Kyrgyzstan, Lao People's Democratic Republic, Mongolia, Nepal, Tajikistan, Turkmenistan, Uzbekistan, Macedonia, Moldova, Bolivia, Paraguay.

Of the 30 landlocked developing countries, the following 12 are located in Asia: Afghanistan, Armenia, Azerbaijan, Bhutan, Kazakhstan, Kyrgyzstan, Lao People's Democratic Republic, Mongolia, Nepal, Tajikistan, Turkmenistan and Uzbekistan. LLDCs do not have territorial access to the sea, thus suffering remoteness and isolation from world markets and higher transportation. Out of 30 landlocked developing countries in the world, 16 including Afghanistan, Bhutan, Lao People's Democratic Republic and Nepal are least developed. Thus strategies for such countries are highly region specific.

3.1 Changing Environment and Impact on Biodiversity in Asian Context

Biodiversity is the degree of variation of life forms within a boundary. The boundary could be as big as a biome or entire planet.

According to the Biological Diversity Act, 2002 of India,

"Biological diversity" means the variability among living organisms from all sources and the ecological complexes of which they are part and includes diversity within species or between species and of eco-systems;

Natural ecosystems provide a variety of service useful for mankind. Natural ecosystem regulates water flows; flood control; food and shelter during disasters; decontamination; carbon sequestration; nutrient and hydrological cycling. Good ecosystem management provides numerous benefits to human societies and other species.

Climate and natural ecosystems are closely coupled, and the stability of that coupled system is an important ecosystem service. An ecosystem is an interdependent, functioning system of plants, animals, and microorganisms. All living things have been classified into five kingdoms out of which only two kingdoms Animals and Plants receive greatest attention. Out of other three kingdoms monera, protesta are microscopic and unicellular where as some fungi species are multi-cellular organism.

Of all the species greater effort for conservation is given to those species which are either helpful to human beings or not detrimental to human beings. Table 3.2

Table 3.2 Types of species and effort of conservation

Type of species	Efforts of conservation	Description
Domestic animals and cultivated plants	High	High efforts of conservation are made as it provides food, recreation, medicine and other benefits. Survival of humans is not possible or highly difficult in the absence of these species. Economically important
Wild animals which does not interfere with human settlement	Moderate	Some countries protect some of the species by legislation. Even if efforts are not made to conserve, search and kill may not happen. If found may be killed for free meet, hide and recreation
Wild animals which interferes with human settlements	Low to moderate	Some countries protect some of the species by legislation. Some time the local community may kill the animals like snake, carnivorous animals to save themselves and domestic animals
Rodents, vectors and weeds	Low	Some times high efforts are put to search and kill to safeguard humans, domestic animals and crop. Not conserved by any laws

explains the types of species and efforts of conservation followed conventionally through out the world.

Ecosystems are of basic importance to environmental functioning and sustainability, and they provide many goods and services crucial to individuals and societies. These goods and services comprise: (1) providing food, fibre, fodder, shelter, medicines and energy; (2) processing and storing carbon and nutrients; (3) absorbing waste; (4) purifying water; (5) reducing soil degradation; (6) providing opportunities for recreation and tourism; and (7) housing the Earth's reservoir of genetic and species diversity; (8) providing cultural, religious, and aesthetic value as well as an intrinsic existence value; (9) regulating water runoff and moderating floods. Living organism will have intricate independence and without the support of the other organisms within their own ecosystem, life forms would not survive.

Table 3.3 gives stages of tree and impact of climate which is typical for trees. *Forest a* complete ecosystem itself is provides fiber, fuel, food, other non-wood products, fresh water, and genetic resources in addition to regulating services like air quality, climate regulation, water regulation, erosion regulation, water purification and waste treatment, disease regulation, pest regulation, pollination, and natural hazard regulation.

Animal life cycles differ from species to species. Life cycles of animal varies whether it is egg lying or not. Impact of climate on animals also depends on whether is aquatic or terrestrial. Both animals and plants have hugely adapted to climates they live-in. Young and old organism are more prone to climate change as young ones will be in the stage of acclimatization and yet to fully develop resistance to diseases where as old organisms might have lost immunity to diseases or in the process of loosing immunity.

The region's ecosystems are unique and represent several of Earth's major biodiversity hotspots. Important ecosystem of Asia includes the eastern Himalayas,

Table 3.3 Stages of tree and impact of climate

Stage	Description	Impact of climate	Description
Seed	Part of fruit which germinates into new plant	High	Stagnate water due to heavy rain may degrade the seeds. Imbalance in season leading late rains may lead to attack by insects
Seedling	The above-ground part of the embryo that sprouts from the seed	High	Tender body parts are sensitive to temperature and water. High temperature will dry the seedling leading to death and heavy rain may degrade the seedling. Less immune to disease and attack by micro organism. Absence of water due to absence of rain after sprouting will lead to death
Sapling	After seedling reaches 1 m tall and until it reaches 7 cm in stem diameter	Moderate	Will tolerate moderate climate changes
Pole	Young trees form 7–30 cm diameter	Moderate	Will tolerate climate change up to certain extent. Depletion of water table beyond root zone will have greatest impact
Mature tree	Over 30 cm diameter, reproductive years begin	Low	Will tolerate climate change to great extent
Old tree	Height growth slows greatly, with majority of production in seed production	Moderate	Will tolerate climate change to certain extent. Heavy rain may cause damage to branches
Over mature tree	Decay of braches and roots become common	Moderate	Will tolerate climate change up to certain extent. Increase in temperate and rain will augment the decay
Snag	Standing dead wood	Low	Will accelerate the spreading of forest fire
Log/debris	Fallen dead wood	Low	Will accelerate the spreading of forest fire

Indo-Malayan archipelago, Japan, mountains of southwest china, Western Ghats and Sri Lanka.

Temperate forests in Asia are cleared by anthropogenic activity and global warming is sufficient to trigger structural changes in remaining temperate forests. The nature and magnitude of these changes depend on connected changes in water availability, water-use efficiency and other human activities in the area. Shifts in temperature and rain may result in altered growing seasons and boundary shifts amongst grasslands, forests, and shrub lands (IPCC 1996). Gobi would change to warm temperate desert scrub, and the area of cool temperate desert scrub would expand to the Khangai Mountains, displacing forest areas due

Table 3.4 Major threatened regions with respect to biodiversity in Asia

Threat due to	Important biodiversity in the area affected
Sea level rise	Philippines, Japan, Indonesia, the Philippines, Malaysia, Papua New Guinea, Solomon Islands and Timor Leste
Ice melting	Himalaya, Mountains of Central Asia
Rise in ambient temperature	Western Gahts and Sri Lanka
Sea water intrusion	Ayeyarwady delta (Myanmar), Chao Phraya delta (Thailand), Mekong delta (Vietnam), Song Hong delta (Vietnam), Zhujiang delta (China), Changjiang delta (China), Huanghe delta (China), Sundarban (India and Bangladesh), Rann of Kutch (India), Lena delta (Russia), Ganges-Brahmapura delta (India and Bangladesh), Indus delta (Pakistan), Shatt-el-Arab delta (Iraq and Iran)

to projected shifts to warmer and drier conditions in Mongolia (Ulziisaikhan 1996).

Table 3.4 shows major threatened regions with respect to biodiversity in Asia. Major habitats of some of the sensitive and endemic species in Asia is hosted by mountains of Central Asia, Himalaya, tropical forest of Indo-Burma region, islands of Japan, mountains of southwest China, Philippines, Western Ghats of India and Sri Lanka.

The amount of snowfall may decrease due to global resulting in spring and summer drought. Changes in snowfall amounts depend on changes in snow and freezing precipitation in the winter in the mountain regions. Increases in diseases, insect damages, and other meteorological hazards result in a shortened life for forest trees (Tsunekawa et al. 1996). When the original species in a forest decline, other species would dominate. Forests of the Hokuriku region on the Japan are likely to respond to climatic change (Kojima 1996).

The Mountains of Central Asia consists of two of Asia's major mountain ranges, the Pamir and the Tien Shan stretching southern Kazakhstan, most of Kyrgyzstan and Tajikistan, eastern Uzbekistan, western China, northeastern Afghanistan, and a small part of Turkmenistan. Mountains of southwest China hosts golden monkey, giant panda, and red panda.

Stretching over 3,000 km of northern Pakistan, Nepal, Bhutan and India, some of the world's major rivers, Ganges, Indus, Brahmaputra, Yangtze, Mekong, Salween, Red River (Asia), Xunjiang, Chao Phraya, Irrawaddy River, Amu Darya, Syr Darya, Tarim River and Yellow River, rise in the Himalayas. These rivers flow through Afghanistan, Bangladesh, Bhutan, People's Republic of China, India, Nepal, Burma, Cambodia, Tajikistan, Uzbekistan, Turkmenistan, Kazakhstan, Kyrgyzstan, Thailand, Laos, Vietnam, Malaysia and Pakistan. The hotspot hosts numerous large birds and mammals, including vultures, tigers, elephants, rhinos and wild water buffalo.

Japan's biodiversity is unique due to its location at the intersection of three of the Earth's tectonic plates. The slippage of these plates generates forces that result in numerous volcanoes, hot springs, mountains and earthquakes. Japan's vegetation includes subtropical broadleaf evergreen forests and mangrove swamps.

Philippines which is a cluster of more than 7,100 islands hosts 6,000 plant species and many birds which is under constant thread due to pressure from anthropogenic activity.

The world's most diverse mountain forests are found among the peaks and valleys of the Eastern Himalayas (CEPF 2007). Rapid variations of altitude over relatively short horizontal distances have resulted in wealthy biodiversity with a wide variety of plants and animals, frequently with sharp change in vegetation sequences (ecotones), ascending into barren land, snow, and ice. The abundance of terrestrial ecosystems in the region is grouped into 25 ecoregions (WWF 2004). Eastern Himalayas are vulnerable to climate change due to its ecological fragility and economic marginality (Tse-ring et al. 2010).

The migration route and migration pattern such as ribbon fish, small and large yellow croakers, could be affected by climate change (Su and Tang 2002; Zhang and Guo 2004). Change in rain pattern could affect the migration of fish from the river to the floodplains for spawning, dispersal and growth (FAO 2003). A great number of plant and animal species are reported to be moving to higher latitudes and altitudes as a due to climate change in parts of Asia (Yoshio and Ishii 2001; IUCN 2003). Changes in the flowering date of Japanese Cherry; change in alpine flora in Hokkaido and other high mountains; and the change in the distribution of southern broad-leaved evergreen trees have been reported (Oda and Ishii 2001; Ichikawa 2004; Kudo et al. 2004; Wada et al. 2004). Further the coral reefs of Southeast Asia are the most seriously threatened (Hoegh-Guldberg et al. 2009).

Changing climate is increasing the risk of floods, landslides and severe storms in some parts of in other areas. Rising sea levels are resulting in storm surge and inundation of fresh water supplies. Damage to coastal vegetation from storms and wild fires are damaging barriers to erosion. The current rate of sea-level rise in Asia is reported to be between 1 and 3 mm/yr which is slightly greater than the global average (Dyurgerov and Meier 2000; Nerem and Mitchum 2001; Antonov et al. 2002; Arendt et al. 2002; Rignot et al. 2003; Woodworth et al. 2004).

Changing climate has also brought changes in active layer thickness (ALT) over permafrost in cold regions of Asia. Monitoring of the ground thermal regime in Central Asia has revealed that the permafrost has been undergoing considerable change due to climate change (Lin et al. 2010). Studies conducted by Wu and Zhang (2010) revealed that ALT had no or very limited change from 1956 to 1983 and a sharp increase of ~39 cm from 1983 to 2005.

Global warming is likely to alter the distribution, reproduction and life history of living organism. Long-term studies have shown that higher surface air temperatures are associated with a lower body size of ungulates (Post et al. 1997). Warmer temperature will result in earlier spawning by amphibians (Beebee 1995). Studies have shown increase in effects of climate change in bird populations (Crick et al. 1997; Montevecchi and Myers 1997; Winkel and Hudde 1997; McCleery and Perrins 1998; Brown et al. 1999). Climate change has resulted in abnormalities in pollination, fertilization and difficulty in regeneration in Uttrakhand Himalyan region. Tourism has been blamed for this negative effect on biodiversity.

The Mediterranean region comprises of enclosed sea bounded by Europe, Africa and Asia. The area is characterized by a relatively high degree of biological diversity comprising rocky intertidal estuaries, and sea grass meadows. The continental shelf is very narrow and the coastal marine areas are rich ecosystems characterized by many endemic species higher than that of the Atlantic Ocean.

Increase in temperature is likely to augment growth of species up to certain extent due to increase metabolism. This in crease in metabolism demand more nutrients due to increase in metabolism. Changes in metabolism are species dependent and beyond certain level of temperature species may contract decease due to increased activity of pathogens. The warming up of climate leads to dehydration in flora and drying in fauna within leading to death and change in population at different levels of food chain. Population of microbial species which is critically depends on optimum temperature may perish or adopt the new regime.

Warming of water bodies is likely to augment growth of plankton up to certain extent due to increase metabolism. Increase in plankton population may decrease visibility and affect respiration of higher aquatic animals. Change in population of micro organism and plankton will affect food web. Algal blooms may cause negative impacts to other organisms via production of natural toxins, mechanical damage to other organisms. High winter flows associated with early melting can scour the streambed and destroy eggs of some aquatic species.

Change in raining pattern is likely to increase growth up to certain extent. But increase growth is species dependent. Some species like Pepper may yield more up to certain extent of rain due to increase in pollination. Some species like mango may shed the flowers due to heavy rains in flowering season. Fauna population may react differently depending on species. Some vectors and insects may bread more due to availability of good breeding places. There would be increase in vector born decease and change in population at different levels of food chain. Excessive rain will result in death of species due to injury and difficulty in access to food as well as shelter. The precipitation leads to variation of ground water table and surface water levels thus leading to flood or draught and subsequent impact on species depending on these water sources. The increase in surface water will increase population of aquatic animals some extent depending on the region but too much rain may disrupt aquatic ecosystem.

In addition to climate driven factors any area will have certain pollution and the effect of pollution depends on type of species of living organism and pollutants. Pollutants may add to fertility for some extent due to increase in oxides of nitrogen in air and subsequent entry into soil. Land contamination and water pollution may lead to decease and death of flora fauna depending on type of pollution. Noise pollution will have grater impact on fauna due to sensitivity of fauna to vibration and energy. Radiation due to radio active pollutants may lead to destruction/mutation of cells in flora and fauna. Thermal pollution may deter quality of water affecting flora and fauna.

The environment will not get affected discretely due to pollution or climate change. It gets affected by region specific factors like wild life poaching and hunting, deforestation, conflicts, corruption, misuse of land and resources. Earthquake may cause destruction. Emission form volcano may cause disruption of activities of birds. Volcanic ash may disrupt animal health and increase temperature of water and add suspended particles along with heavy metal to water. Tsunamis in the region will have negative impact by destruction and increase in salinity of freshwater. Tsunami will increase salinity of soil and affect growth of flora.

Thus policy legislation and coping mechanism can not be formulated discretely to climate as the disaster and climate impact is always associated with regional factors and pollution.

3.2 Climate Change and Asian Economics

Economics is the study of production, distribution and consumption of goods as well as services. An economic system of a country or region is referred as economy. The economy of Asia differs widely between, and within, states due to range of differing cultures, environments and government systems.

About 6,000 million lives in Asia comprising of 60% of the world's population with roughly 25% of the global domestic product. Asia witnessed extraordinary economic development in the region with China, India, Malaysia, the Philippines, Indonesia, Thailand and Vietnam posting healthy rates of economic growth in last decade. Asia includes mature economic centre as well as the native cultures of developing countries which live in isolation from the rest of the world. Increasing population, globalization and the information technology revolution has resulted in higher demands for energy, mobility, communications and other goods as well as services. The scope and magnitude of the environmental consequences of this change has lead to increase in air pollution at local and regional levels. The problem of air pollution is not confined only at local scale and encompasses complex interlinkages with respect to haze, smog, ozone as well as global warming.

International Panel on Climate Change (IPCC 2007) 4th Assessment Report indicated that seas may rise by at least one meter by the end of this century. Sea level rise could have devastating impacts on coastal ecosystems such as coral reefs, mangroves and sea grass beds, as well as inundating coastal communities.

People around the world are affected by climate change and urban areas are particularly vulnerable. Climate change is expected to affect through rising sea levels, inland flooding, temperature extremes, and other climatological, meteorological disasters. China and India witnessing rapid economic growth are also the leading producers of rice in the world. Most of the rice harvest comes from irrigated agriculture in the Ganges, the Yangtze, and the Yellow River basins. However, Stern Report on 'The Economics of Climate Change' indicates "China's human development could face a major 'U-turn' by this mid-century unless urgent measures are now taken to 'climate proof' development results" (Stern 2006).

Due to extent of illiteracy, poverty and low standards of living in numerous Asian countries, huge efforts have to be made to increase average living conditions.

Climate change is a potentially disastrous global externality and one of the world's greatest combined action problems. The distribution of causes and effects is highly uneven across regions and across generations. Massive ambiguity surrounds existing estimates of future indemnity that may result from climate change, but these possible damages are to a considerable level irreversible and may be calamitous if global warming is unimpeded. The costs of climate change reduction also have a depressed component – that is, cannot be fully recovered – and are dependent on a huge number of factors, including the rate at which the global economy grows over the long term and the rapidity at which low-emission technologies come out and spread across the global economy.

Table 3.5 shows impact of climate change on different sectors and possible adaptation measures. Damages from climate change and the expenses to reduce them across generations have significant implications on policy options.

A reasonable rise in temperature increases agricultural productivity in region with low initial temperatures, but decreases it in hotter region. Likewise, warming reduces deaths in region with initially colder climates, but increases mortality and morbidity in region with warmer climates. Increase in temperature is likely to increase in electricity consumption due to increased demand by Air Conditioner and refrigeration. The warming also increases consumption of ice cream, beer and cool drinks. Expansion of railway lines, electric cables and metal structure may require additional maintenance to overcome changing environment. Melting of bituminous roads is likely to demand additional maintenance and new layer of pavement. The warning of globe will impact sericulture and crops which are sensitive to temperature.

Although warming reduces expenditures on winter heating in regions with an initially cooler climate, such region may invite added expenditures on summer cooling. Regions with initially warmer climates also incur additional costs for cooling. Cost of heating and cooling depends on affordability of citizens and extent of temperature rise. Citizens with lower income prefer to burn some waste on streets and warm themselves generating some more pollution but eventually with direct economic burden on country. The subsequent impact on health on poor due to air pollution will have to be born by countries.

Change in precipitation will result in loss of crop/yield depending on intensity. Change in raining pattern will affect duration and timings with respect to crop cycle due to change in sowing date and other phonological reasons. Washout of pesticides may incur additional burden. Increase in commutable disease in the areas of poor sanitation may demand additional medicine and medical attention. Increase in intensity of rain will result in overflow of sewers and disruption of, solid waste management. Extended rains in harvest season may create hard ship in drying of food grains in sun. Rains during harvest of fruits may result in rotting and damage to crops. Low-income and small countries are more vulnerable to natural disasters because of more limited geographical and economic diversification.

Coastal zones in semi arid region of Asia are not significantly affected by sea-level changes (IPCC 1996). Many economic activities of countries in the region are

Table 3.5 Impact of climate change on different sectors and possible adaptation measures

Sector	Description	Examples	Impact	Adaptation measures
Primary sector	Changing natural resources into primary products	Agriculture, agribusiness, fishing, forestry and all mining and quarrying industries	Even though total output in terms of currency may increase due to price adjustment, artificial increase in prices of goods due to middlemen, overall productivity in terms of quantity may come down with respect to Agriculture, Agribusiness, fishing, forestry, etc. Open mining and quarrying may get affected by heavy rains	• Use of more tolerant crop varieties with respect to heat/drought, disease, pest, salt • Cold regions shall introduce higher yielding, earlier maturing crop varieties in cold regions • Improve application of agrochemicals • Change sowing dates effectively • Breed livestock for greater tolerance and productivity • Encourage vegetarianism • Breed fish tolerant to climate change • Improve irrigation systems and their efficiency • Adopt rain and snow water harvesting • Improve information exchange system • Improve preparedness and management system of natural calamities like, draught, flood, hurricane, and cyclone • Shift cultivation from non-food crops like tobacco, flowers to food items

Secondary sector	Create a finished, usable product: production and construction	Aerospace manufacturing, automotive industry, brewing industry, chemical industry, textile industry, consumer electronics, energy industry, steel production, tobacco industry	Some manufacturing activity like ice-cream, beer, cool drinks, air conditioner, and refrigerator may get boost due to increased demand. Activities like solar cells and machinery for non conventional energy will get uplift due to inclinations to adopt this technology to avoid carbon emission	• Be prepared for the demand • Increase efficiency of power transmission • Minimise waste
Tertiary sector	Servicing	Government, healthcare/hospitals, public health, waste disposal, education, banking, insurance, financial services, legal services, consulting, news media, hospitality industry (e.g. restaurants, hotels, casinos), tourism, retail sales, franchising, real estate	Some sectors like health care will get a boost due to increased diseases; Waste disposal may get attention to avoid carbon emission. Hospitality industry will perform depending on the location and service. Heavy rains and temperature increase may affect tourism	• Improve information exchange system • Improve sea defense and flood management • Prepare disaster management plans and train personnel • Evolve vaccination for new diseases • Improve waste management practice

oil production which is mostly land-based and is not as vulnerable to the impacts of climate change. Transportation of oil also is less affected in these areas.

In addition to climate change impacts, Asian countries are daunted by pollution which demands more medicine and medical attention to citizens. The pollution has lead to acid rain and atmospheric brown cloud in many areas. Economy is already affected due to more cost for water treatment and increased consumption of packaged water.

Many parts of Asia are affected by regional factors like communal riots and international conflict where local disasters will bring in additional disruption to economy. Corruption may increase in economic activities resulting in sort term benefits but will have long term negative consequences due to construction of poor quality infrastructure which demands more resources to repair and maintenance.

The gross effect of climate change, pollution regional factors will result in short term benefits due to increased economic activity resulting in sale of medicine; maintenance of infrastructure; sale of liquor; and ice cream, etc., but in long term, its impact would be negative due to decrease in manpower; demand for additional infrastructure; and ill health of people due to increased alcoholic and cholesterol rich food.

Unusual snowfall in February 2007 and hailstorms in summer destroyed potato crops in Nepal twice in a row (Inforesources focus 2008). Climate impacts have become more frequent in other potato growing regions of the world. While in some regions the crop is destroyed by hails and snows in other regions it is being destroyed by drought. This crop is of invaluable importance to diets and livelihoods of people worldwide with China as main having a crop yield of 71 million tonnes, accounting to over 20% of global production.

The rise in temperature will result in forced maturity and poor harvest index because of limited water supply (Yadav et al. 1987). Further rise in temperature will decrease in crop life span and grain yield in food grains (Mavi and Chaurasia 1974; Bagga and Rawson 1977; Singh et al. 1991; Wardlaw 1970; Wardlaw et al. 1989; Hundal et al. 1993).

Yield of major crops are expected to reduce in China whereas rice production may likely to be affected in India due to heat stress. Crop diseases for major cereals could become more widespread in whole Asia due to warmer and wetter Asian climate. Studies indicated that a 0.5°C rise in winter temperature will reduce wheat yield by 0.45 tonnes/ha in India (Kalra et al. 2003).

The health of communities will be deteriorated due to climate change. Riverside community will be affected during floods due to increase in diarrhea and malaria. People would be affected during post-cyclone due to transfer of pathogens and damage to water treatment facility. The other health problem due to change in climate will be stress due to hot waves and cold waves. The changing climate will also spread communicable diseases faster. The extreme events will cause injury to people and animals.

The impacts due to climate change are superimposed on a range of other environmental and social stresses, many already recognised as severe (Ives and Messerli 1989). Himalayan region is the source of ten major rivers – Amu Darya,

Brahmaputra, Ganges, Indus, Irrawaddy, Mekong, Salween, Tarim, Yangtze and Yellow. The flow of these rivers will vary due to change in climate in Himalayas.

3.3 Transcending the Tensions Between Climate and Development

The meaning of word transcending means surpassing. There is always a conflict between the people whether to choose safer climate or development. Climate change in recent years has been result of development. But this does not mean that one has to stop developing for sake of safer climate. The climate has been sever and detrimental prior to industrial revolutions as well and many civilizations perished due to climate change.

Development includes good governance, health care, education, disaster preparedness, infrastructure, gender equality and environment protection. Disasters bring in loss of human lives, livelihoods, infrastructure, and environment. More than 70 of death due to disaster occur is Asia. Natural hazards have disastrous in the due to unplanned settlements, unsafe building practices, poor economic conditions, corruption, etc.

The processes of adaptation are highly complex and dynamic. Development need not always end up in air conditioned buildings and cars everywhere. Development can also address increased life expectancy; decrease in social and economic imbalance; descent disaster resistant safe house; access to better health care; access to better education; and access to better food. Disparity in income has been source of crime in many places and disparity in income is associated with poor education and discrimination of greater number of people by few people in power and wealth. Forests are greatly affected by climate, soil, fire, industry, tourism, and construction. Hence expected impact of global warming on forests would vary from region to region.

Surpassing problems due to climate change not only needs technology but also need strategic intervention from policy makers. Planning without considering possible climate change will result in costs beyond initial estimated cost of project. Adaptation to climate changes is not a new concept. Human societies have adapted to different climates in the past. Success of adaptation depends on the availability of finance, knowledge and technical capability.

One school of thought is that population in poor countries is typically more vulnerable to climate change because of lower per capita income, limited availability of public services (such as health care). However, this argument is not always true as cost of commodity vary from country to country. The population near equator whose ancestors lived for long periods in the regions has larger quantities of *eumelanin* pigment in their skins making their skins dark and protects them against high levels of exposure to the sun. Figure 3.2 shows natural disaster summary 1900–2009 and Fig. 3.3a, b shows average annual damage caused by

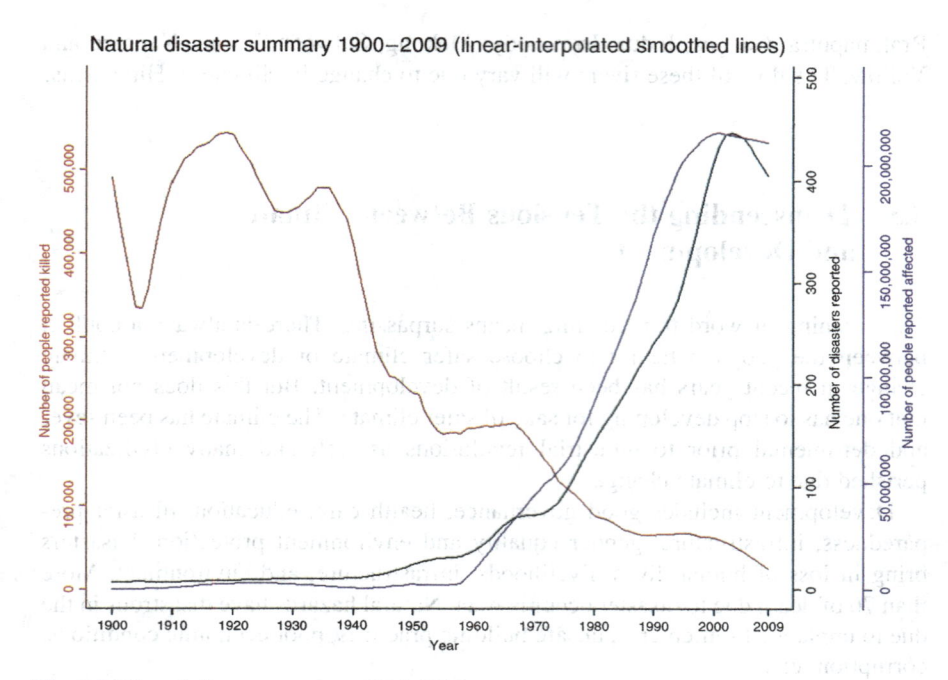

Fig. 3.2 Natural disaster summary 1900–2009

reported natural disaster 1900–2009. As evident, the Africa has the least economic losses in the period 1900–2009 possibly due to low cost of buildings and infrastructure created in the continent, where as Asia recorded highest damage followed by America and Europe.

Two main areas of uncertainty persist still today. The first area is the limitation of current scientific knowledge about climate change and the second is how best to quantify the economic impact of climate change (WEO 2008).

Recent studies imply that considerable decreases in cereal production potential in Asia could be likely by the end of this century due to climate change (Cruz et al. 2007). However, regional differences with respect to wheat, maize and rice yields due to climate change could likely be significant (Parry et al. 2004; Rosenzweig et al. 2001). Studies shows that crop yields could increase up to 20% in East and South-East Asia while it could decrease up to 30% in Central and South Asia (Cruz et al. 2007).

The climate changes that could affect precipitation distribution will eventually affect runoff to rivers and lakes in Philippines (Aida and Nathaniel 1999). The Philippine archipelago comprising of 7,100 islands, clustered in three major island groups with a total land area of 300,000 km^2 is influenced substantial amounts of rains almost all year round. Uneven distribution of rain has resulted in extreme events such as floods and droughts, the country experienced imbalances in water supply and demand (Jose et al. 1993).

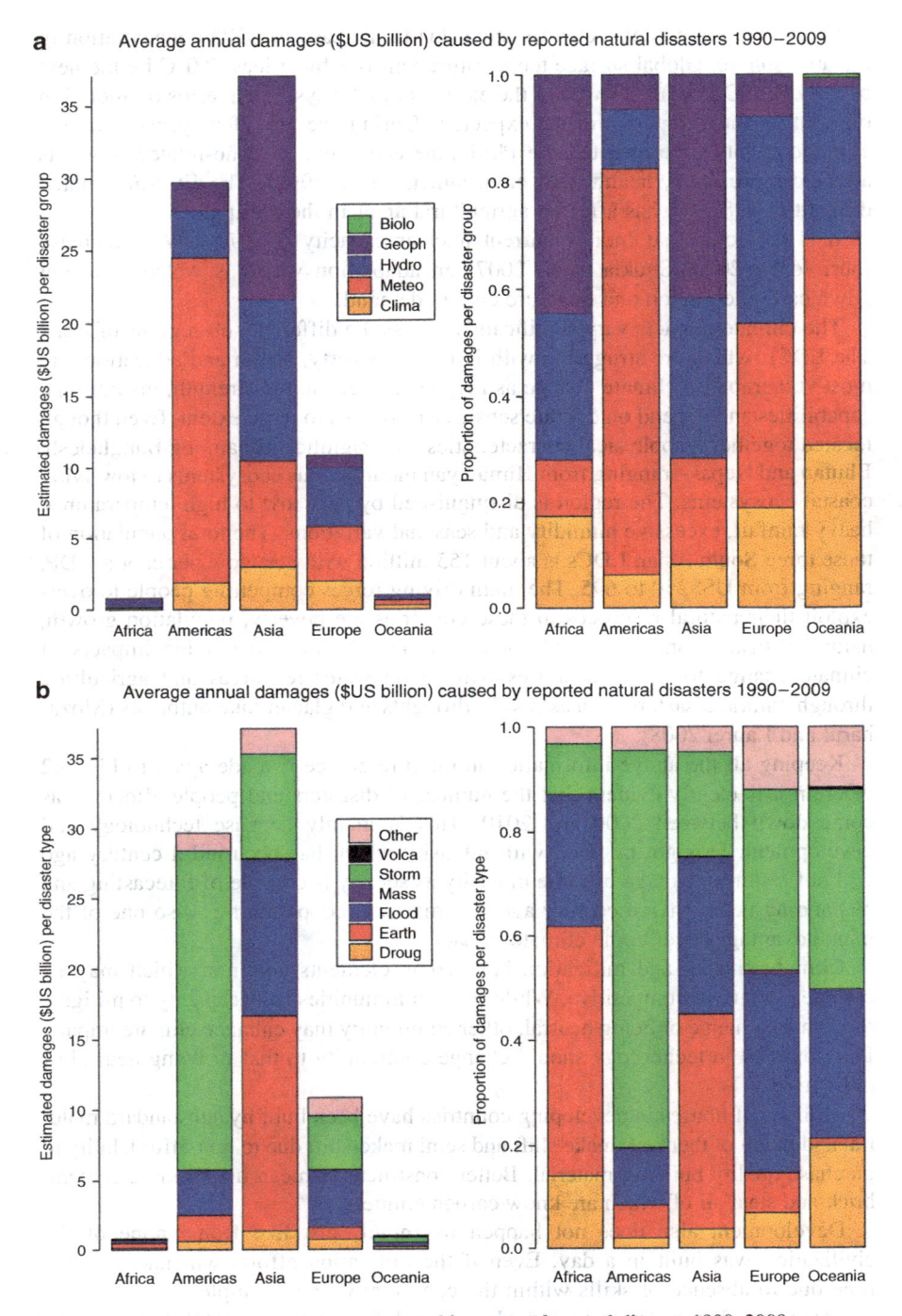

Fig. 3.3 (**a**, **b**) Average annual damage caused by reported natural disaster 1900–2009

Global warming is expected to occur due to increase in GHG concentration in the atmosphere. Global surface temperature will rise by at least 2.0°C by the next century (IPCC 1996). Changes in the earth's climatic system in terms of alteration of rainfall and temperature are expected. During the past few years, extreme climatic events have affected the Philippine economy. El Niño-related droughts affected agriculture, health, and environment (Jose 1992). The El Niño-related drought of 1982–83, has affected agricultural areas in the Philippines.

It is projected that India's current nuclear capacity of 2.72 GW in increase fourfold by 2020 (Shukla et al. 2007) an adaptation which is welcome to cut down carbon emission and to assure energy demand.

The climate impacts vary significantly across the different sub-regions of Asia. The LDCs, which are struggling with issues of poverty, health and education, are most vulnerable to climate change as they lack of economic strength, institutional capabilities and depend on climate sensitive resources to large extent. Even though located together, geophysical characteristics vary significantly among Bangladesh, Bhutan and Nepal – ranging from Himalayan mountainous ecosystems to low-lying coastal ecosystems. The region is distinguished by very low to high temperatures, heavy rainfall, excessive humidity and seasonal variations. The total population of these three South Asian LDCs is about 155 million with very low per capita GDP, ranging from US$250 to 695. The main driving forces compelling people to over-exploit their natural resources in these countries are poverty, population growth, natural disasters, and low levels of education. The most damaging impacts of climate change for these countries will be on water resources and agriculture through natural disasters such as floods, droughts and glacier lake outbursts (Mozaharul and Laurel 2008).

Keeping all the above information in mind reference is made again to Fig. 3.2 wherein it is clearly evident that the number of disasters and people affected has come down between 2000 and 2010. This is mainly because technology and development. Drought or flood with intensity which has occurred a century ago will not result in damage of same intensity as society is capable of forecasting and responding better than a century ago. International cooperation is also one of the main advantages over come climate impact.

Climate change and mitigation have some elements common which may be complementary or competitive. While some communities have capacity to mitigate the climate change or being neutral, other community may enhance climate impact. Development in technology should change community to that of being neutral or mitigative.

Millions of houses in developing countries have been built by substandard material and many of them are make shift and semi make-shift due to non-affordability to purchase quality building material. Better constructions mean use of more cement; brick and steal all of which are know carbon emitters.

Development also does not happen overnight. Just like Rome none of the civilization was built in a day. Even if the mitigation efforts will take its own time due to absence of skills within the community. For example even though it s known fact that metro train and Mass Rapid Transit System (MRTS) do bring down the emissions, the country can't implement it as it has to depend on external expertise. Other hindrance includes non agreement of data, pacts, and soft laws.

There is urgent need to large emission reduction, but national circumstances and performance of individual nations vary considerably. Future GHGs emission and driving forces influencing them are highly uncertain; hence all mitigation scenarios do assume certain technical change in coming years.

In summary, as highlighted by Umesh (2009), the increasing consumption of earth resources like fossil fuel, mining, pumping out ground water, etc. is creating disturbances inside earth and in earth's atmosphere. These disturbances in earth systems will attract vulnerable events like earthquakes, tsunami, cyclones, etc. Regions prone to these vulnerable events will be devastated which will force people to migrate to safer places. On the other hand, due to global warming and melting of ice, regions like Scandinavia and Russia will get more agricultural land and greenery. This will attract migration of people towards northern parts of these countries. Even people from other parts of Europe, Africa and Asia may shift towards these areas in due course of time. Another factor affecting migration will be sinking of islands due to sea level rise. Depletion of ground water will also be major factor affecting migration to other places within region or in some cases out of the region (if severity increases). In tropical region like south Asia, glaciers are melting; rivers are not having enough water to recharge the ground water. The uncertainty of monsoon rains is increasing. The rain water is not properly trapped to recharge the ground water; most of it goes as run off. Very soon in these regions, conditions will be so that people will be forced to leave agriculture and migrate to some other places. This will also affect the agricultural patterns and population distribution.

References

Aida M. Jose, Nathaniel A. Cruz, 1999: Climate change impacts and responses in the Philippines: water resources, Climate research, Vol. 12, PP 77–84

Antonov J.I, Levitus S and Boyer T.P, 2002: Steric sea level variations during 1957–1994: Importance of salinity. J. Geophys. Res., **107**, 8013, doi:10.1029/2001JC000964.

Arendt A.A, Echelmeyer K.A, Harrison W.D, Lingle C.S and Valentine V.B, 2002: Rapid wastage of Alaska glaciers and their contribution to rising sea level. Science, **297**, 382–386.

Bagga A.K. and Rawson H.M. 1977: Contrasting response of morphologically similar wheat cultivars to temperature appropriate to warm temperature climates with hot summers : A study in controlled environments. Australian Journal of Plant Physiology **4**: 877–887.

Brown J.L, Li S.H, Bhagabati N, 1999: Long-term trend toward earlier breeding in an American bird: a response to global warming ? Proc.NatlAcad.Sci.USA 96, 5565–5569.

Beebee T.J.C, 1995: Amphibian breeding and climate. Nature 374, 219–220.

Bryceson D.F, 2004: Agrarian vista or vortex: African rural livelihood policies. Review of African Political Economy, **31**, 617–629.

CCCD (Commission on Climate Change and Development), 2009: Closing the Gaps: Disaster risk reduction and adaption to climate change in developing countries.

CEPF, 2007: Ecosystem profile: Indo-Burman Hotspot, Indo-China Region. Arlington: WWF, US-Asian Program/CEPF

Crick H.Q.P, Dudley C, Glu D.E, Thomson D.L, 1997: UK birds are laying eggs earlier. Nature 388, 526.
Cruz R.V, Harasawa H, Lal M, Wu S, Anokhin Y, Punsalmaa B, Honda Y, Jafari M, Li C, Huu Ninh N, 2007: Asia. Climate Change 2007: Impacts, Adaptation and Vulnerability. Contribution of Working Group II to the Fourth Assessment Report of the Intergovernmental Panel on Climate Change, [Parry M.L, Canziani O.F, Palutikof J.P, van der Linden P.J and Hanson C.E, Eds.], Cambridge University Press, Cambridge, UK, 469–506.
Dyurgerov M.B. and Meier M.F, 2000: Twentieth century climate change: evidence from small glaciers. P. Natl. Acad. Sci. USA., 97, 1406–1411.
FAO (Food and Agriculture Organization), 2003: World Agriculture: Towards 2015/2030. An FAO Perspective. Bruinsma, Ed., FAO, Rome and Earthscan, London, 520 pp
Hoegh-Guldberg O, Hoegh-Guldberg H, Veron J.E.N, Green A, Gomez E. D, Lough J, King M, Ambariyanto, Hansen L, Cinner J, Dews G, Russ G, Schuttenberg H.Z, Peñafl or EL, Eakin C. M, Christensen T.R.L, Abbey M, Areki F, Kosaka R.A, Tewfi k A, Oliver J, 2009: The Coral Triangle and Climate Change: Ecosystems, People and Societies at Risk. WWF Australia, Brisbane, 276 pp.
Hundal S.S, Kaur P, Singh G, Singh R, 1993: Simulated rice and wheat yields in Punjab (India) under changing climate scenarios. Proceedings of the Indo-German Conference on Impact of Global Climatic Changes on Photosynthesis and Plant Productivity held at HAU, Hisar on Dec. 1–3,
Ichikawa A, 2004: Global Warming – The Research Challenges: A Report of Japan's Global Warming Initiative. Springer, 160 pp.
Inforesources Focus, 2008: Potatoes and Climate Change, No. 1/08
IPCC (Intergovernmental Panel on Climate Change), 1996, Climate change 1995: the IPCC second assessment report, Vol 2: scientific-technical analyses of impacts. Adaptations and mitigation of climate change. Watson R.T, Zinyowera M.C, Moss R.H (eds). Cambridge University Press, Cambridge
Ives J.D, Messerli B, 1989: The Himalayan Dilemma: Reconciling Development and Conservation. London: John Wiley and Sons
IPCC, 2001: Climate Change 2001: Impacts, Adaptation, and Vulnerability. Contribution of Working Group II to the Third Assessment Report of the Intergovernmental Panel on Climate Change, McCarthy J.J, Canziani O.F, Leary N.A, Dokken D.J and White K.S, Eds., Cambridge University Press, Cambridge, 1032 pp.
IPCC, 2007: Synthesis Report. Contribution of Working Groups I, II and III to the Fourth Assessment Report of the Intergovernmental Panel on Climate Change. In: Team C.W, Pachauri R.K, Reisinger A (eds). Intergovernmental Panel on Climate Change, 2008, Geneva, Switzerland 104
IUCN (International Union for Conservation of Nature), 2003: Indus Delta, Pakistan: economic costs of reduction in freshwater flows. Case Studies in Wetland Valuation No. 5, Pakistan Country Office, Karachi, 6 pp. Accessed 24.01.07: www.waterandnature.org/econ/CaseStudy05Indus.pdf.
Jariwala C.M, 1980: The Constitution 42nd Amendment Act and the Environment, in Legal Control of Environmental Pollution, N.M. Tripathi Private Ltd., Bombay, p. 2.
Jose A.M, 1992: Preliminary assessment of the 1991–1992 ENSO-related drought event in the Philippines. PAGASA, Quezon City
Jose A.M, Francisco R.V, Cruz N.A, 1993: A preliminary study on the impact of climate variability/change on water resources in the Philippines. PAGASA, Quezon City
Kalra N, Aggarwal P.K, Chander S, Pathak H, Choudhary R, Choudhary A, Mukesh S, Rai H.K, Soni U.A, Anil S, Jolly M, Singh U.K, Owrs A and Hussain M.Z, 2003: Impacts of climate change on agriculture. Climate Change and India: Vulnerability Assessment and Adaptation. Shukla P.R, Sharma S.K, Ravindranath N.H, Garg A and Bhattacharya S, Eds., Orient Longman Private Ltd., Hyderabad, pp.193–226.

Kojima S, 1996: Global climatic warming and possible responses of vegetation of Hokuriku region. Japan. Jour. of Phytogeography and Taxonomy, 44, 9–18.

Kudo G, Nishikawa Y, Kasagi T and Kosuge S, 2004: Does seed production of spring ephemerals decrease when spring comes early? Ecol. Res., 19, 255–259.

Leatherman S.P, 1997: Beach ratings: a methodological approach. Journal of Coastal Research, 13, 1050–1063.

Lin Zhao, Qingbai Wu, Marchenko S.S, Sharkuu N, 2010: Thermal State of Permafrost and Active Layer in Central Asia During the international polar year. Volume 21, Issue 2, Pages 198–207, April/June 2010. DOI:10.1002/ppp.688

Maul G.A, 1996: Marine Science and Sustainable Development. Coastal and Estuarine Studies, America Geophysical Union, Washington, DC, US 467 pp.

Mavi H.S & Chaurasia R, 1974: Influence of meteoro-logical factors on the phenology of wheat in the Punjab. Indian Journal of Ecology 1: 17–23.

McCleery R. H. & Perrins C. M. 1998: Temperature and egg-laying trends. Nature 391, 30–31.

Mozaharul Alam and Laurel A. Murray, 2008: Facing Upto Climate Change In South Asia, SAARC Workshop Climate Change and Disasters, PP 29–40

Montevecchi W.A. & Myers R.A, 1997: Centurial and decadal oceanographic influences on changes in northern gannet populations and diets in the northwest Atlantic: implications for climate change. ICES J. Mar. Sci. 54, 608–614.

Nerem R.S. and Mitchum G.T, 2001: Observations of sea level change from satellite altimetry. Sea Level Rise: History and Consequences, B.C. Douglas, M.S. Kearney and S.P. Leatherman, Eds., Academic, San Diego, California, 121–163.

Oda K. and Ishii M, 2001: Body color polymorphism in nymphs and ntaculatus (Orthoptera: Tettigoniidae) adults of a katydid, Conocephalus. Appl. Entomol. Zool., 36, 345–348.

Parry M.L, Rosenzweig C, Iglesias A, Livermore M and Fischer G, 2004: Effects of climate change on global food production under SRES emissions and socio-economic scenarios. Global Environ. Chang., 14, 53–67.

Post E, Stenseth N.C. & Fromentin J.M. 1997: Global climate change and phenotypic variation among red deer cohorts. Proc. R. Soc. Lond. B 264, 1317–1324.

Rosenzweig C, Iglesias A, Yang X.B, Epstein P.R and Chivian E, 2001: Climate change and extreme weather events: implications for food production, plant diseases and pests. Global Change and Human Health, 2, 90–104.

Rignot E, Rivera A and Casassa G, 2003: Contribution of the Patagonia ice fields of South America to sea level rise. Science, 302, 434–437.

Singh D.P, Singh P, Pannu R.K & Sharma H.C, 1991: Carbon dioxide enrichment, climate change and Indian agriculture : A preliminary analysis. Proceedings of Symposium on Impact of Global Climatic Changes on Photosynthesis and Plant Productivity. ICAR, New Delhi pp. 279–296.

Shukla P.R, Garg A, Dhar S and Halsnaes K, 2007: Balancing Energy, Development and Climate Priorities in India: Current trends and future Projections. UNEP Risoe Centre on Energy, Climate and Sustainable Development, Roskilde, Denmark. ISBN:378-87-550-3627-7. September 2007. 39

Stern N, 2006: The Economics of Climate Change. London: Cambridge University Press

Su J.L and Tang Q.S, 2002: Study on Marine Ecosystem Dynamics in Coastal Ocean. II Processes of the Bohai Sea Ecosystem Dynamics. Science Press, Beijing, 445 pp.

Tse-ring K, Sharma E, Chettri, N, Shrestha, A (eds), 2010: Climate change vulnerability of mountain ecosystems in the Eastern Himalayas; Climate change impact an vulnerability in the Eastern Himalayas – Synthesis report. Kathmandu: ICIMOD

Tsunekawa A, Ikeguhi H, and Omasa K, 1996: Prediction of Japanese potential vegetation distribution in response to climatic change. In: Climate Change and Plants in East Asia [Omasa, K., K. Kai, H. Toda, Z. Uchijima, and H. Yoshino (eds.)], Springer-Verlag, Tokyo, Japan, pp. 57–65

Ulziisaikhan V, 1996: Impact Assessment of Climate Change on Forest Ecosystem in Mongolia. Paper presented at the Regional Workshop on Climate Change Vulnerability and Adaptation in Asia and the Pacific, January 15–19, 1996, Manila, Philippines, pp. 1–10.

Umesh Kulshrestha, 2009: <http://www.icsu-visioning.org/tag/extreme-events> retrieved on 25, September 2010.

UNESCO, 1988: Man Belongs to the Earth: UNESCO's Man and the Biosphere Programme

UN-OHRLLS (Office of the High Representative for the Least Developed Countries, Landlocked Developing Countries and Small Island Developing States), 2009: The Impact Of Climate Change On The Development Prospects Of The Least Developed Countries And Small Island Developing States.

Wada N, Watanuki K, Narita K, Suzuki S, Kudo G and Kume A, 2004: Climate change and shoot elongation of Alpine Dwarf Pine (Pinus Pumila): comparisons among six Japanese mountains. 6th International Symposium on Plant Responses to Air Pollution and Global Changes, Tsukuba, 215.

Wardlaw I.F, 1970: The early stages of grain development in wheat : response to light and temperature in a single variety. Australian Journal of Biological Sciences 23: 765–774.

Wardlaw I.F, Dawson I.A & Munibi P, 1989: The tolerance of wheat to high temperatures during re-productive growth. II. Grain development. Australian Journal of Agricultural Research 40: 15–24.

WEO (World Economic Outlook), 2008: International Monitory Fund, 133–184

Woodworth P.L, Gregory J.M and Nicholls R.J, 2004: Long term sea level changes and their impacts. The Sea, A. Robinson and K. Brink, Eds., Harvard Univ. Press, Cambridge, Massachusetts, 717–752.

Winkel W & Hudde H, 1997: Long-term trends in reproductive traits of tits (Parus major, P. caeruleus) and pied flycatchers Ficedula hypoleuca. J. Avian Biol. 28, 187–190.

Wu Q, and Zhang T, 2010: Changes in active layer thickness over the Qunghai-Tibetan Plateau from 1995 to 2007. J. Geophys. Res., 115, D09107, doi:10.1029/2009JD012974

WWF, 2004: Terrestrial ecoregions of the world, Version 2.0. Vector digital data. Washington DC: WWF-US

Yadav S.K, Singh D.P, Singh P & Kumar A, 1987: Diurnal pattern of photosynthesis, evapotranspiration and water use efficiency of barley under field conditions. Indian Journal of Plant Physiology 30: 233–238.

Yoshio M and Ishii M, 2001: Relationship between cold hardiness and northward invasion in the great mormon butterfly, Papilio memnon L. (Lepidoptera: Papilionidae) in Japan. Appl. Entomol. Zool., 36, 329–335.

Zhang Q and Guo G, 2004: The spatial and temporal features of drought and flood disasters in the past 50 years and monitoring and warning services in China. Science and Technology Reviews, 7, 21–24.

Chapter 4
Greenhouse Gas Inventory

The "greenhouse effect" is the warming of the Earth due to the presence of GHGs. The name "green house" is borrowed from phenomenon used in greenhouses to raise temperature capturing long wave radiation within the green house. Solar radiation from the sun absorbed by the surface of the Earth and then radiated back to the atmosphere in the form of long wave infrared radiation.

The atmosphere is fundamental to life on earth as it regulated temperature, provides air to breath, it is part of bio-geo-chemical cycles. Green house effect is a key mechanism of temperature regulation and without it earth's average temperature would not be 15°C but −6°C. Global warming is expected to result in changes in weather patterns, including an increase in global precipitation and changes in frequency as well as magnitude of storms, flood and droughts.

Ecosystems are sustained by processes that keep the Earth's atmosphere suitable for life. Greenhouse effect responsible for trapping part of the sun's heat in the atmosphere makes earth favorable for life. But, unnaturally high amounts of GHGs can trap more heat than is required; resulting in global warming, change in precipitation patterns, and sea level rise.

GHG sources can be tracked to following major categories

- Energy generation (thermal power plant, heating, etc.)
- Industry (manufacturing and use of GHGs, fertilizers, cement, brick making, etc.)
- Transport (air, water, land, space)
- Residential and commercial building (heating, cooling, cooking, etc.)
- Land use change and deforestation (cutting trees, changing location of water bodies, urbanization, etc.)
- Agriculture (rice cultivation, animal husbandry, etc.) and
- Waste and wastewater (incineration, anaerobic digestion, etc.)

The naturally occurring GHGs in the atmosphere are water vapor, CO_2, CH_4, N_2O, and O_3. Anthropogenic activity has not only increased the concentration of above GHGs but also contributed gases like Hydroflurocabons (HCFs) and Perfluorocarbons.

Incorporation of carbon into biosphere through photosynthesis and diffusion into oceans are about 110 and 2.5 billion tones carbon per year respectively resulting in increase in atmospheric carbon at the rate of 5–6.5 billion tones carbon per year in recent years.

Many of the GHGs have been phased out as part of exercise carried out during execution of Montreal protocol. For example carbon tetrachloride (CTC) an Ozone Depleting Substance (ODS) and GHG is a solvent and cleaning agent was used in many industries due to its high solvency power, low cost and non-flammable property. But it was substituted by other chemicals some of which is listed in Table 4.1 to reduce release of CTC. Due to phasing out of CTCs it manufacturing and imports were restricted and hence became costly compelling the users to switch over to substitutes.

But the practice cannot be adopted in all activities that result in emission of GHGs. Figure 4.1 shows animals waiting to die in butcher house, Fig. 4.2 depicts animal after butchering, and Fig. 4.3 shows slaughterhouse waste. The animal husbandry in India is mostly a method to convert waste into food. The animals are fed low quality food waste and then tortured in slaughter house and killed un-hygienically generating methane and CO_2 in every stage. In such scenarios methane and CO_2 from animal husbandry and slaughtering can't be phased out or reduced as long as old practice continues.

GHG inventory is required to run mathematical models. Quality of emission inventory depends on the quality of statistics available. Figure 4.4 shows relation between GHG Inventory and Policy. The output of sophisticated models operated by scientific community will become the basis for policy development and implementation. Error in GHG inventory leads to erroneous output from mathematical models and ends in wrong policy like building infrastructure where it may not be needed.

Likewise the error in mathematical model will also lead to improper decisions and may lead to wrong resource allocation. Improper policy implementation due to corruption, negligence of policy and improper allocation of resources (human, monitory, land etc) will results in consequences of likely disasters.

GHG inventory is still in evolutionary stage. 122 of 148 non-Annex I Parties of United Nations Framework Convention on Climate Change (UNFCCC) which submitted their initial national communications by 1 April 2005 had inconsistency in methodology and reporting in spite of availability of *Revised 1996 IPCC Guidelines for National Greenhouse Gas Inventories*.

With the intense understanding of the GHG inventory, IPCC has published *2006 IPCC Guidelines for National Greenhouse Gas Inventories* which can be referred for calculation of GHG.

The 2006 IPCC guidelines has five volumes – (1) General Guidance and Reporting, (2) Energy, (3) Industrial Processes and Product Use (IPPU), (4) Agriculture, Forestry and Other Land Use (AFOLU), and (5) Waste.

Respiration is not considered in carbon inventory as the amount of carbon dioxide released during respiration of living organism will not be greater than the carbon organisms put into their bodies by eating plants or animals that eat plants. The plants get the carbon from atmosphere. But burning fossil fuels is a concern because it is not a closed loop in time scales. Extracting fossil fuel and burning them

Table 4.1 List of viable alternate solvent for CTC in different industries

Sl. No	Industry	Viable alternative solvent
1	Metal degreasing	Chloroform
		Cyclohexane
		Dichloroethane
		Ethyl acetate
		Hexane
		Isopropyl alcohol
		d-Limonene
		Methanol
		Methyl ethyl ketone
		Methylene chloride
		Mineral turpentine
		N-Methyl pyrrolidone
		Perchloroethylene
		Toluene
		White petrol
		Xylene
2	Precision cleaning	Methylene dichloride
		Perchloroethylene
		Trans-1,2-dichloroethylene (trans)
		Toluene
		White petrol
3	Particulate analysis of components	Perchloroethylene (PCE)
		Isopropyl alcohol (IPA)
		Methylene Dichloride (MDC)
4	Sampling of raw material foundries	Perchloroethylene (PCE)
5	Jewellery casting	Perchloroethylene (PCE)
		Toluene
		Xylene
		Isopropyl Alcohol (IPA)
		Hexane
		Methylene Dichloride (MDC)
		Perchloroethylene (PCE)
		Toluene
6	Refrigerator system repair	White petrol
		Methylene Dichloride (MDC)
7	Oxygen system	Perchloroethylene (PCE)
		Methylene Dichloride (MDC)
		Perchloroethylene (PCE)
8	Electrical system	Isopropyl alcohol
9	Solvents in motor and generator cleaning	Mineral turpentine
		Perchloroethylene (PCE)
		White petrol
		White petrol
		n-Hexane
		Isopropyl alcohol
10	Offset printing	Methylene Dichloride (MDC)
		Perchloroethylene (PCE)
		White petrol
		Hexane
		Mineral turpentine oil

(*continued*)

Table 4.1 (continued)

Sl. No	Industry	Viable alternative solvent
11	Textiles-Stain removal	Perchloroethylene (PCE)
		Mineral turpentine oil
		NC thinner
		Ethyl acetate
12	Spinning mills	Methyl ethyl ketone
		Xylene
13	Activated carbon testing	n-Butane

Fig. 4.1 Animals waiting to die

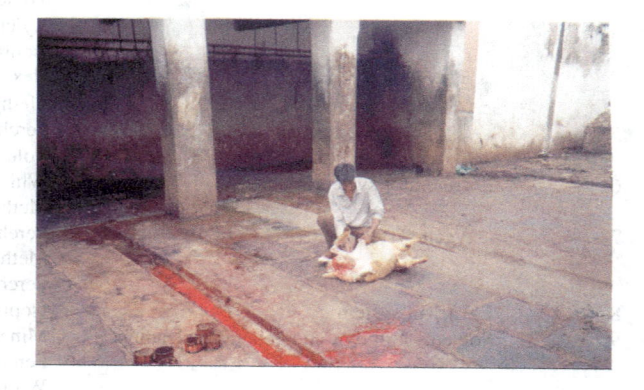

Fig. 4.2 Animal after butchering

puts back carbon into the atmosphere which was removed by plants millions of years ago. On the other hand it has been observed that cattle-raising and meat production contribute for up to one fifth of releases of reactive GHG. No inventory data is available for GHG generation from wild life some of which has similar intestinal track and digestive system similar to that of domestic animals which are known produce methane (see Box 4.1).

Fig. 4.3 Slaughterhouse
waste

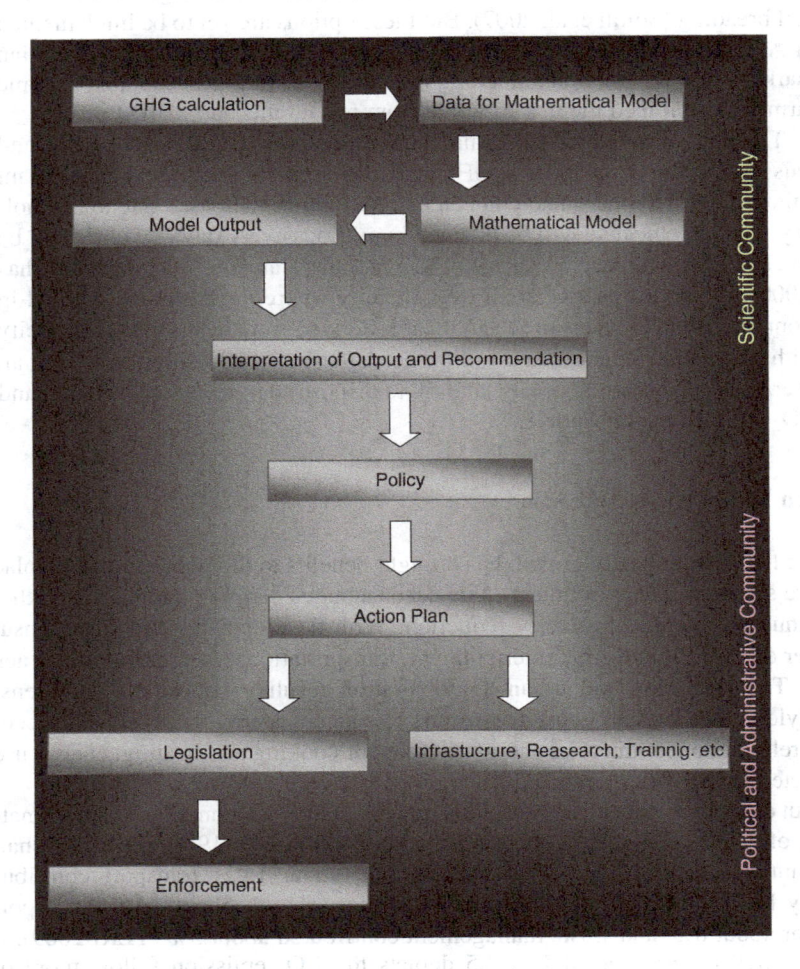

Fig. 4.4 GHG Inventory and policy

Box 4.1 Livestock and GHG

Most of the world's livestock are ruminants-such as sheep, goats, camel, buffalo and cattle which have four-chambered stomach. A ruminant is a mammal that digests food by initially softening it in one of the four chambers, then regurgitating the semi-digested mass called cud, and chewing it again. Re-chewing the cud is called "ruminating". Ruminating mammals include cattle, goats, deer, camels, antelope, sheep, alpacas, llamas, wildebeest, giraffes, bison, yaks, water buffalo, pronghorn, and nilgai. The bacteria in the stomach will break down food and generate methane a known GHG.

The mitigation options available to emission of CH_4 from ruminant are improved feeding practices; specific agents and dietary additives; vaccine against methanogenic bacteria; and long term animal management and animal breading (Smith et al. 2007). But these options are yet to be implemented in Asia as additives and vaccines are not available commercially in Asian markets. On the other hand feeding practices may not be acceptable to some farmers which feed them with waste from restaurants and market places.

There is considerable worry that there is increase in meat consumption and thus there could be increase in GHG due to increase in population of ruminant animals. On the other hand chicken cultivation has increased due to technology wherein chickens will be hatched in millions every day and birds will be butchered when it reaches an age of few months. Bangalore in India alone has 3,000 chicken shops spread all over the city wherein birds are butchered in front of costumers to ensure them meat is derived from healthy bird. This city with population of about six million needs flesh of 500,000 birds/day on an average. In the absence of any land fill to dispose such waste the methane and CO_2 escapes to atmosphere.

Green House Gases and Asia

While the Asia's speedy growth has brought benefits to the poor, it has also placed severe stress on its environment. Asia needs a huge energy to maintain growth and continue the record of poverty reduction. With development, countries consume higher energy for industry, entertainment, transportation, research and other activities. The migration and urban agglomeration results in more energy-intensive lifestyles. As the Asia gains the means to extend energy services to its people who rely on noncommercial energy sources for cooking and heating, energy use is expected to increase dramatically.

Out of total global emission in 2004, energy supply accounted for approximately 26% of GHG emission; industry accounted for about 19%; land use change accounted around 17%; agriculture released about 14%; transport contributed nearly 13%; residential, commercial and service sectors collectively was responsible for about 8%; and waste management contributed about 3% (TERI 2009).

Detailed observation of Fig. 4.5 depicts the CO_2 emission follow more of a growth curve used for population studies. Population of species in an ecosystem

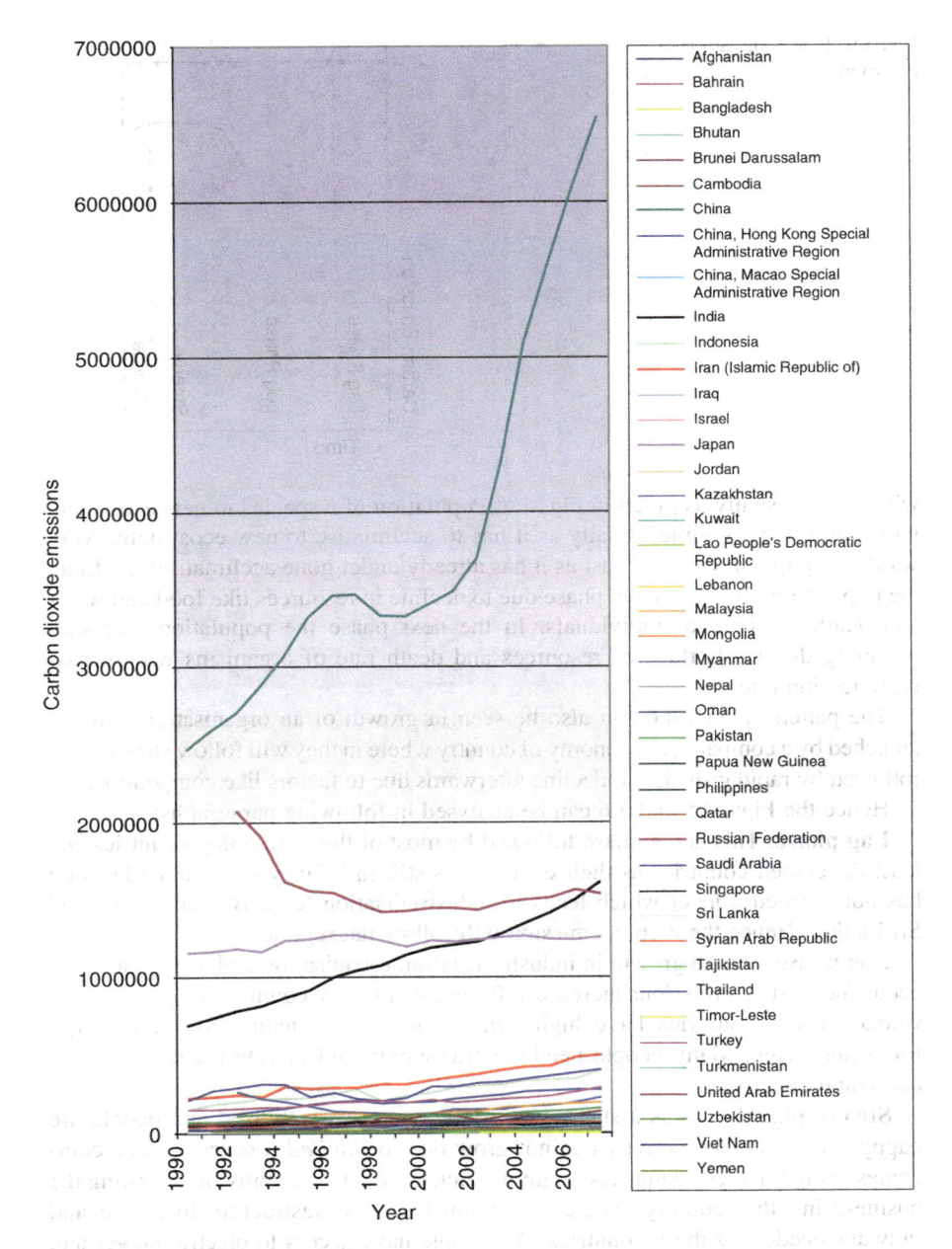

Fig. 4.5 Carbon dioxide emissions, total, per capita and per $1 GDP (PPP)
Source: Based on data downloaded from http://mdgs.un.org/unsd/mdg/SeriesDetail.aspx?srid=749&crid on 15.7.2010

Fig. 4.6 Phases of carbon emission

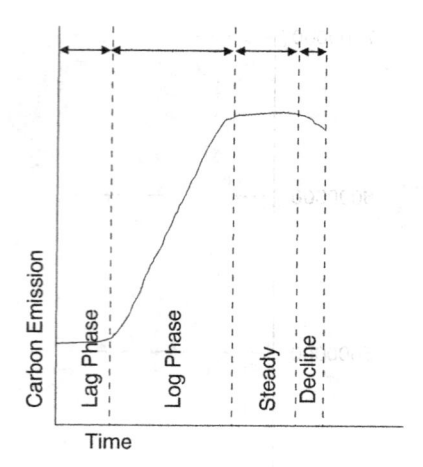

will behave usually as given in Fig. 4.6. Population of a species in new ecosystem will follow a flat regime initially as it has to acclimatise to new ecosystem. After wards the population grows fast as it has already under gone acclimatisation. Later the population reaches steady phase due to decline in resources like food and water and death of some of individuals. In the next phase the population will start declining due to shortage of resources and death rate of organisms will surpass reproduction rate.

The pattern in Fig. 4.6 can also be seen in growth of an organisation; product launched by a company; or economy of country where in they will follow steady stage followed by rapid growth and decline afterwards due to factors like competition.

Hence the Figs. 4.5 and 4.6 can be analysed in following paragraphs:

Lag phase: This is the phase followed by most of the developing countries and least developed countries as their economy is still in infancy stage and education has not reached a level which leads to industrialisation (e.g., Bhutan, Nepal and Sri Lanka). Hence the carbon emission will follow flat regime.

Log phase: Due to growth in industrialisation, urbanisation and use of carbonaceous fuel carbon emissions increase at faster rate in some countries (e.g., India and China). These countries have high demand for infrastructure, job and energy for better living. Many people need electric supply and private transportation to overcome poverty.

Steady phase: This is a stage where in economy is stabilised and people are happy. The industrial activity can not grow due to '*limited resources*'. The companies started in these countries might be contributing to economy by operating the business in other country. There is not much new infrastructure like road and railways needed by these countries. All people have access to electric power and no demand for further personal transportation. These reasons coupled with mitigation measure make the emission regime 'flat' (e.g., Japan and UAE).

Decline phase: This is the stage when decline in carbon emission can be observed due to efforts by the country to curb carbon emission and slowing of industrial activity.

4.1 Green House Gases and Asia

Asia is undergoing rapid development resulting in levels of GHGs and air pollution in many cities to that of Europe in the first decades of the twentieth century. About two thirds of the estimated 800,000 deaths worldwide caused by exposure to urban air pollution in 2000 occurred in the developing Asia (WHO 2002).

Most common Anthropogenic GHG sources in Asia are industrial, domestic stationary sources, mobile sources, marine emissions, solid-waste burning, Agriculture (Including Animal Husbandry) rural area sources, and terrestrial and transport sources. Apart from these anthropogenic sources natural sources include, emission from marsh, water evaporation, forest fires and wild life.

Major industrial sources of GHG include coal fired power plants, large-scale boilers, GHG manufacturing units, incineration of waste generated within the industry, Diesel Generators sets (DG Sets) installed in industry for power generation during power failure and emergency.

Domestic sources include cooking, building heating, DG sets installed in apartments and some houses for power generation during power failure and emergency. Fuels used in house hold varies from location to location depending mainly on the affordability and reliability of the fuel supply. The fuels used include agricultural waste; dried animal dung; combustible waste; wood; liquefied petroleum gas (LPG); coal; coal gas; and natural gas.

Commercial sources like hotels, restaurant include cooking fumes, DG sets installed for power generation during power failure and emergency and fuel-combustion.

Power supply is not continuous in many Asian countries. During shortage of power generated in power plants distribution will made according to discretion of power supply companies or as per instruction of government. Due to frequent power failures usually all industries, commercial establishments and wealthy residents install DG sets for generation of power during power failure. Capacity of DG Sets varies from 5 to 1,000 kVA and above.

Mobile sources include emissions from air, water and land locomotives. Open burning of solid waste is the major source of GHG.

Unlike in developed countries emissions from rice production and burning of biomass are concentrated in the group of developing countries. CH_4 emissions from rice occurred mostly in South and East Asia (Smith et al. 2007). CH_4 and N_2O emissions from wastewater are generally higher than in developed countries due to rapid population growth and urbanization without concurrent development of wastewater infrastructure. Decentralized treatment processes and septic tanks might also result in comparatively large emissions of CH_4 and N_2O, particularly in China, India and Indonesia due to increase in wastewater volumes as a result of economic development (Scheehle and Doorn 2003).

Coal continues to make up the bulk of China's energy consumption (70% in 2008), making China the largest producer and consumer of coal in the world. China also became a large-scale coal importer in the year 2009(BP 2010).

As China's economy continues to grow, China's coal demand is projected to rise significantly. Even though coal's share of China's overall energy consumption will decrease, coal consumption will maintain to rise in absolute terms. China's continued and increasing reliance on coal as a power source has contributed appreciably to China's emergence as the world's largest emitter of acid rain-causing sulphur dioxide and green house gases, including carbon dioxide.

In developing countries of Asia, controlled incineration of waste is not practiced frequently because of high capital and operating costs. Indiscriminate burning of waste for volume reduction is still a common practice that contributes to urban air pollution as well as CO_2.

Incineration is not the choice for wet waste, and municipal waste in several developing countries (Bogner et al. 2007) but there is rise in number of incinerators for disposal of biomedical waste to combat spread of infection especially in India.

Iran saw the world's highest volumetric consumption increase with respect to Natural Gas, while India recorded growth of 25.9% which was the highest in percentage terms (BP 2010).

In 2004, energy accounted for 26% of GHG emissions; industry accounted for 19%; land use change and deforestation accounted for 17%, agriculture accounted for 14%, transportation accounted for 13%, Residential commercial and service sector accounted for 8%; and waste accounted for 3% (Depicted pictorially in Fig. 4.7) gives world GHG emissions by sector in 2004. Emissions of GHGs covered under the Kyoto protocol increased by about 70% during 1970–2004. A CO_2 emission has been major emission from the power generation and road transport sectors. Emissions of CH_4 and N_2O have increased by 40 and 50% respectively from the 1970 levels. Agriculture is the main contributor of CH_4 emissions. Fertilizer use is major reason for rise in N_2O emissions.

China's total GHG emissions in 1994 are 4,060 million tons of CO_2 eq. and as per tentative estimates by experts from China, China's total GHG emission in 2004 is about 6,100 million t CO_2 eq. (NDRCPRC 2007). As on 2004 per capita CO_2 emissions (in metric tons) of USA, European Union, Japan, China, Russia,

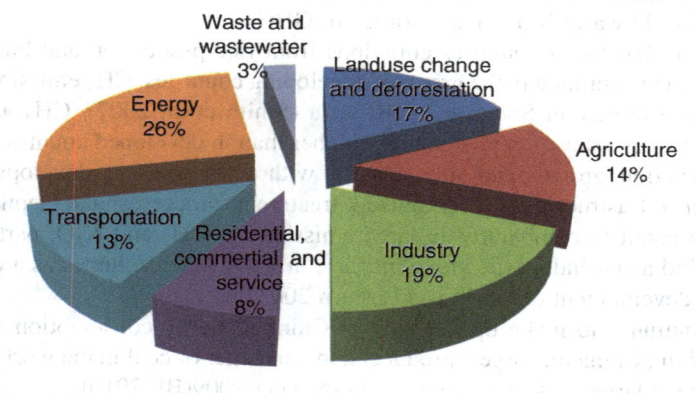

Fig. 4.7 World GHG gas emissions by sector in 2004

and India and world average were 20.01, 9.40, 9.87, 3.60, 11.71, 1.02 and 4.25 respectively.

The investigation by Tek et al. (2008) revealed that during time period from 1990 to 2005 Annex I countries reduced emission by 4.6%; the Economic in Transition (EIT) decreased their emission by 36.2% and non-EIT parties increased their emissions by 10%. Amongst the major carbon dioxide emitters in the world (IEA 2009), China has been responsible for two thirds of the global raise in anthropogenic carbon dioxide emissions of 3.1% in 2007 (Yan and Yang 2010). Studies conducted by Chen and Zhang (2010) revealed that China's contribution to GHG to be 7,456.12 $MtCO_2$-eq in the year 2007.

Studies on non agricultural open fires in Asia from 2000 to 2009 by Song et al. (2010) revealed that the lowest emissions usually occurred in the fire year 2000, while high emissions were in the fire year 2002 or 2004. The total emissions as per the studies for CO_2, CO, $NMHC_s$, NO_x, BC, and OC were 83 (69–103), 6.1 (4.6–8.2), 0.64 (0.36–1.0), 0.085 (0.074–0.10), 0.023 (0.020–0.028), and 0.27 (0.22–0.33) Tg yr^{-1}, respectively with Indonesia being the biggest contributor mostly due to peat burning.

4.2 Countrywise Scenario in Asia

Developing Asia is most dynamic part of the world in many aspects. Even though millions of Asians still live in poverty, the continent is observing steady growth of per capita income. In the coming decades, if current development continues, developing Asia may well be approaching the per capita income of middle-income Latin America (United Nations Economic and Social Affairs 2004).

Countries report their GHG emissions to UNFCCC according to the IPCC Guidelines. The Annex I parties to the Convention report annually but Non-Annex I countries do not report on a regular basis.

Figure 4.8 shows GHG emissions per capita throughout the world as updated on 2009 and Table 4.2 shows latest available data with respect to GHG as on July 2010. In the Asia largest emitters are China (51%), India (15%) and the Islamic Republic of Iran (5%).

Distribution of wealth and of GHG is unbalanced among different world regions and the growth of population higher in the present developing countries than in the industrialized countries (Tom et al. 2000).

Bio-fuels contributed, about 28% of the total carbon emissions from energy use in Asia with China and India being largest emitting countries (David and Stephanie 1999). Biomass energy accounted for the 46% of energy requirement in Pakistan resulting in loss 14.7% of its forest habitat between 1990 and 2005 interval (Tahir et al. 2010).

South East Asia was one of the fastest growing regions in the world until the Asian crisis. The use of cheap outdated technologies to keep up with the flourishing demand in every sector of the economy created stern environmental concerns in the

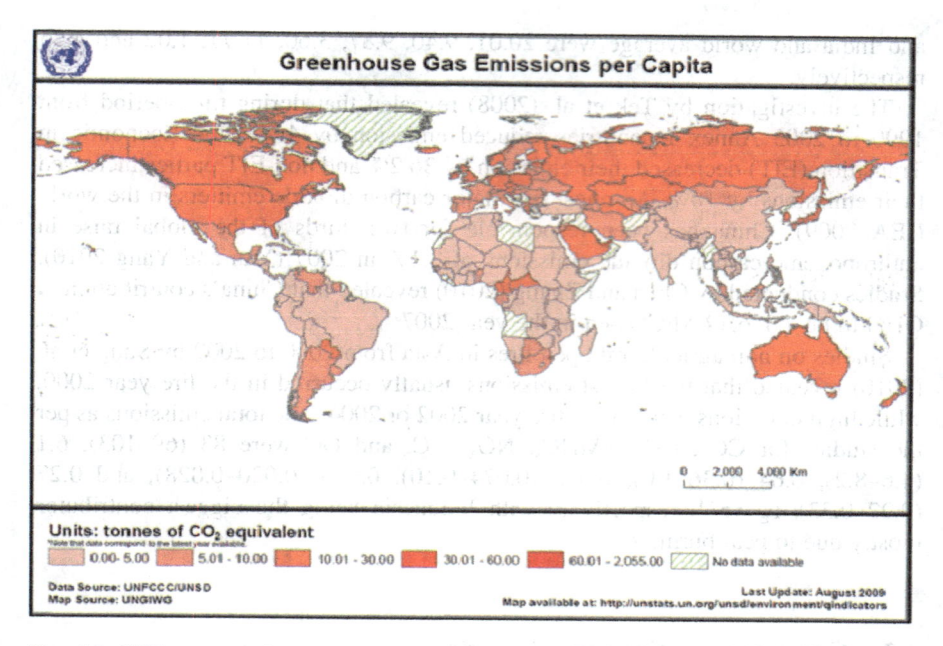

Fig. 4.8 GHG gas emissions per capita (http://unstats.un.org/unsd/environment/air_greenhouse_emissions.htm downloaded on 18.7.2010)

area (Clemencon 1997; Chen 1995). The biomass resources in this area such as rain forests have been robustly affected and the governments now comprehend the need for sustainable management of these resources (Brookfield and Byron 1993). At the same time, the energy demand of the developing economies of the ASEAN (Association of South East Asian Nations); countries have augmented steadily (Bretz 1997; Brandon 1996).

More than 30% of agricultural N_2O emission was produced by animals and CH_4 emission from livestock waste in the world was about 28 Gt (Li et al. 2003; OECD 1991). Approximately 50% of GHG can be reduced if all the livestock manure in China were treated by anaerobic process (Hou et al. 2006). With an annual production of about 2.6 million cars in 2009, India became seventh largest vehicle producing country in the world. This is a milestone the country attained 6 years ahead of the set target. The Indian automobile sector is one of the fastest growing markets globally. In 2008 India became the second-largest producer of small cars. Studies by Streets et al. (2003) estimate total Asian emissions as: 34.3 Tg SO_2, 26.8 Tg NO_x, 9,870 Tg CO_2, 279 Tg CO, 107 Tg CH_4, 52.2 Tg NMVOC, 2.54 Tg black carbon (BC), 10.4 Tg organic carbon (OC), and 27.5 Tg NH_3.

Although urban growth is most often recognized to migration from rural to urban areas, natural increases in the urban population and reclassification of city boundaries also contribute to the increase in city populations (HEI International Scientific Oversight Committee 2010). With urbanization the living style energy consumption patterns also change. In spite of many Asian countries does not have electric

Table 4.2 Green house gases for Asian and non Asian countries

Country	Latest year available	Total GHG emissions (million tonnes of CO_2 equivalent)	% Change since 1990 (%)	GHG emissions per capita (tonnes of CO_2 equivalent)	Total GHG emissions including LULUCF/ LUCF (million tonnes of CO_2 equivalent)
Albania	1994	5.53	−22.5	1.74	7.06
Algeria	1994	91.76	−	3.31	100.34
Antigua and Barbuda	1990	0.39	−	6.30	0.29
Argentina	2000	282.00	22.0	7.63	238.70
Armenia	1990	25.31	−	7.14	24.70
Australia	2006	536.07	28.8	25.99	549.85
Austria	2006	91.09	15.1	11.01	72.94
Azerbaijan	1994	43.17	−29.0	5.62	42.09
Bahamas	1994	2.20	14.9	7.97	2.20
Bahrain	1994	19.60	−	34.90	19.60
Bangladesh	1994	45.93	−	0.37	53.76
Barbados	1997	4.06	24.8	15.90	4.05
Belarus	2006	81.00	−36.4	8.29	55.00
Belgium	2006	136.97	−5.2	13.08	135.91
Belize	1994	6.34	−	29.63	2.31
Benin	1995	39.35	−	6.88	−8.18
Bhutan	1994	1.29	−	2.51	−2.26
Bolivia	2000	21.46	40.1	2.58	49.94
Botswana	1994	9.29	−	6.15	−29.44
Brazil	1994	658.65	11.1	4.14	1,476.73
Bulgaria	2006	71.34	−38.9	9.28	53.12
Burkina Faso	1994	5.97	−	0.61	4.58
Burundi	1998	2.00	−	0.32	−1.00
Cambodia	1994	12.76	−	1.15	−5.15
Cameroon	1994	165.73	−	12.11	187.91
Canada	2006	720.63	21.7	22.09	751.97
Cape Verde	1995	0.29	−	0.73	0.22
Central African Republic	1994	37.74	−	11.61	−101.58
Chad	1993	8.02	−	1.20	−38.18
Chile	1994	54.89	−	3.87	45.69
China	1994	4,057.62	−	3.39	3,650.14
Colombia	1994	137.49	22.9	3.84	152.08
Comoros	1994	0.52	−	1.08	−0.38
Congo	1994	1.38	−	0.51	−68.49
Cook Islands	1994	0.08	−	4.33	−0.07
Costa Rica	1996	10.79	76.9	3.03	10.07
Cote d'Ivoire	1994	24.73	−	1.71	4.88
Croatia	2006	30.83	−5.2	6.95	23.34
Cuba	1996	40.19	−36.8	3.67	18.54
Czech Republic	2006	148.20	−23.7	14.49	144.83
Dem. Rep. of the Congo	1994	44.64	−	1.03	−176.84
Denmark	2006	71.91	2.2	13.24	70.11
Djibouti	1994	0.51	−	0.84	−0.09
Dominica	1994	0.15	−98.8	2.18	−0.22
Dominican Republic	1994	20.44	61.7	2.56	13.94
Ecuador	1990	30.77	−	2.99	77.72
Egypt	1990	116.74	−	2.02	106.84
El Salvador	1994	11.72	−	2.07	15.66

(*continued*)

Table 4.2 (continued)

Country	Latest year available	Total GHG emissions (million tonnes of CO_2 equivalent)	% Change since 1990 (%)	GHG emissions per capita (tonnes of CO_2 equivalent)	Total GHG emissions including LULUCF/LUCF (million tonnes of CO_2 equivalent)
Eritrea	2000	0.76	–	0.21	0.76
Estonia	2006	18.88	−54.6	14.04	15.40
Ethiopia	1995	47.74	11.0	0.84	37.87
Fiji	1994	1.39	–	1.83	−6.31
Finland	2006	80.29	13.2	15.26	46.85
France	2006	546.53	−3.5	8.90	476.63
Gabon	1994	6.52	–	6.19	−494.35
Gambia	1993	4.26	–	4.24	−45.73
Georgia	1997	12.89	−71.6	2.62	14.04
Germany	2006	1,004.79	−18.2	12.20	968.39
Ghana	1996	13.14	17.8	0.74	−5.88
Greece	2006	133.11	27.3	12.01	127.91
Grenada	1994	1.61	–	16.21	1.51
Guatemala	1990	14.74	–	1.65	−24.80
Guinea	1994	5.06	–	0.70	−12.54
Guinea-Bissau	1994	1.69	–	1.49	–
Guyana	1998	3.07	40.8	4.05	−27.80
Haiti	1994	5.13	–	0.67	6.06
Honduras	1995	10.83	–	1.94	15.46
Hungary	2006	78.62	−20.0	7.82	72.72
Iceland	2006	4.23	24.2	14.05	5.36
India	1994	1,214.25	–	1.30	1,228.54
Indonesia	1994	334.19	25.2	1.77	498.31
Iran (Islamic Republic of)	1994	385.43	–	6.30	417.01
Ireland	2006	69.76	25.6	16.33	69.27
Israel	2005	73.41	–	10.97	73.02
Italy	2006	567.92	9.9	9.63	455.71
Jamaica	1994	116.31	–	47.58	116.15
Japan	2006	1,340.08	5.3	10.51	1,248.58
Jordan	1994	21.94	–	5.33	18.37
Kazakhstan	2004	210.21	−34.7	13.93	202.77
Kenya	1994	21.47	–	0.80	−6.53
Kiribati	1994	0.03	–	0.39	0.03
Korea, Dem. People's Rep.	1990	201.93	–	10.02	187.30
Korea, Republic of	2001	542.89	87.6	11.62	508.25
Kyrgyzstan	2000	15.05	−58.3	3.04	14.08
Lao People's Dem. Rep.	1990	6.87	–	1.63	−97.44
Latvia	2006	11.62	−56.1	5.10	−6.19
Lebanon	1994	15.70	–	4.63	15.91
Lesotho	1994	1.82	–	1.07	3.08
Liechtenstein	2006	0.27	19.0	7.79	0.27
Lithuania	2006	23.22	−53.0	6.85	15.27
Luxembourg	2006	13.32	1.0	28.37	13.03
Madagascar	1994	21.93	–	1.72	−217.04
Malawi	1994	7.07	−12.1	0.71	24.59
Malaysia	1994	136.68	–	6.81	75.60
Maldives	1994	0.15	–	0.62	0.15
Mali	1995	8.67	–	0.91	−1.08
Malta	2000	2.85	28.4	7.33	2.61

(*continued*)

Table 4.2 (continued)

Country	Latest year available	Total GHG emissions (million tonnes of CO_2 equivalent)	% Change since 1990 (%)	GHG emissions per capita (tonnes of CO_2 equivalent)	Total GHG emissions including LULUCF/ LUCF (million tonnes of CO_2 equivalent)
Mauritania	1995	4.33	–	1.91	3.58
Mauritius	1995	2.06	–	1.83	1.84
Mexico	2002	548.50	29.2	5.38	638.35
Micronesia, Federated States of	1994	0.25	–	2.36	0.25
Monaco	2006	0.09	−13.0	2.88	0.09
Mongolia	1998	15.90	−17.7	6.82	15.60
Morocco	1994	44.37	–	1.67	39.88
Mozambique	1994	8.19	20.8	0.53	15.97
Namibia	1994	5.60	–	3.54	−0.11
Nauru	1994	0.04	–	4.06	0.03
Nepal	1994	31.19	–	1.48	39.31
Netherlands	2006	207.48	−2.0	12.66	210.05
New Zealand	2006	77.87	25.7	18.75	55.12
Nicaragua	1994	7.65	–	1.68	−5.40
Niger	1990	4.86	–	0.61	10.96
Nigeria	1994	242.63	–	2.25	347.64
Niue	1994	4.42	–	2,052.95	4.51
Norway	2006	53.51	7.7	11.44	25.68
Pakistan	1994	160.60	–	1.26	167.12
Palau	2000	0.09	–	4.67	0.09
Panama	1994	10.69	–	4.08	34.40
Papua New Guinea	1994	5.01	–	1.09	4.60
Paraguay	1994	140.46	149.9	29.94	159.96
Peru	1994	57.58	–	2.45	98.80
Philippines	1994	100.87	–	1.47	100.74
Poland	2006	400.46	−11.7	10.49	359.95
Portugal	2006	82.74	40.0	7.81	78.58
Republic of Moldova	1998	10.51	−68.4	2.49	9.07
Romania	2006	156.68	−36.7	7.27	119.19
Russian Federation	2006	2,190.24	−34.2	15.37	2,478.03
Rwanda	2002	2.38	–	0.28	−4.63
Saint Kitts and Nevis	1994	0.16	–	3.77	0.07
Saint Lucia	1994	0.89	–	6.12	0.54
Samoa	1994	0.56	–	3.36	0.48
Sao Tome and Principe	1998	0.12	–	0.89	−1.42
Saudi Arabia	1990	165.27	–	10.16	150.03
Senegal	1995	9.57	–	1.11	3.57
Seychelles	1995	0.26	–	3.43	−0.58
Singapore	1994	26.86	–	7.96	26.86
Slovakia	2006	48.90	−33.6	9.07	45.87
Slovenia	2006	20.59	10.8	10.27	15.86
Solomon Islands	1994	0.29	–	0.82	0.29
South Africa	1994	379.84	9.4	9.38	361.22
Spain	2006	433.34	50.6	9.94	400.34
Sri Lanka	1995	29.13	–	1.60	408.21
St. Vincent and the Grenadines	1997	0.41	4.6	3.80	0.28

(continued)

Table 4.2 (continued)

Country	Latest year available	Total GHG emissions (million tonnes of CO_2 equivalent)	% Change since 1990 (%)	GHG emissions per capita (tonnes of CO_2 equivalent)	Total GHG emissions including LULUCF/LUCF (million tonnes of CO_2 equivalent)
Sudan	1995	54.24	–	1.76	71.97
Suriname	2003	3.34	–	6.85	4.87
Swaziland	1994	2.64	–	2.78	−0.62
Sweden	2006	65.75	−8.7	7.21	27.74
Switzerland	2006	53.21	0.7	7.11	50.98
Tajikistan	1998	4.29	−81.9	0.71	2.80
Thailand	1994	223.98	–	3.76	285.84
The Former Yugoslav Rep. of Macedonia	1998	15.07	−2.4	7.56	12.79
Togo	1998	6.28	–	1.28	34.41
Tonga	1994	0.23	–	2.37	−0.08
Trinidad and Tobago	1990	16.01	–	13.14	14.51
Tunisia	1994	25.14	–	2.85	23.37
Turkey	2006	331.76	95.1	4.60	255.66
Turkmenistan	1994	52.31	–	12.77	51.93
Tuvalu	1994	0.01	–	1.09	0.01
Uganda	1994	41.55	–	2.05	49.80
Ukraine	2006	443.18	−51.9	9.51	410.56
United Arab Emirates	1994	130.44	–	56.71	126.21
United Kingdom	2006	655.79	−15.1	10.83	653.83
United Rep. of Tanzania	1994	39.24	0.6	1.35	952.80
United States	2006	7,017.32	14.4	22.96	6,170.53
Uruguay	2000	29.73	7.5	8.95	17.19
Uzbekistan	1994	153.89	−5.7	6.85	153.49
Vanuatu	1994	0.30	–	1.79	0.30
Venezuela	1999	192.19	–	8.03	177.90
Viet Nam	1994	84.45	–	1.18	103.84
Yemen	1995	17.87	–	1.15	8.20
Zambia	1994	32.77	–	3.70	36.23
Zimbabwe	1994	27.59	–	2.40	−34.65

Source: http://unstats.un.org/unsd/environment/air_greenhouse_emissions.htm downloaded on 18.7.2010

connection to rural poor, their no dearth of electricity when it comes to lighting the stadium (Fig. 4.9) by flood light to play games. Even though Asian countries receive good sunlight huge energy is wasted towards lighting stadiums in night. The divide between rich and poor is extremely high in many Asian countries. Many rich people waste the energy for playing games, casino, ornamental lighting to enhance beauty of buildings and other luxury. On the other hand children of poor people have to study in kerosene lamps and sometimes suffer snake/insect bites due to poor visibility.

Increase in epidemics swine flu and avian flu in recent past has resulted in increased activity in solid waste management. Many countries like India which did not had landfill sites a decade ago is now having landfill sites all over the

Fig. 4.9 In spite of sufficient day light in many Asian countries energy is wasted by using floodlights in stadiums contributing to GHGs. Mean while many people in rural area struggle without electric connection

Fig. 4.10 Combustion is still an important option in many developing countries and developed countries

country and number is still growing to tackle accumulation of hazardous waste, incineration ash and non-incinerable waste. Incineration is still an option with respect to incinerable waste (Fig. 4.10) as technologies like pyrolisis and microwave is yet to prove its economics in Asia.

As could be seen in Fig. 4.11 vehicles in Asia have multiplied over last two decades both in terms of variety and numbers. Transport sector contributes about 14% of global GHG emissions as on date. Carbon dioxide represents the largest proportion of GHG gas emissions and increased faster than those from all other sectors and projected to increase more speedily in future. Less than 1% of the population of India currently owns automobiles so as many developing countries, which is a much smaller proportion than the developed countries. The total motor vehicles in India increased from about 310,000 in 1951 to about 8,900,000 at the last part of fiscal year 2005–06 (CPCB 2010). But being second large country with a

Fig. 4.11 Vehicles in Asia has multiplied over last two decades both in terms of variety and numbers

Fig. 4.12 Cement industry is one of the many industry which generates GHGs

population of one billion, the numbers would be significant with respect to countries contribution to GHG.

Gasoline and diesel buses consume 40–50 l per 100 km traveled where as motorbikes, consume less than 10 l per 100 km. Diesel and gasoline cars consume about 10–20 l per 100 km. On the other hand energy intensity range order will be reversed if we consider per capita energy requirement for fleet. Gasoline motorbike consume 7–9 l per person per100 km traveled where as requirement of fuel per capita per 100 km in bus is 2 l. Diesel and gasoline cars are require 4–7 l per person per 100 km.

Major industrial activities contributing to GHGs by non energy generating industry are iron and steel; cement (Fig. 4.12); paper and pulp; fertilizes; petroleum; Metal processing; lime; Metallurgical coke production; chemical industry etc. Cement and steel production has increased in the area due to tremendous in construction activities.

In spite of scope for reduction in emission by energy efficiency and technology modification the manufacturers still have to switchover to climate friendly technologies.

In conclusion the increase in energy usage and GHG emission is not happening by economic growth alone but also due to other reasons also which could be easily avoided.

References

Bogner J, Abdelrafie Ahmed M, Diaz C, Faaij A, Gao Q, Hashimoto S, Mareckova K, Pipatti R, Zhang T, Waste Management, In Climate Change 2007: Mitigation. Contribution of Working Group III to the Fourth Assessment Report of the Intergovernmental Panel on Climate Change [Metz B, Davidson O.R, Bosch P.R, Dave R, Meyer L.A(eds)], Cambridge University Press, Cambridge, United Kingdom and New York, NY, USA.

BP (British Petroleum), 2010: BP Statistical Review of World Energy, BP Oil Company Ltd., London

Brandon C, 1996: Confronting the growing problem of pollution in Asia. Journal of Social, Political and Economic Studies;21(2):199–204.

Bretz E, 1997: Developing Asia energy demand outpaces supply. Electrical World.

Brookfield H, Byron, 1993: South-East Asia's environmental future. United Nations University Press, Tokyo, Japan.

Clemencon R, 1997: Economic integration and the environment in Southeast Asia: securing gains from open markets while preventing further environmental degradation. Journal of Environment and Development; 6(3):317.

Chen X, 1995: Industrialization, energy efficiency and environmental protection in Asian industrializing countries: the role of technological change. World Resource Review;7(2):289–93.

Chen G.Q, Zhang B, 2010: Greenhouse gas emissions in China 2007: Inventory and input–output analysis. Energy Policy, doi:10.1016/j.enpol.2010.06.004

CPCB (Central Pollution Control Board, India) 2010: Status of the Vehicular Pollution Control Programme in India

David G. Streets, Stephanie T. Waldhoff, 1999: Greenhouse-gas emissions from biofuel combustion in Asia, Energy 24, 841–855

HEI International Scientific Oversight Committee, 2010: Outdoor Air Pollution and Health in the Developing Countries of Asia: A Comprehensive Review. Special Report 18. Health Effects Institute, Boston, MA.

Hou Jicong, Xie Yanhua, Wang Fengde, Dong Renjie, 2006: Greenhouse gas emissions from livestock waste: China evaluation, International Congress Series 1293, 29–32

IEA, 2009: CO_2 emissions from fuel combustion 2009-highlights.

Li M.F. et al., 2003: Research on greenhouse gas emission from agricultural producing, Journal of Shandong Agricultural University (Natural Science), 34 (2), 311–314.

NDRCPRC (National Development and Reform Commission People's Republic of China) 2007: China's National Climate Change Programme

OECD, 1991: Estimation of greenhouse gas emissions and sinks, Final Report, OECD/OCDE, Paris

Scheehle E. and Doorn M, 2003: Improvements to the U.S. wastewater CH4 and N2O emissions estimates. Proceedings 12th International Emissions Inventory Conference, "Emissions Inventories – Applying New Technologies", San Diego, California, April 29-May 3, 2003, published by U.S. EPA, Washington, DC., <www.epa.gov/ttn/chief/conference/ei12>, accessed 12/08/07.

Smith, Martino P.D, Cai Z, Gwary D, Janzen H, Kumar P, McCarl B, Ogle S, O'Mara M, Rice C, Scholes B, Sirotenko O, 2007: Agriculture. In climate Change 2007: Mitigation Contribution of Working Group III to the Fourth Assessment Report of the Intergovernmental panel on Climate Change [Metz B, Davidson O.R, Bosch P.R, Dave R, Mayer L.A(eds)], Cambridge University Press, Cambridge, United Kingdom and New York, NY, USA.

Streets D.G, Bond T.C, Carmichael G.R, Fernandes S.D, Fu Q, He D, Klimont Z, Nelson S.M, Tsai N.Y, Wang M.Q, Woo J.H, and Yarber K.F, 2003: An inventory of gaseous and primary aerosol emissions in Asia in the year 2000: *J. Geophys. Res.*, <http://www.cgrer.uiowa.edu/EMISSION_DATA/publications/Streets_TRACE-P_revised.pdf> accessed on 29.12.2010

Tek N. Maraseni, Jerry Maroulis & Mehryar Nooriafshar, 2008: An analysis of anthropogenic greenhouse gas emissions status of Annex I countries: can they meet Kyoto targets?, Volume 3 Issue 2, AJBBS, 3–15

Tahir S.N.A, Rafique M, Alaamer A.S, 2010: Biomass fuel burning and its implications: Deforestation and greenhouse gases emissions in Pakistan, Environmental Pollution 158, 2490–2495

TERI (The Energy and Resources Institute), 2009: Simplifying Climate Change, based on the findings of the IPCC Fourth Assessment report, pp140

Tom Kram, Tsuneyuki Morita, Keywan Riahi, R. Alexander Roehrl, Sascha Van Rooijen, Alexei Sankovski, And Bert De Vries, 2000: Global and Regional Greenhouse Gas Emissions Scenarios, Technological Forecasting and Social Change 63, 335–371

United Nations Economic and Social Affairs. 2004: World Economic Situation and Prospects 2004. Development Policy and Analysis Division, Department of Economic and Social Affairs, United Nations, New York, NY.

World Health Organization, 2002: World Health Report 2002: Reducing Risk, Promoting Healthy Life. WHO, Geneva, Switzerland.

Yan Y.F, Yang L, 2010: China's foreign trade and climate change: a case study of CO2 emissions. Energy Policy 38(1),350–356.

Yu Song, Di Chang, Bing Liu, Weijie Miao, Lei Zhu and Yuanhang Zhang, 2010: A new emission inventory for non agricultural open fires in Asia from 2000 to 2009, Environ. Res. Lett. 5 (January-March 2010) 014014 doi:10.1088/1748-9326/5/1/014014, <http://iopscience.iop.org/1748-9326/5/1/014014/fulltext> accessed on 29.12.2010

Chapter 5
Climate Change and Hydro-Meteorological Risks in Asia

Meteorology is the study of weather processes and short term forecasting. The branch of meteorology that deals with the occurrence, motion, and changes of state of atmospheric water is called hydro-meteorology. Disasters related to hydro meteorology are called hydro meteorological risk. Extreme hydro-meteorological events are mounting concern across the world. The relations with climate change hydro-meteorological events are evident from scientific literature.

Global land precipitation has raised by about 2% since the start of the twentieth century (Jones and Hulme 1996; Hulme et al. 1998). The increase is neither spatially nor temporally uniform (Karl and Knight 1998; Doherty et al. 1999). Annual snow-cover extent (SCE) has declined by about 10% in Northern Hemisphere since 1966 largely due to decreases in spring and summer since the mid-1980s over both the Eurasian and American continents (Robinson 1997). Global warming can lead to change in the global water cycle due to feedbacks between rising temperatures and hydrologic processes. Water resources in Asia are vulnerable to climate change.

Climate-change and environmental alterations have augmented the risks of hydro meteorological risks. Such phenomenon is closely connected to each other. In some cases they may be induced, triggered or amplified by human activities. Climate change is likely to alter the hydro-meteorological patterns of an area, favoring such incidents.

There are huge variations with respect to hydrometeorological risks within Asia. While some parts are receiving heavy rains other parts are facing drought. The ground water level is affected by huge withdrawals. Unlike few decades ago where in people took few months to dig well, people drill bore wells within hours to reach ground water. There has been tremendous drop in groundwater table due to drilling bore well and drawing water using submersible pumps.

Rise in temperature leads to early melt of snow. But there could be heavy snow fall like the one witnessed by many countries in January 2010 which was not observed for several years. Water quality which is already polluted due to industrialisation and urbanisation will deteriorate further due to floods which will pick chemicals and sewage from urban pockets.

Globally, demand for water is increasing. But availability of desired water quantity with required quality is decreasing. Many cities like Bangalore are

R. Chandrappa et al., *Coping with Climate Change*,
DOI 10.1007/978-3-642-19674-4_5, © Springer-Verlag Berlin Heidelberg 2011

Fig. 5.1 Commercial establishment buying water in tankers. Water stress in many Asian urban area make difficult for urban authority to supply water

depending on tanker to fetch water of desirable quality for restaurant, industries and apartments. Even though the ground water in the city is just few meter below ground it can't be used due to contamination from leaky sewers.

The water shortage in many areas is also due to un-prudent use for cultivating crops. In some places rice was grown using ground water due to shortage of surface water resources. Some part of Karnataka state of India is now facing water shortage due to such unconventional, unscientific use of water. Water in many of the Asian cities are supplied by tankers instead of pipe lines as urban authority responsible for supplying water can't cater explosion in demand. Hence as depicted in Fig. 5.1 many people buy water in tankers who will fetch water from rural area located in outskirts of the urban area.

Global warming has severe impact on mountainous region. Nepal witnessed 0.6°C per decade, compared with a global average of 0.74°C over the last 100 years (Mats et al. 2009). Such occurrences coupled with extreme weather events such as high intense rainfalls will result in leading to flash floods, landslides and debris flows.

5.1 Introduction to Hydro-Meteorological Disasters

Hydro-meteorological disasters include floods and cyclones, as much as slow onset disaster types, such as droughts and desertification. Table 5.1 gives time scale and potential risk of hydrometeorological disasters. These types of phenomena are connected to each other, and may be induced, triggered or amplified by human activities. Climate change may alter the hydro-meteorological patterns of an area, which in turn may affect each other. Due to climatic change, this tendency is expected to be enhanced in the future (IPCC 2007). Temperature changes bring in changes in moisture, precipitation and atmospheric circulation. These effects change the hydrological cycle, particularly characteristics of precipitation (amount, frequency, intensity, duration, type) and extremes (Trenberth et al. 2003).

Table 5.1 Time scale and potential risk of hydro-meteorological disasters

Hazard	Description	Time scale	Area affected	Potential risk
Tropical cyclone	Storm system characterized by a large low pressure centre and numerous thunder storms that produce strong winds and heavy rain. Based on location and strength it is called hurricane, typhoon, tropical storm, cyclonic storm, tropical depression or cyclone	2–5 days	Few km to several thousand km^2	Very high with Strong winds and heavy rainfall
Storm surge inundation	Abnormal rise in water elevation caused by severe storms approaching the coast. In other words it is large wave that moves with the storm	One day to several days	Few to several km	Storm surge of 4–10 m height
Strong winds	Wind blowing at very high speed due to meteorological variations	Several days in tropical cyclone and strong monsoon conditions	Several km^2	Very high to moderate (20–80 m/s)
Heavy rainfall	Rainfall of several cm/day	2–3 days (over 6 cm/day). Can be 30–50 cm/day	Few km^2 to several hundred km^2	Very high, flooding
Severe local thunderstorms	Storm with lighting and thunder	6–12 h	Few km^2	High under tornado
Squall lines	Line of severe thunder storms that can form along and/or before cold front (leading edge of cooler mass of air replacing a warmer mass of air at ground level)	6–12 h	Few hundred km^2	Moderate
Hail storms	Storms that produce hail stones	6–12 h	Few hundred km^2	High for crops
Heat waves	Prolonged period of excessive hot weather. Daily maximum temperature is more than five consecutive days exceeds the average maximum temperature by 5°C	3–5 days	Several hundred km^2	High
Cold waves	Rapid fall in temperature within 24 h	3–5 days	Several hundred km^2	High
Land slides	Sliding of earth surface	1–2 days	Several hundred km^2	Very high
Floods	Overflow of a huge quantity of water that submerges land	Few hours to several days	Few km^2 to several thousand km^2 depending on flood type	Very high
Drought	Extended period of deficiency in water supply	Weeks to season	Few hundred km^2 to several thousand km^2 depending on flood type	Moderate

Unlike floods and storms which give no time to escape, drought provides sufficient time for people in the region. The people in drought hit area can adopt themselves provided they have additional skills and lower dependency on agriculture. Diversification of economic activity helps people to cope with disasters by exporting goods/services out of the region and importing food/water into the region.

Human exposure to heat extremes (heat waves and cold waves) on the other hand can be handled by staying in the buildings with temperature regulation like airconditioning and room heating. But protection of crop and animals due to heat extremes needs higher investments. Other hindrance in protecting population due to heat extreme is poverty wherein poor without home have to warm themselves by burning waste materials available to them. Many poor people in Delhi also hire blankets on daily basis to cope with the temperature. Table 5.2 shows climatological disaster (drought, extreme temperature, wild fire) during the period 2000–2009.

Table 5.2 Climatological disaster (drought, extreme temperature, wild fire) during the period 2000–2009 in Asia

Sl. no	Country	Number
1	China P Rep	22
2	India	19
3	Bangladesh	10
4	Pakistan	8
5	Turkey	8
6	Afghanistan	6
7	Indonesia	5
8	Korea Rep	4
9	Thailand	4
10	Nepal	4
11	Cyprus	4
12	Tajikistan	3
13	Japan	3
14	Cambodia	3
15	Viet Nam	3
16	Mongolia	2
17	Philippines	2
18	Iran Islam Rep	2
19	Jordan	2
20	Syrian Arab Rep	2
21	Kazakhstan	1
22	Kyrgyzstan	1
23	Uzbekistan	1
24	East Timor	1
25	Malaysia	1
26	Sri Lanka	1
27	Armenia	1
28	Azerbaijan	1
29	Georgia	1
30	Iraq	1
31	Israel	1
32	Lebanon	1

Created on: Sep-24-2010 – Data version: v12.07
Source: "EM-DAT: The OFDA/CRED International Disaster Database http://www.emdat.be – Université Catholique de Louvain – Brussels – Belgium"

As can be seen China and India which are top two nations in terms of population are also in the same order with respect to climatological disasters.

Each disaster is unique in its own way with respect to duration, impact and potential risk. The floods which can occur only for few hours can take away many lives and huge property depending on the location. The floods witnessed by Delhi in 2010 last for few days and city was in total chaos. The floods witnessed at Ladak in Himalayas in the same year due to cloud burst last only few minutes but left many dead. The coping mechanism in flood prone area often involves providing of early warning system, emergency preparedness plan and disaster management plan. Table 5.3 shows hydrological disaster (flood; mass movement wet) during the

Table 5.3 Hydrological disaster (flood; mass movement wet) during the period 2000–2009 in Asia

Sl. no	Country	Number
1	China P Rep	125
2	India	109
3	Indonesia	88
4	Philippines	51
5	Afghanistan	45
6	Viet Nam	41
7	Pakistan	41
8	Thailand	33
9	Malaysia	24
10	Bangladesh	23
11	Iran Islam Rep	22
12	Turkey	22
13	Sri Lanka	17
14	Nepal	16
15	Yemen	16
16	Japan	12
17	Korea Rep	11
18	Korea Dem P Rep	9
19	Myanmar	9
20	Cambodia	8
21	Saudi Arabia	7
22	Lao P Dem Rep	5
23	Iraq	5
24	Georgia	4
25	Mongolia	3
26	East Timor	3
27	Azerbaijan	3
28	Taiwan (China)	2
29	Timor-Leste	2
30	Bhutan	2
31	Syrian Arab Rep	2
32	Hong Kong (China)	1
33	Maldives	1
34	Armenia	1
35	Lebanon	1
36	Palestine (West Bank)	1

Created on: Sep-24-2010 – Data version: v12.07
Source: "EM-DAT: The OFDA/CRED International Disaster Database http://www.emdat.be – Université Catholique de Louvain – Brussels – Belgium"

period 2000–2009. The ranking of China and India has remained the same as that of climatological disasters.

Table 5.4 enlist meteorological (storm) disaster during the period 2000–2009 in Asia indicating Bangladesh as the most vulnerable country for storm followed by India. But not entire India is vulnerable to storms. The coping mechanism in storm prone area demands for early warning system, emergency preparedness plan and disaster management plan.

Frequent occurrence of floods and storms discourages investments in such area. Further the disaster prone areas are not suitable for industries considering the environmental damage such events can cause.

Table 5.4 Meteorological (storm) disaster during the period 2000–2009 in Asia

Sl. no	Country	Number
1	Bangladesh	40
2	India	30
3	Turkey	22
4	Yemen	16
5	Saudi Arabia	7
6	Iraq	5
7	Iran Islam Rep	5
8	Pakistan	5
9	Iran Islam Rep	5
10	Pakistan	5
11	Turkey	5
12	Georgia	4
13	Afghanistan	4
14	Afghanistan	4
15	Azerbaijan	3
16	Oman	3
17	Syrian Arab Rep	2
18	Sri Lanka	2
19	Cyprus	2
20	Jordan	2
21	Syrian Arab Rep	2
22	Yemen	2
23	Armenia	1
24	Lebanon	1
25	Palestine (West Bank)	1
26	Bhutan	1
27	Georgia	1
28	Israel	1
29	Lebanon	1
30	Kyrgyzstan	1
31	Tajikistan	1

Created on: Sep-24-2010 – Data version: v12.07
Source: "EM-DAT: The OFDA/CRED International Disaster Database http://www.emdat.be – Université Catholique de Louvain – Brussels – Belgium"

5.2 Effect of Climate Change on Hydro-Meteorological Risk

Asia is located in a disaster prone area because of its geography, monsoon flood systems, tropical cyclone strike, and cold waves in some countries and heat wave in some other countries. Mountain region is affected by land slides and avalanches, etc. The natural disaster profile of constituent countries differs from each other. To formulate disaster management policies, develop preparedness and mitigation plans and allocate human and financial resources each country needs to prepare its own profile.

Tables 5.5–5.10 clearly indicate the severity with which Asia is affected by climatological, hydrological and Meteorological Disasters in the world.

As could be seen in Table 5.5 top seven climatological disasters in the world – based on number killed occurred in Asia due to drought. Such occurrence before 1970 was mainly due to the countries affected had not adapted to new agricultural practices. Currently the agricultural practices like proper irrigation, use of high

Table 5.5 Top ten climatological disasters in the world – based on number killed

Disaster	Date	No. of people killed
China P Rep, drought	1928	3,000,000
Bangladesh, drought	1943	1,900,000
India, drought	1942	1,500,000
India, drought	1965	1,500,000
India, drought	1900	1,250,000
Soviet Union, drought	1921	1,200,000
China P Rep, drought	1920	500,000
Ethiopia, drought	May-83	300,000
Sudan, drought	Apr-83	150,000
Ethiopia, drought	Dec-73	100,000

Created on: Jul-17-2010 – Data version: v12.07
Source: "EM-DAT: The OFDA/CRED International Disaster Database http://www.em-dat.net – Université Catholique de Louvain – Brussels – Belgium"

Table 5.6 Top ten climatological disasters in the world – based on number affected

Disaster	Date	No. of people affected
India, drought	May-87	300,000,000
India, drought	Jul-02	300,000,000
India, drought	1972	200,000,000
India, drought	1965	100,000,000
India, drought	Jun-82	100,000,000
China P Rep, drought	Jan-94	82,000,000
China P Rep, extreme temperature	10/1/2008	77,000,000
China P Rep, drought	Apr-02	60,000,000
China P Rep, drought	Oct-09	51,000,000
India, drought	Apr-00	50,000,000

Created on: Jul-17-2010 – Data version: v12.07
Source: "EM-DAT: The OFDA/CRED International Disaster Database http://www.em-dat.net – Université Catholique de Louvain – Brussels – Belgium"

Table 5.7 Top ten hydrological disasters in the world – based on number killed

Disaster	Date	No. of people killed
China P Rep, flood	Jul-31	3,700,000
China P Rep, flood	Jul-59	2,000,000
China P Rep, flood	Jul-39	500,000
China P Rep, flood	1935	142,000
China P Rep, flood	1911	100,000
China P Rep, flood	Jul-49	57,000
Guatemala, flood	Oct-49	40,000
China P Rep, flood	Aug-54	30,000
Venezuela, flood	15/12/1999	30,000
Bangladesh, flood	Jul-74	28,700

Created on: Jul-17-2010 – Data version: v12.07
Source: "EM-DAT: The OFDA/CRED International Disaster Database http://www.em-dat.net –
Université Catholique de Louvain – Brussels – Belgium"

Table 5.8 Top ten hydrological disasters in the world – based on number affected

Disaster	Date	No. of people affected
China P Rep, flood	1/7/1998	238,973,000
China P Rep, flood	1/6/1991	210,232,227
China P Rep, flood	30/06/1996	154,634,000
China P Rep, flood	23/06/2003	150,146,000
India, flood	8/7/1993	128,000,000
China P Rep, flood	15/05/1995	114,470,249
China P Rep, flood	15/06/2007	105,004,000
China P Rep, flood	23/06/1999	101,024,000
China P Rep, flood	14/07/1989	100,010,000
China P Rep, flood	8/6/2002	80,035,257

Created on: Jul-17-2010 – Data version: v12.07
Source: "EM-DAT: The OFDA/CRED International Disaster Database http://www.em-dat.net –
Université Catholique de Louvain – Brussels – Belgium"

Table 5.9 Top ten meteorological disasters in the world – based on number killed

Disaster	Date	No. of people killed
Bangladesh, storm	12/11/1970	300,000
Bangladesh, storm	29/04/1991	138,866
Myanmar, storm	2/5/2008	138,366
China P Rep, storm	27/07/1922	100,000
Bangladesh, storm	Oct-42	61,000
India, storm	1935	60,000
China P Rep, storm	Aug-12	50,000
India, storm	14/10/1942	40,000
Bangladesh, storm	11/5/1965	36,000
Bangladesh, storm	28/05/1963	22,000

Created on: Jul-17-2010 – Data version: v12.07
Source: "EM-DAT: The OFDA/CRED International Disaster Database http://www.em-dat.net –
Université Catholique de Louvain – Brussels – Belgium"

Table 5.10 Top ten meteorological disasters in the world – based on number affected

Disaster	Date	No. of people affected
China P Rep, storm	14/03/2002	100,000,000
China P Rep, storm	20/04/1989	30,007,500
China P Rep, storm	16/07/2006	29,622,000
China P Rep, storm	1/9/2005	19,624,000
Bangladesh, storm	11/5/1965	15,600,000
Bangladesh, storm	29/04/1991	15,438,849
China P Rep, storm	8/9/1996	15,005,000
China P Rep, storm	1/7/2001	14,998,298
India, storm	12/11/1977	14,469,800
India, storm	28/10/1999	12,628,312

Created on: Jul-17-2010 – Data version: v12.07
Source: "EM-DAT: The OFDA/CRED International Disaster Database http://www.em-dat.net – Université Catholique de Louvain – Brussels – Belgium"

yielding variety, use of agrochemicals have increased crops and brought down number of death due to hunger. It is interesting to see the top ten climatological disasters based on number of affected occurred only in China and India (Table 5.6). One possible reason is that these two countries top the highly populated nations with China being highest populated country followed by India. Even though occurrence of drought in recent years has affected millions of people, the events have not taken lives as it did prior to 1970 due to international cooperation nations increased capability to cope with such disasters.

A few regions have cooled since 1901. Increase in temperature during this time has been strongest in the continental interiors of Asia and northern North America (Trenberth et al. 2007).

Since the 1950s, studies in parts of North America and southern South America, Europe, northern and eastern Asia, southern Africa and Australasia show a decrease in the number of very cold days and nights and an increase in the number of extremely hot days and warm nights. Precipitation has raised over land north of 30°N over the period 1900–2005 but downward drift dominate the tropics since the 1970s (Trenberth et al. 2007).

Although there was no remarkable trend from 1880 to 1998 during summer in eastern China, precipitation for 1990–1998 was the highest on record for any period of comparable length (Gong and Wang 2000). Increases in drought were found in northern China but not in northwest China (Zou et al. 2005).

The Asian monsoon is of two types – South Asian (or Indian monsoon systems) and East Asian (Ding et al. 2004). Guo et al. (2003) found a reduction in the East Asian summer monsoon between 1951 and 2000 with stronger monsoon prevailing in the first half of the period and a weaker monsoon in the other half. Even though there exists a weakening trend beginning in 1920s, it is not seen in the longer record extending back to the 1850s, which shows decadal-scale variability before the 1940s (Allan and Ansell 2006; Trenberth et al. 2007).

Relation between monsoon-related events viz., rainfall over South and East Asia, tropical pacific circulation decreased between 1890 and 1930 but increased during 1930–1970 (Kripalani and Kulkarni 2001) The linear trends of rainfall decreases for 1900–2005 were 7.5% in both the western Africa and southern Asia regions (Trenberth et al. 2007).

Rainfall in boreal Asia is highly variable both seasonal and spatial scales. Annual mean precipitation in Russia is diminishing but these tendencies have augmented during 1951–1995 (Rankova 1998). In long-term mean precipitation has decreased 4.1 mm/month/100 years. As a result of this augment in precipitation, water storage in a 1-m soil layer has grown by 10–30 mm (Robock et al. 2000). The rising trends in soil moisture have created favorable situation for infiltration into groundwater.

In tropical Asia, hills and mountain ranges are main cause for spatial variations in rainfall. Approximately 70% of the total annual precipitation over the Indian subcontinent is restricted to the southwest monsoon season (June–September). The western Himalayas receive more snowfall than the eastern Himalayas during winter but eastern Himalayas receive more precipitation than in the western Himalayas during the monsoon season (Kripalani et al. 1996). Rainfall during the Indian monsoon season shows decadal variability. Long-term time series of summer monsoon precipitation have no noticeable trends, but decadal departures are observed above and below the long time average alternatively for three successive decades (Kothyari and Singh 1996). Studies have revealed that the impact of El Niño is more severe during the below-normal period, while the impact of La Niña is more severe during the above-normal period (Kripalani and Kulkarni 1997; Kripalani et al. 2001, 2003). Recent decades have shown an increase in extreme rainfall events over northwest India during the summer monsoon (Singh and Sontakke 2002) but the number of rainy days during the monsoon in east coastal stations has decreased in the past decade.

As can be seen in Table 5.7 seven out of top ten floods based on number killed occurred in China and nine out of top ten floods based on number affected occurred in China (Table 5.8)

Top ten Storms in world has occurred only in Asia that to in India, China, Bangladesh and Myanmar (Tables 5.9 and 5.10). Possible coping mechanism in these areas shall include acquiring new skills and migration to non disaster prone area apart from installation of early warning system with emergency preparedness plan, and disaster management plan.

Annual mean rainfall is significantly low in the majority parts of the arid and semi-arid region of Asia. Moreover, temporal variability is quite high: some time, as much as 90% of the annual total is observed in just 2 months of the year at a few places in the region. Precipitation during the past 50 years in some countries in the northern parts of this region has shown growing trend on an annual mean basis. Declining trend is observed in Kazakhstan with respect to annual rainfall for the period 1894–1997 (Mohammed 2005). In Pakistan, 7 of 10 stations have shown increasing rainfall during monsoon season (Chaudhari 1994).

In temperate Asia, the East Asian monsoon significantly influences temporal and spatial changes in rainfall. Annual mean rainfall in Mongolia is 100–400 mm and is restricted mainly to summer. Summer rainfall seems to have decreased over the period 1970–1990 in Gobi; the number of days with comparatively heavy rainfall events has dropped notably (Rankova 1998).

In China, annual precipitation is diminishing continuously since 1965; this decline has become serious since the 1980s (Chen et al. 1992). The summer monsoon is stronger in northern China during globally warmer years (Ren et al. 2000). Drier conditions prevailed over most of the monsoon-affected region during globally colder years (Yu and Neil 1991).

The annual mean rainfall in Sri Lanka is trendless. A positive trends in February and negative trends in June have been reported, however (Chandrapala and Fernando 1995) a long-term declining trend in rainfall in Thailand is reported (OEPP 1996). In Bangladesh, decadal variations were below long term averages until 1960; afterward, they have been much above normal (Mirza and Dixit 1997).

Mountains are unique areas for detecting climate change and for assessing climate-related impacts (Whiteman 2000). Infrastructure in the mountains and downstream could be threatened. Impacts in valley interact with one another in unexpected ways, in some cases resulting in greater impacts (Eamer et al. 2007). In the Eastern Himalayas significant amounts of snowfall at high altitudes is rapidly lost by melting in the summer (Owen et al. 1998).

Study of the glaciations of the of Pamir-Alay mountain system located on the territory of Uzbekistan, Kyrgyzstan and Tajikistan, showed that losses are especially high in Bartang, Muksu, and the systems of the Fedchenko Glacier and in the center and in the south of the region. Area losses are less in the South of the Fergana Valley, Surkhandarya River and Kashkadarya River. The glacial area has been reduced by 1,233 km^2 compared to area in 1957. The glaciers of the region have lost 126 km^3 of ice between 1957 and 1980-year periods (NCRUCC 1999).

The coping mechanisms which often adopted are migration. Migration is a common response to calamities such as floods and famines (Mortimore 1989). Wilbanks et al. (2007) have drawn noticeable links to amplified flood hazard due to climate change. Recent trends in many Asian countries show that such migrations are multidirectional and some times intercontinental. Rural population is entering urban settlement and people in urban areas are migrating to developed countries. Reasons for migration are often multiple but ultimately to live better which include escaping disasters, overcome hunger and poverty.

References

Allan R.J, and Ansell T, 2006: A new globally complete monthly historical gridded mean sea level pressure data set (HadSLP2); 1850–2003. J. Clim., 19, 5816–5842.

Chandrapala L, Fernando T.K, 1995: Climate variability in Sri Lanka: a study of air temperature, rainfall and thunder activity. Proceedings of the International Symposium on Climate and Life in the Asia–Pacific, April 10–13. Darussalam University of Brunei

Chaudhari Q.Z, 1994: Pakistan's summer monsoon rainfall associated with global and regional circulation features and its seasonal prediction. Proceedings of the International Conference on Monsoon Variability and Prediction, May 9–13, Trieste, Italy.

Chen S.J, Kuo Y.H, Zhang P.Z, Bai Q.F, 1992: Climatology of explosive cyclones off the East Asian coast. Mon Weather Rev;1202: 3029–35.

Ding Y.H, Li C.Y and Liu Y.J, 2004: Overview of the South China Seas monsoon experiment. Adv. Atmos. Sci., 21, 343–360.

Doherty R.M, Hulme M, Jones C.G, 1999: A gridded reconstruction of land and ocean precipitation for the extended tropics from 1974–1994. Int J Climatol; 19:119–42.

Eamer J; Lambrechts C; Prestrud P; Young O, 2007: 'Policy and perspectives.' In Eamer, J (ed) Global outlook for ice and snow, pp 215–228. Nairobi: UNEP

Gong D.Y, and Wang S.W, 2000: Severe summer rainfall in China associated with enhanced global warming. Clim. Res., 16, 51–59.

Guo Q, Cai J, Shao X, Sha W, 2003: Interdecadal variability of East-Asian summer monsoon and its impact on the climate of China. Acta Geogr. Sin., 4, 569–576.

Hulme M, Osborn T.J, Johns T.C, 1998: Precipitation sensitivity to global warming: comparison of observations with HadCM2 simulations. Geophys Res Lett ;25:3379–82.

IPCC, 2007: Climate Change 2007: The Physical Science Basis. Contribution of Working Group I to the Fourth Assessment Report of the Intergovernmental Panel on Climate Change [Solomon S, Qin D, Manning M, Chen Z, Marquis M, Averyt K.B, Tignor M and Miller H.L (eds.)]. Cambridge University Press, Cambridge, United Kingdom and New York, NY, USA, 996 pp.

Jones P.D, Hulme M, 1996: Calculating regional climatic time series for temperature and precipitation: methods and illustrations. Int J Climatol; 16:361–77.

Karl T.R, Knight R.W, 1998: Secular trends of precipitation amount, frequency, and intensity in the USA. Bull Am Meteorol Soc; 79:231–41.

Kripalani R.H, Inamdar S.R, Sontakke N.A, 1996: Rainfall variability over Bangladesh and Nepal: comparison and connection with features over India. Int J Climatol;16:689–703.

Kripalani R.H and Kulkarni A, 1997: Climatic impact of El Niño/La Niña on the Indian monsoon: A new perspective. Weather, 52, 39–46.

Kripalani R.H, and Kulkarni A, 2001: Monsoon rainfall variations and teleconnections over South and East Asia. Int. J. Climatol., 21, 603–616.

Kripalani R.H, Kulkarni A, and Sabade S.S, 2001: El Niño Southern Oscillation, Eurasian snow cover and the Indian monsoon rainfall. Proc. Indian Nat. Sci. Acad., 67A, 361–368.

Kripalani R.H, Kulkarni A, Sabade S & Khandekar M, 2003: Indian monsoon variability in a global warming scenario. Natural Hazards, 29, 189–206.

Kothyari U.C, Singh V.P, 1996: Rainfall and temperature trends in India. Hydrol Process;10: 357–72.

Mirza M.Q, Dixit A, 1997: Climate change and water management in the GBM [Ganges–Brahmaputra–Meghna] basins. Water Nepal; 5:71–100.

Mats Eriksson, Xu Jianchu, Arun Bhakta Shrestha, Ramesh Ananda Vaidya, Santosh Nepal, Klas Sandström, 2009: The Changing Himalayas, Impact of climate change on water resources and livelihoods in the greater Himalayas, International Centre for Integrated Mountain Development (ICIMOD).

Mohammed H.I. Dore, 2005: Climate change and changes in global precipitation patterns: What do we know?, Environment International,31, 1167–1181

Mortimore M.J, 1989: Adapting to Drought: Farmers, Famines, and Desertification in West Africa. Cambridge University Press, Cambridge, 299 pp.

NCRUCC (National Commission of the Republic of Uzbekistan on Climate Change), 1999: Initial Communication of the Republic of Uzbekistan under the United Nations framework convention on climate change.

OEPP, 1996: Report on Environmental Conditions of the Year 1994. Bangkok, Thailand Office of Environmental Policy and Planning, Ministry of Science, Technology and Energy

Owen L.A, Derbyshire E, Fort M, 1998: 'The Quaternary glacial history of the Himalaya'. Journal of Quaternary Science 13:91–120

Rankova E, 1998: Climate change during the 20th century for the Russian Federation. In the abstract of the Book of the 7th International Meeting on Statistical Climatology, May 25–29, Whistler, BC, Canada; p. 98.

Robock A, Konstantin Y.V, Srinivasan G, Entin J.K, Hollinger S.E, Speranskaya N.A, Liu S and Namkhai A, 2000: The global soil moisture data bank, Bull Am Meteorol Soc; 81: 1281–99.

Ren G, Wu H, Chen Z, 2000: Spatial pattern of precipitation change trend of the last 46 years over China. J Appl Meteorol; 11(3):322–30.

Robinson D.A: Hemispheric snow cover and surface albedo for model validation. Ann Glaciol 1997;25:241–5.

Singh N, Sontakke N.A, 2002: Natural and anthropogenic environmental changes of the Indo-Gangetic Plains, India. Clim Change; 52:287–313.

Trenberth K.E, Dai A, Rasmussen R.M, and Parsons D.B, 2003: The changing character of precipitation. Bull. Am. Meteorol. Soc., 84, 1205–1217.

Trenberth K.E, Jones P.D, Ambenje P, Bojariu R, Easterling D, Klein Tank A, Parker D, Rahimzadeh F, Renwick J.A, Rusticucci M, Soden B and Zhai P, 2007: Observations: Surface and Atmospheric Climate Change. In: Climate Change 2007: The Physical Science Basis. Contribution of Working Group I to the Fourth Assessment Report of the Intergovernmental Panel on Climate Change [Solomon, S., D. Qin, M. Manning, Z. Chen, M. Marquis, K.B. Averyt, M. Tignor and H.L. Miller (eds.)]. Cambridge University Press, Cambridge, United Kingdom and New York, NY, USA.

Wilbanks T.J, Romero Lankao P, Bao M, Berkhout F, Cairncross S, Ceron J.P, Kapshe M, Muir-Wood R and Zapata-Marti R, 2007: Industry, settlement and society. Climate Change 2007: Impacts, Adaptation and Vulnerability. Contribution of Working Group II to the Fourth Assessment Report of the Intergovernmental Panel on Climate Change, [Parry M.L, Canziani O.F, Palutikof J.P, van der Linden P.J and Hanson C.E, Eds.,] Cambridge University Press, Cambridge, UK, 357–390.

Whiteman D, 2000: Mountain meteorology. Oxford: Oxford University Press

Yu B, Neil D.T, 1991: Global warming and regional rainfall: the difference between average and high intensity rainfalls. Int J Climatol;11: 653–61.

Zou, X.K, Zhai P.M, and Zhang Q, 2005: Variations in droughts over China: 1951–2003. Geophys. Res. Lett., 32, L04707, doi:10.1029/2004GL021853.

Chapter 6
Asian Geological Settings: Its Effect on Climate Change and Disaster

The total population of Asia as on 2005 is approximately 3,879 million making most populated continent with about 60% of the World's population and growing at a rate of approximately 2%. Eastern, southern and south-east Asia are the most populated where as the desert, mountain and tundra regions are the least populated.

Asia is the World's largest continent with 43,810,582 km^2 of land area covering approximately 30% of the Earth's land and 8.66% of the Earth's surface.

The climate of Asia varies depending on position and physical geography. The climate varies significantly throughout Asia due to variation in:

(a) Elevation
(b) Altitude
(c) Distribution of land and water
(d) Latitude
(e) Storms
(f) Monsoons
(g) El Niño and
(h) La Niña

In addition to above the water bodies and diverse biodiversity also plays major role in changing the climate. Forest control humidity by evapotranspiration and water cycle, the water bodies do control temperature by absorbing heat.

There are eight different types of climate depending on the geographical region and biological setting:

Deciduous forest – Area with deciduous forest is characterized by four distinct seasons with warm summers and cold wet winters. Plant life in this region includes tall and short trees, shrubs, small plants and mosses.

Coniferous forest – Also known as Taiga, these areas are, cold and dry with snowy winters and warmer summers. The coniferous forest contains coniferous trees such as pine, fir and spruce.

Alpine/mountain – These areas are cold, windy and snowy. It is characterized by winter from October to May with temperatures below freezing. Summer in this area will be from June to September where the temperature can reach 15°C. Only short grasses and shrubs grow in the tundra and alpine/mountain regions.

R. Chandrappa et al., *Coping with Climate Change*,
DOI 10.1007/978-3-642-19674-4_6, © Springer-Verlag Berlin Heidelberg 2011

Rainforest – These are characterized by high temperatures and high rainfall throughout the year. The rainforest are characterized by jungles of dense, wet forests.

Desert – These areas are characterized by warm to high temperatures with very little rainfall.

Tundra – This area is characterized by a layer of permafrost. Winters are very cold, summers are warm and there is little rainfall.

Grassland – These area witness hot summers and cold winters with above average rainfall. These areas are characterized by large open areas of tall or short grass.

Savanna – These areas are characterized by very high temperatures all year and rain during the summer season. These areas are characterized by large open areas of tall or short grass.

6.1 Geological Setting and Common Disasters in Asia

Asia is the central and eastern part of Eurasia, joined to Europe by a long border generally following Ural Mountains.

Asia lies between $81°13'N$ to $12°04'S$ and $26°4'E$ to $169°0'W$. Northernmost point lies in Arctic Cape, Komsomolets Island, Severnaya Zemlya, Russia and southern most point lies in Pamana Island, a small island off Rote Island, Indonesia. If the Cocos (Keeling) Islands is considered as part of Asia, then South Island is the southernmost point. Cape Baba, Turkey is the westernmost point of the Asian part. Big Diomede, Russia forms easternmost point.

The endpoints of mainland are:

- Northernmost point – Cape Chelyuskin, Russia ($77°43'N$)
- Southernmost point – Cape Piai, Malaysia ($1°16'N$)
- Westernmost point – Cape Baba, Turkey ($26°4'E$)
- Easternmost point – Cape Dezhnev (East Cape), Russia ($169°40'W$)

The Asia can be further divided into Central Asia, East Asia, North Asia, Southeast Asia, West Asia, and South Asia. The classification is not rigid as different literature use the terms depending on the subject and issues discussed. Table 6.1 gives different regions of Asia used in this book.

Asia lies almost completely in the northern hemisphere and hence the seasons of northern Asia, central Asia and eastern Asia are similar where as the seasons of southern and south-east Asia differ slightly due to monsoons as indicated below.

Northern Asia, Central Asia, Eastern Asia

- Spring – March, April, May
- Summer – June, July, August
- Autumn – September, October, November
- Winter – December, January, February

Table 6.1 Different regions of Asia

Asian regions	Description
Central Asia	This area includes republics of Kazakhstan (excluding its small European territory), Uzbekistan, Tajikistan, Turkmenistan and Kyrgyzstan, Afghanistan, Mongolia, Iran and the western regions of China, Former Soviet states in the Caucasus region
East Asia	This area includes Japan; North and South Korea on the Korean Peninsula; Eastern regional of China; Taiwan; Mongolia
North Asia	This term refers to the larger Asian part of Russia, also known as Siberia
Southeast Asia	This region includes Malay Peninsula, Indochina and islands in the Indian Ocean and Pacific Ocean. The countries it contains are: countries Burma (Myanmar), Thailand, Laos, Cambodia, Vietnam, Malaysia, Brunei, the Philippines, Singapore and Indonesia. East Timor
West Asia	This area includes island nation of Cyprus; Syria; Israel; Jordan; Lebanon; Iraq and the Asian portion of Egypt; Saudi Arabia; United Arab Emirates; Bahrain; Qatar; Oman; Yemen; Kuwait; Transcaucasia; Georgia; Armenia; Azerbaijan; Iranian Plateau; Iran and Afghanistan
South Asia	This region includes India, Sri Lanka, Pakistan, Nepal, Maldives Bangladesh, Bhutan

Southern Asia, South-East Asia

- Spring – December, January, February
- Summer – March, April, May
- Autumn – June, July, August
- Winter – September, October, November

Medieval Europeans considered Asia as a distinct landmass but modern discovery no longer consider Europe and Asia to be separate continents. It is rather defined in terms of geological landmasses (physical geography) or tectonic plates (geology). In terms of physical geography, Europe is a western peninsula of Eurasia or the Africa-Eurasia landmass. In terms of tectonic plates, Europe and Asia are parts of the Eurasian plate.

6.1.1 Main Geographical Features

The mean elevation of the continent is 950 m (3,117 ft), the highest among the continents. The plateau and mountainous areas broadly sweep SW-NW across Asia, rising to the highest peaks in the world in the Himalaya Mountains. To the northwest of Asia is plains, while in the south lie the geologically distinct areas of the Arabian Peninsula, Indian subcontinent and Malay Peninsula. Large numbers of islands are scattered in southeast of the continent.

Besides its mainland, Asia includes a large number of islands. Indonesia, Brunei, East Timor, Singapore, Japan, Philippines, Taiwan, Sri Lanka, Maldives and Cyprus are solely made up of one or more islands, and have no territory on the mainland. Other prominent islands include Borneo; Sumatra; Java; Honshū; Bali, Madura and Sulawesi of Indonesia; Hokkaidō, Shikoku, Kyūshū and Okinawa of Japan; the Andaman and Nicobar of India; Luzon, Visayas and Mindanao of the Philippines; Ko Pha Ngan and Ko Samui of Thailand.

Southeast Asia has the utmost relative rate of deforestation among any major tropical region, and likely to lose about 75% of its original forests by 2100 along with up to 42% of its biodiversity (Navjot et al. 2004).

The climate in South Asia is characterized by a tropical monsoon climate with two monsoon systems: the Southwest or summer monsoon (June-September) and the Northeast or winter monsoon (December-April). The rainfall for the duration of the summer monsoon mainly accounts for the total annual rainfall over most of South Asia exception being Sri Lanka). The monsoon rainfall in South Asia is characterized by huge spatial and temporal variability. The arid, semi-arid region around Pakistan and Northwest India receive monsoon rainfall as low as 50 mm whereas parts of Northeast India and the west coast receive over 1,000 mm. The region shows huge year-to-year variations in the rainfall resulting in floods/ droughts over large areas.

Table 6.2 Gives geological settings in Asia which influence the climate and get influenced by climate. But the anthropogenic activities are destroying forest for wood and land; destroying hills and mountains for minerals and rocks; natural water bodies are greatly disturbed by building dams and destroying lakes to make way to building.

Drought and Flood are two common disasters in Asia followed by other disasters like storm, cyclone and earthquake. The combination these would further lead to land movement disasters like rock fall or wild fires or tsunamis. The droughts and floods are often occurring due to result of *El Niño*. *El Niño* is linked with a variation in atmospheric pressure known as the Southern Oscillation, and the overall phenomenon is often referred as the *El Niño* Southern Oscillation (ENSO). Figure 6.1 shows pictorial representation of development of *El Niño*. ENSO consists of *El Niño* – a "warm phase" in the equatorial Pacific Ocean – and *La Niña* – a "cool phase" in the central Pacific Ocean. These phases results in warmer and colder than normal surface water temperature.

El Niño is an intermittent oceanographic occurrence in which a strong and wide warming occurs in the upper ocean in the tropical eastern Pacific. The *El Niño* effect results in strengthening of a warm ocean current in the mid-Pacific, disrupting entire weather mechanism. ENSOs lead to irregular pattern of rainfall and temperature and there will be drop in fish catch; crop loss; and wild fire. ENSO leads to dry conditions in Australia and Southeast Asia, where as more rain and flood occur in Ecuador and Peru. During *La Niña*, rains are enhanced over Australia and Southeast Asia, but the central equatorial Pacific becomes drier.

Twenty ENSOs have been recorded in the 120 years since 1877. Occurrence of *La Niña is less frequent* and is about half as often as does *El Niño*. In 1982–1983, *El Niño* caused severe flooding in Latin America and droughts in parts of Asia. During 1997–1998, it caused severe droughts in Australia, Indonesia, Papua New

Table 6.2 Geological settings in Asia

Region	Landforms	Hydrologic features	Climatic variations	Biotic communities
Central Asia	Gobi desert, Kara Kum desert, Lop desert, Ordos desert, Taklamakan desert, Badain Jaran desert, Himalaya mountains, Kunlun mountains, Mount Kailash, Tian Shan ranges, Yin mountains, Tien Shan mountains, Kunlun-Shan mountains, Eastern Pamir, Western Pamir and Pamir-Alai mountains	Yellow river (Huang He), Yangtze river (Chang Jiang), Pearl river (Zhu Jiang), Hai river	Huge differences in latitude, longitude, and altitude give rise to variations in precipitation and temperature within Central Asia. Precipitation varies regionally even more than temperature	The dominant vegetation in the area is desert, semi-desert and steppe on lower slopes and foothills and in some of the outlying ranges. Riverine woodland survives in few places. At higher altitudes, steppe communities, dominated by various species of grasses and herbs occur. Coniferous forest type occurs on the moist northern slopes of the Tien Shan. At the very highest and coldest elevations, there are cushion plants, snow-patch plants and tundra-like vegetation
East Asia	Himalayan mountains, Greater Hinggan Range, Plateau of Tibet, Gobi desert, Kunlun Shan mountains, North China plain, Altun Shan mountains, Sichlian basin, Tarim basin, Yunnan plateau, Taklimakan desert, Japan trench, Tian Shan mountains, Ryukyu trench, Altai mountains, Izu trench, Khangai mountains	Hwang Ho (Yellow) river, Brahmaputra river, Yangtze river, Indus river, Hsi river, South China sea, Irrawaddy river, East China sea, Salween river, Yellow sea, Mekong River, sea of Japan, Ganges river	Region is characterized by varying climate. The major climate observed is dry climates characterized by evaporation exceeding precipitation; humid mesothermal climate with no distinct dry season; Humid microthermal climate with no distinct dry season; vertical climates go through as many variations as those ranging from the tropics to the poles	The region has monsoon forests, mid-latitude forests (mostly deciduous), coniferous forests (boreal) and temperate grasslands

(continued)

Table 6.2 (continued)

Region	Landforms	Hydrologic features	Climatic variations	Biotic communities
South Asia	Himalaya mountain, Karakorum mountain, Hindu Kush mountain, Western Ghats Hill Range, Western Ghats hill ranges, Easter Ghats hill ranges, Deccan plateau, Vindya hill ranges, Knuckles mountain range, Thar desert, Aravalli hills	Indus rivers, Ganges rivers, Irrawaddy (Ayeyarwady) river, Brahmaputra river, Godavary river, Kaveri river, Krishna river, Narmada river, Taapi river, Mahanadi river, Arabian sea, Indian ocean, Bay of Bengal, Cholistan desert	The climate of south Asia varies from area to area from tropical monsoon in the south to temperate in the north. The variation is influenced by altitude, proximity to the sea coast and the seasonal variations of the monsoons. Southern parts of south Asia are mostly hot in summers and receive rain during monsoon period(s). The northern belt of Indo-Gangetic plains experiences extremes of weather with hot summer due to heat waves from Thar desert and cool in winter due to cold wave from Himalayas. The Himalayas receives snowfall at higher altitudes of Himalayan ranges. Since the Himalayas block the north-Asian bitter cold winds, the temperatures are moderate in the plains down below	The region is characterised by tropical forest found in the Andaman and Nicobar Islands; the Western Ghats. Small patches of rain forest are found in some parts of south Asia. Semi-evergreen rain forest is more extensive than the evergreen forests The tropical vegetation of north-east India typically occurs at elevations up to 900 m Apart from above the region is characterised by Evergreen and semi-evergreen rain forests, moist deciduous monsoon forests, riparian forests, swamps and grasslands. Evergreen rain forests are found in the foothills of the eastern Himalayas
West Asia	Anatolian plateau, An Nafud desert (*part of the Arabian desert*), Ar Rub" Al Khali desert (part of *Arabian desert*), Asir and Hejaz mountains, Caucasus	Major rivers in the region are Tigris and Euphrates	Western Asia is dominantly arid and semi-arid, and is subject to frequent drought. There are two wind phenomena in Western Asia: the *sharqi* and the *shamal*. These winds are	Main biotic communities include Mediterranean forests, rangelands and deserts Apart from terrestrial eco system, marine ecosystems include mudflats, mangrove swamps,

	mountains, Dash-E Lut (*Lout desert*), Dasht-E Kavir (*Kavir desert*), Elburz mountains (*Or Alborz*), Hindu Kush mountains, Kara Kum desert, Pamir, Syrian desert, Taurus mountains, Tien Shan (*Also Tian*), Zagros mountains		dry and dusty, with gusts up to 80 km/h and carry sand a few thousand meters high	sea grass and coral reefs. Biotic community in rivers and springs represent freshwater ecosystems
South East Asia	Southeast Asia is divided into to two geographic regions: (1) the Asian mainland and island arcs and (2) archipelagos	Irrawaddy river, Chao Pya river, Mekong river, Red river	The climate in Southeast Asia is dominantly tropical-hot and humid all year round with plentiful rainfall due to shift in winds or monsoon	Southeast Asia's coral reefs and rain forest have the highest degree of biodiversity

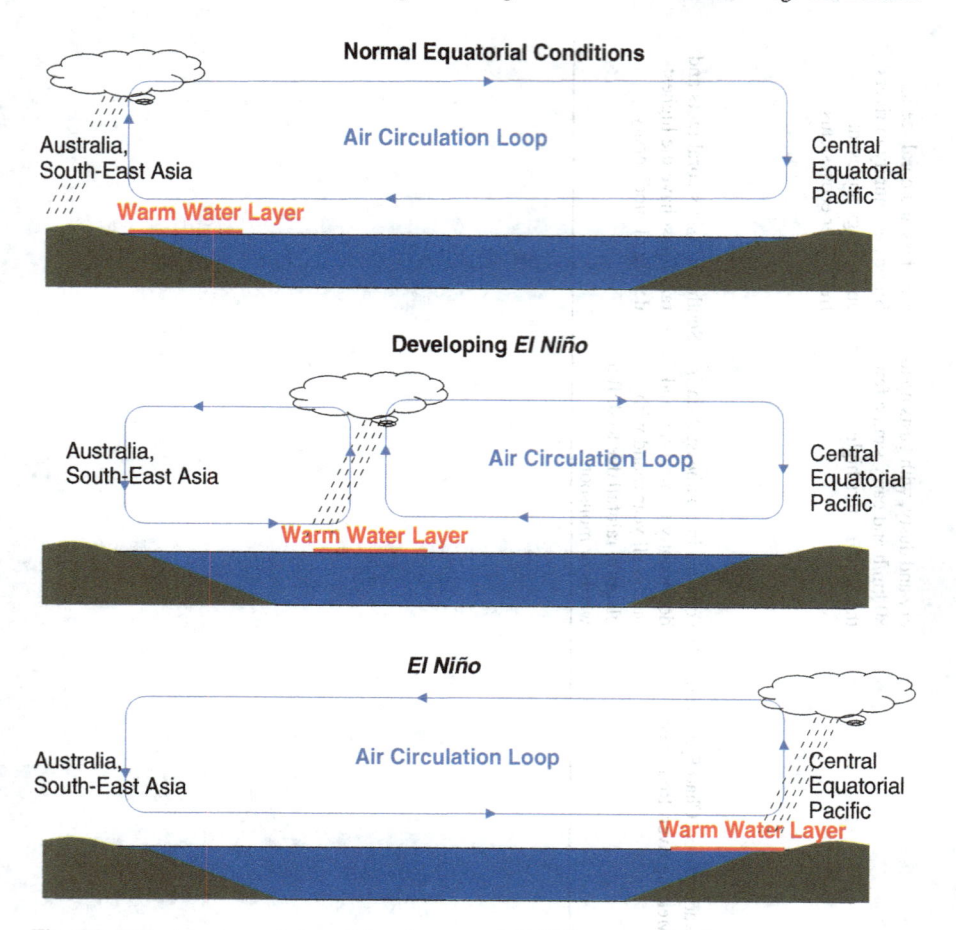

Fig. 6.1 Pictorial representation of development of *El Niño* warm water layer

Guinea, and Philippines; forest fires in Indonesia, Malaysia, Thailand, and Amazon (ADB 2001).

Past studies indicate that there is an ongoing trend to warmer climate over Central Asia. Current precipitation pattern in the upstream areas are changing, on the other hand tendencies are less clear due to the complex mountainous topography (Krysanova et al. 2007).

A change in precipitation at different altitudes is anticipated. Glaciers in mountain are very much sensitive to climate warming. Small glaciers are disappearing, and large glaciers are decreasing in the Amudarya basin. The glacier area in the Amudarya basin has been shrunk from 7,144 km^2 in 1957 to 6,205 km^2 in 1980 (Agaltseva 2005).

The runoff of the major Central Asian rivers is fed by snow and glacier melt in the high altitudes of the Hindukush, Pamir and Tienshan mountains. Climate variation will change the relative runoff contributions from snow, glacier melt and rain (Ososkova et al. 2000). Such changes will alter the flow patterns of the river making people depending on these rivers to change their options.

The Himalayas display immense climatic variability. The mountains act as a barrier to atmospheric circulation for summer monsoon and winter westerlies. The summer monsoon lasts for 8 months (March-October) in the eastern Himalayas, 4 months (June-September) in the central Himalayas, and 2 months (July-August) in the western Himalayas (Chalise and Khanal 2001). The monsoon enter northwards along the Brahmaputra river into the southeast Tibetan Plateau, but seldom as far as the Karakoram (Hofer and Messerli 2006; Rees and Collins 2006). The highest annual rainfall in the region amounting to more than 12,000 mm occurs in Cherapunjee in India. Natural resources in these basins offer the basis for a significant part of the region's economy and important environmental services, which are also of importance beyond the region (Penland and Kulp 2005; Nicholls 1995; Woodroffe et al. 2006; Macintosh 2005; Sanlaville and Prieur 2005).

Bloom in economy has resulted in series of anthropogenic activities as sown in Fig. 6.2 leading to consumption of cement, steel, and wood leading to increase in GHGs.

The Arab region consisting of 22 countries with 10 in Africa and 12 in West Asia is situated in a hyper-arid to arid region with some pockets of semi-arid areas characterized by an extremely harsh environment, with scarcity of water resources, very low precipitation, low biodiversity, and desertification. The region has extended coastal zones on the Mediterranean Sea, the Red Sea, the Gulf and the Atlantic Sea. The coastal area has large percentages of the population living in economic centres.

Physical impact due to climate change in a country need not have to affect only that particular nation. Although Bangladesh accounts for only 7% of the Ganges Brahmaputra Meghna (GBM) basin, due to its geographical location at the tail end of the basin, flooding in Bangladesh depends on the rainfall in the entire GBM basin of India. Hence institutional arrangements with neighbouring countries by international cooperation are as important in managing floods.

Fig. 6.2 Anthropogenic activities are destroying forest for wood and land; destroying hills and mountains for minerals and rocks. Natural water bodies are greatly disturbed by building dams and destroying lakes to make way to building

Box 6.1 Impact of corruption on climate change

Even though array of national and international environmental legislations exists throughout the world, it has not been implemented in the spirit of laws related to human rights. The best example is comparison of adding poison to cup of tea and discharging poison to drinking water source. The person who has added poison to cup of tea will be prosecuted severely when caught with evidences. But in many countries environmental legislations are not implemented seriously due to the presence of corruption making way to discharge of pollutants. There is not much work done to combat adulteration of fuels in corrupt places. This clearly makes way for corruption to dominate in the scene and contribute to emission of GHGs without knowledge of international community. Further corruption also leads to erroneous statistics which will further corrupt the end results of mathematical models used by scientific community to predict climate change.

Corruption will also play major role in energy efficiency as the power theft is often accounted as transmission loss and encourage misuse of power by burning light in day time and running machines when not required.

The corruption will also lead to purchase of unnecessary and inefficient machineries as well as equipments. The officials also misuse cars and official vehicles emitting more of carbon into atmosphere.

The decisions in land use planning is also influenced by corruption leading to unscientific land uses resulting in higher GHGs emissions and reduced carbon sinks.

The corruption in recruitment will also have high impact on the quality of staff incubated to institutions which are responsible for control carbon emission and pollution.

Given the persistent impacts of climate-related risks over time, Bangladesh is also one of the most Climate flexible countries and can provide many lessons on developing climate resilient strategies for other developing countries. But, damages from recent cyclones and floods indicate that substantial risks remain.

Not all countries share same risk in Asia. Countries in central Asia will not have threat of sea level rise as primary risk. But if these countries depend on coastal countries for some of the food and products then effect on economy due disruption of food and material supply cannot be over ruled. Anthropogenic activities reinforced with corruption (Box 6.1) often lead to change in climate at regional and local level thus demanding huge corrective action at international level.

Twenty-one countries of Asia have predominantly low relief with predominantly arid and semi-arid climate. The countries in arid and semi arid region are Afghanistan, Bahrain, Iran, Islamic Republic of Iraq, Israel, Jordan, Kazakstan, Kuwait, Kyrgyz Republic, Lebanon, Oman, Pakistan, Qatar, Saudi Arabia, Syria, Tajikistan, Turkey, Turkmenistan, United Arab Emirates, Uzbekistan and Yemen. Two-thirds of the region is hot or cold desert. Northern part of the region is characterized by cold winters and hot summers. Zone neighboring to the Mediterranean Sea is classified as a Mediterranean zone characterized by wet and moderately warm winters and dry summers. Permafrost zones exist in high mountain areas in the southeast part of this region.

Temperate Asia is composed of three regions: monsoon Asia, with its tropical sub region; the inner arid/semi-arid regions; and Siberia. Tropical cyclones are common in the coastal regions. Inner Siberia is called the "cold pole" due to its mean monthly temperature in January of about $-50°C$ which is the coldest part of the northern hemisphere in winter. Taklamakan Desert in this area experiences extremely dry, hot climate. Temperate Asia comprises of China, Hong Kong, Japan, Democratic People's Republic of Korea, Republic of Korea, and Mongolia. Climate differs extensively within Temperate Asia. It has a humid, cool, temperate climate in the north; a tropical monsoon climate in the remote south; and a desert climate or steppe climate in the west as well as northwest. In the rest of the area a humid, temperate climate prevails. Tropical cyclones are common in the coastal regions.

Tropical Asia includes 16 countries – Bangladesh, Bhutan, Brunei Darussalam, Cambodia, India, Indonesia, Laos, Malaysia, Myanmar, Nepal, Papua New Guinea, Philippines, Singapore, Sri Lanka, Thailand, and Vietnam. Climate in this region is characterized by seasonal weather patterns coupled with the two monsoons and the occurrence of tropical cyclones in the northern Indian Ocean and the north-western Pacific Ocean.

Appendix C gives disasters witnessed by Asian countries from 1900 to 2010. As could be seen from the summary of disasters witnessed in each country, China and India are leading the impacts due to disasters. Countries in south Asia and south east Asia are more vulnerable to climate change with United Arab Emirates being the most safe as per the records since beginning of past century. The absence of climate related disaster in Singapore shows not all Island countries are vulnerable to climate change. This could also one of the reason for its high economic activities. Absence of drought in many of the desert dominated countries may be due to their adaptation strategies to protect life and economy of the region from climate extremes.

6.2 Climate Change and Disasters After Industrialization

Climate change was dramatic after industrialisation in eighteenth and nineteenth century in developed countries and twentieth and twenty-first century in transition and developing countries. Along with industrialisation, world wars in the last century also emitted GHGs emission which is not quantified. In addition to GHG emission the industrialisation and wars destroyed carbon sinks like forests and continued till date.

Irrespective of time, Asia always stood top among the continents disasters due to its high population and vulnerability to climate changes. The continent is attacked by droughts, floods and cyclones with nearly half of the world's disaster affected population.

Infrastructure has accounted for the huge share of adaptation costs. Coastal zones which are home to an ever growing concentration of people and economic activity will be most vulnerable due to changes envisaged with respect to climate. Climate change has already affected the hydrological cycle affecting agriculture by altering yields. Climate change has also resulted in increases in the occurrence of vector-borne disease; water-borne diseases; heat- and cold-related deaths; injuries and deaths due to flooding, and mass movement.

The arid and semi-arid parts comprising of large areas of Pakistan and northwestern Indian states of Rajasthan, Punjab, Haryana and Gujarat experience frequent droughts, and the eastern Himalayan sub-region, fed by the Ganga-Brahmaputra-Meghna river system, are subjected to frequent floods.

During the period 1871–2000 India suffered 22 drought years and 19 flood years with prolonged drought condition in the years 1904–05, 1965–66 and 1985–87. Similarly, country suffered prolonged flood conditions during 1892–94 and 1916–17. Occurrence of droughts and floods in South Asia indicate a clear relationship between the El Nino/La Nina and drought/flood events in the east Pacific Ocean. During the period 1856–1997 there were 30 El Nino years and 16 La Nina years resulting in ten droughts and nine floods. However relationship is weakening in recent years (UNEP and C4 2002).

In addition to climate related disasters industrialisation also brought human induced disasters like fire, industrial accidents, and air/rail/ship accidents. The combination of these disasters occurring at same time poses great challenge in coming days.

References

ADB (Asian Development Bank), 2001: Fire, Smoke, and Haze The ASEAN Response Strategy, PP 246

Agaltseva N, 2005: Climate changes impact to water resources within Amudarya river basin, Report for the NEWATER Amudarya Case Study, NIGMI, Tashkent.

Chalise S.R; Khanal N.R, 2001: An introduction to climate, hydrology and landslide hazards in the Hindu Kush-Himalayan region. In Tianchi, L; Chalise, SR; Upreti, BN (eds) Landslide Hazard Mitigation in the Hindu Kush-Himalayas, pp 51–62. Kathmandu: ICIMOD

Hofer T; Messerli B, 2006: Floods in Bangladesh: History, Dynamics and Rethinking the Role of the Himalayas. New York: United Nations University Press

Krysanova V, Buiteveld H, Haase D, Hattermann F, van Niekerk K. Roest, Santos P.M, Schlüter M, 2007: CAIWA (International Conference on Adaptive & Integrated Water Management) Conference 2007 from 12th to 15th of November 2007.

Macintosh D, 2005: 'Asia, eastern, coastal ecology'. In Schwartz, M (ed) Encyclopedia of Coastal Science. pp 56–67 Dordrecht: Springer,

Navjot S. Sodhi, Lian Pin Koh, Barry W. Brook and Peter K.L. Ng, 2004: Southeast Asian biodiversity: an impending disaster, TRENDS in Ecology and Evolution Vol.19 No.12 December 2004

Nicholls R.J, 1995: Coastal mega-cities and climate change. GeoJournal 37: 369–379

Ososkova T, Gorelkin N, and Chub V, 2000: Water resources of Central Asia and Adaptation Measures for Climate Change. Environmental Monitoring and Assessment 61:161–166.

Penland S, Kulp M.A, 2005: 'Deltas'. In Schwartz, M.L (ed) Encyclopedia of Coastal Science, pp 362–368. Dordrecht: Springer

Rees G.H, Collins D.N, 2006: 'Regional differences in response of flow in glacier-fed Himalayan rivers to climate warming'. Hydrological Processes 20: 2157–2167

Sanlaville P, Prieur A, 2005: 'Asia, Middle East, coastal ecology and geomorphology'. In Schwartz, ML (ed) Encyclopedia of Coastal Science, pp 71–83. Dordrecht: Springer

UNEP and C4, 2002: The Asian Brown Cloud: Climate and Other Environmental Impacts UNEP, Nairobi

Woodroffe C.D, Nicholls R.J, Saito Y, Chen, Z, Goodbred, S.L, 2006: 'Landscape variability and the response of Asian megadeltas to environmental change'. In Harvey, N (ed) Global change and integrated coastal management: the Asia-Pacific region, pp 277–314. New York: Springer

Chapter 7
Climatic and Non Climatic Hazards: Asian Context

Disasters can occur from natural and/or technological hazards and their combinations. The natural hazards fall into: astronomical (solar flares, geomagnetic storms, supernovas, and the collision of celestial matter with the Earth); biological (disease, pestilence, genetic mutations, and conflicts between human and wild animals); hydrometerological (floods, fog, severe storms, hurricanes, storm surges, tornadoes, temperature extremes); and geologic (earthquakes, tsunamis, liquefaction, landslides, rock slides, avalanches, mud flows, volcanoes, tsunamis, lahars, fumaroles, and lava flows).

Technological hazards are of two types: deliberate and accidental. Deliberate acts include industrial/vehicular pollution, pesticide and herbicide use, waste dumping, terrorist acts, war, etc. Accidental acts are accidents at industries/vehicles, etc.

Climatic and non climatic hazards have always been a matter of concern to the people all over the world. Disaster risk is closely connected to human development. Different types of disasters can either occur separately or together. Though there have been noteworthy accomplishments in science and technologies, people still continue to suffer the consequences of hazards on all continents. Hazards endanger human life as well as cause damage to settlements, communication networks. Both climatic and non climatic hazards are harmful for agriculture and may create problems in water supply. Landslides and floods can occur with earthquake, which can end in devastating scenario. In view of the climate change, floods and droughts may become more frequent. Warming of the atmosphere will lead to an increased evaporation from the oceans, which leads to an increased precipitation. But precipitation will not be uniform. Only some parts of the world (in high latitudes and tropical regions) would experience higher precipitation, whereas other regions would get lower amounts of rainfall. Also, regional variations do occur due to local topographical conditions. Climate change therefore has the potential to increase the frequency of disasters in many regions.

Hazards are strongly related to geomorphology of the earth. Hazards are the result of abrupt changes in long-term behavior due to minute changes in the initial conditions (Scheidegger 1994). In this sense, geomorphic hazards can be classified

R. Chandrappa et al., *Coping with Climate Change*,
DOI 10.1007/978-3-642-19674-4_7, © Springer-Verlag Berlin Heidelberg 2011

as endogenous (volcanism and neotectonics), exogenous (floods, avalanche, tsunamis, coastal erosion), and those stimulated by climate and land-use change (desertification, permafrost, degradation, soil erosion, salinization, floods) (Slaymaker 1996).

Figure 7.1 shows number of natural disasters by country from 1976 to 2005; Fig. 7.2 shows number of persons reported affected by natural disasters in 2009; Fig. 7.3 shows reported economic damages from natural disasters in 2009; Fig. 7.4 shows number of persons reported killed by natural disasters in 2009 and Fig. 7.5 shows natural disaster occurrence in 2009. The reader may note that term natural disasters is used loosely in literature. However, no disaster is natural, it is the hazard which can be natural or man-made, and once damage is beyond coping capacity of community, the hazard becomes a disaster.

A quick glance at these figure shows that more people in developing countries get killed due to natural disasters. This is mainly due to lack of early warning systems, delay in disaster response and recovery, absence of emergency preparedness and disaster management plans.

Risk is external fact of life. Vulnerability is inability to manage risk. Irrespective of development all are vulnerable to certain extent to natural disasters. Economic damages are higher in developed countries due to high cost of property. Thus even small number of disasters in developed countries could lead to great economic loss where as developing countries have to suffer large number of disaster to make loss equivalent to that of developed country.

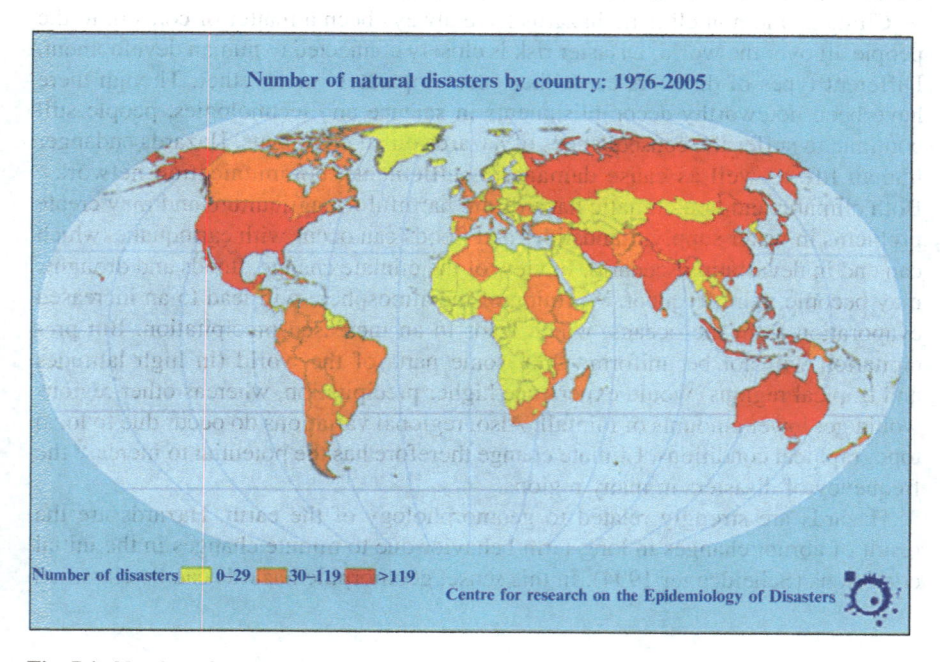

Fig. 7.1 Number of natural disasters by country from 1976 to 2005

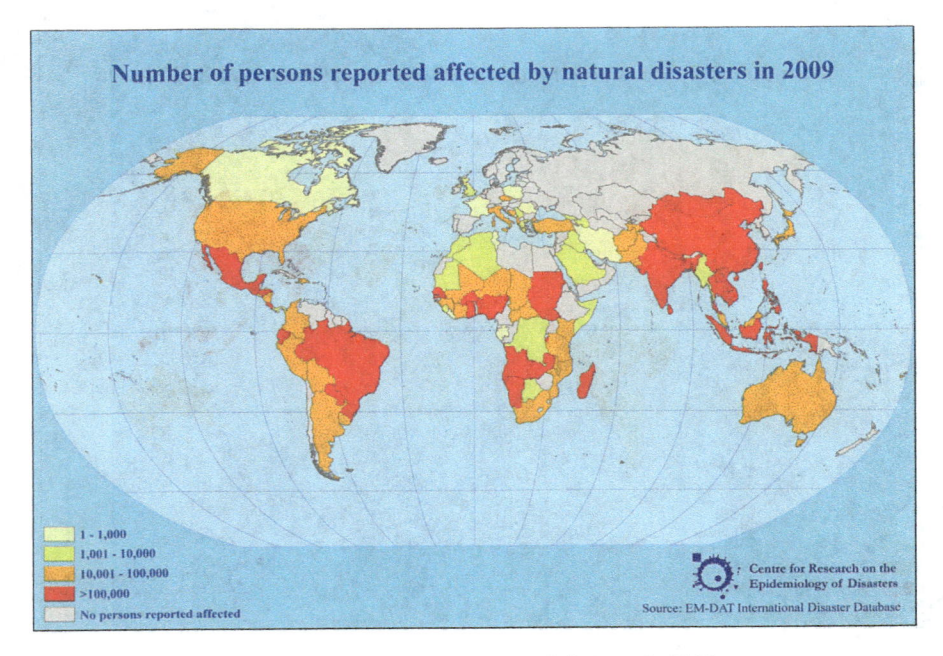

Fig. 7.2 Number of persons reported affected by natural disasters in 2009

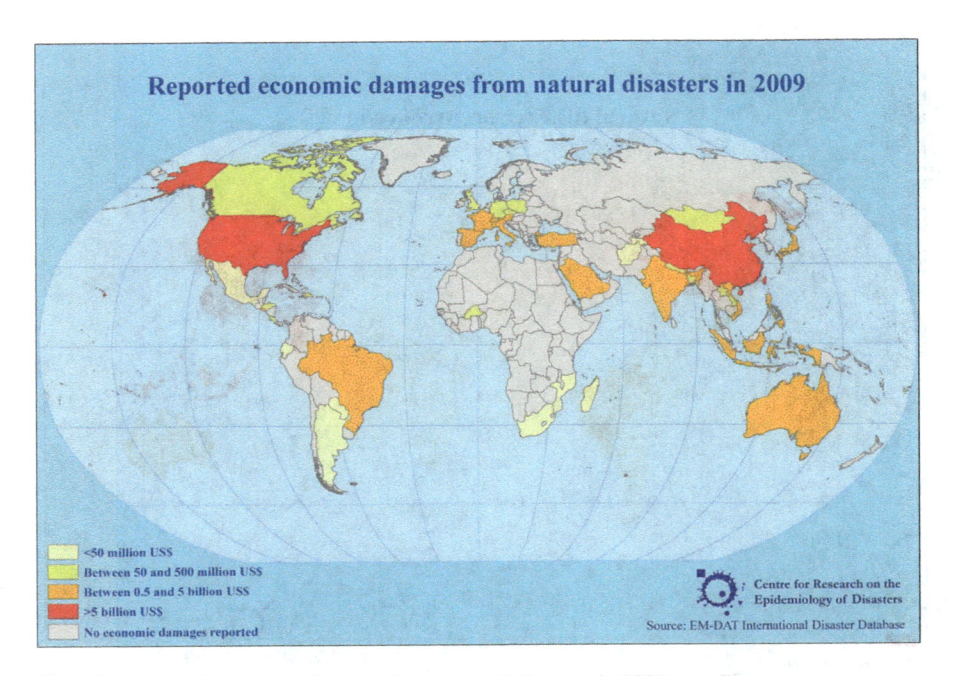

Fig. 7.3 Reported economic damages from natural disasters in 2009

Fig. 7.4 Number of persons reported killed by natural disasters in 2009

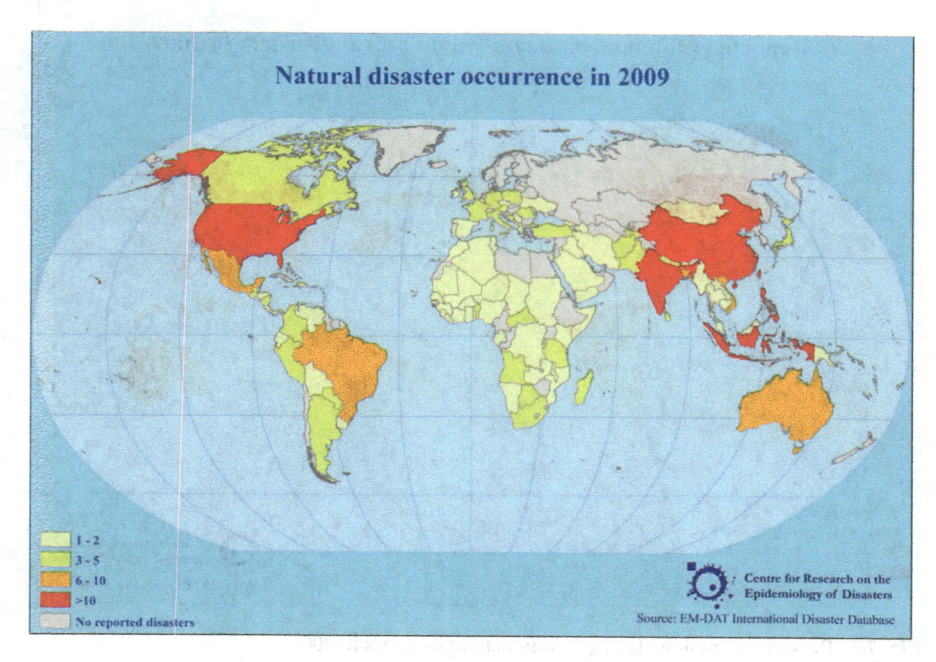

Fig. 7.5 Natural disaster occurrence in 2009

Hazard could be manmade (Box 7.1) or natural (Fig. 7.6). But in either the case damage will bring loss to life, and infrastructure. Such losses are likely to affect economic activities and demand creation of new infrastructure.

Box 7.1 Man made hazards
Man made hazards are two types:
    Technological hazards – Hazard due to unintentional accident cased due to unpredicted technical fault
    E.g. Industrial hazards, structural collapse, fire, radiation contamination, accident during transportation, space disasters, chemical hazards, biological hazards.
    Sociological hazards – Hazard to humans and environment due to intentional sociological reasons.
    E.g. Crime, arson, Civil disorder, terrorism, war, etc.

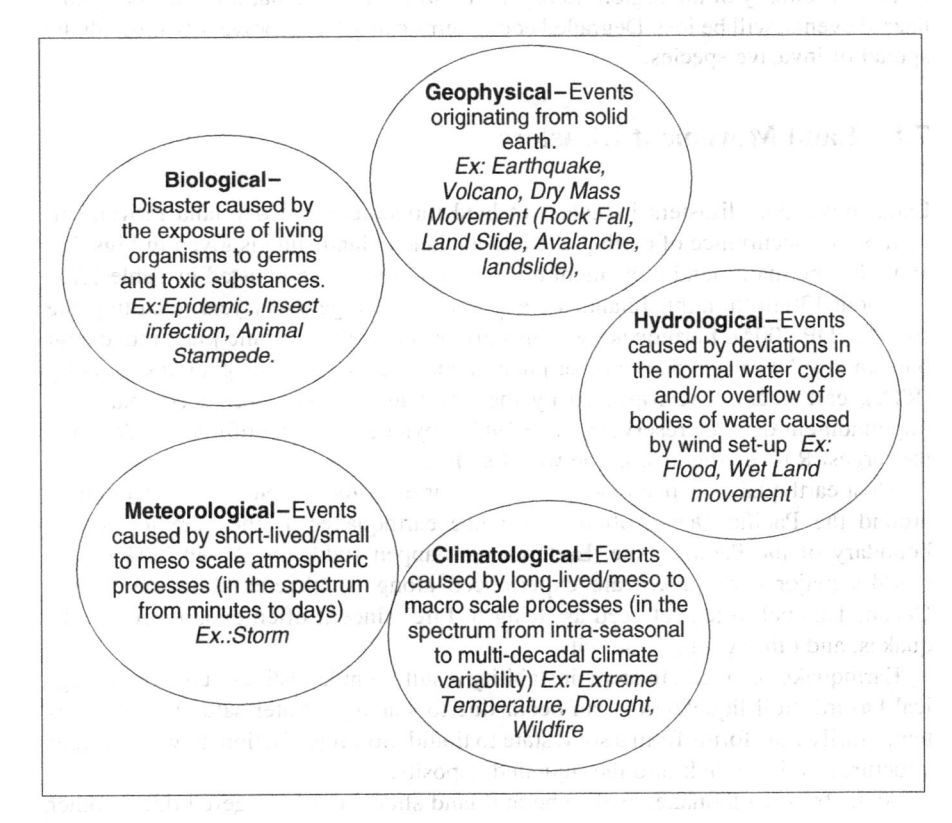

Fig. 7.6 Natural disasters

Some of the manmade disasters can lead to emission of GHGs and cause climate change. While others associated with climate change like floods can cause chemical spillage and damage the environment in the locality.

Social hazards can be curbed by proper negotiation, enforcement of law, awareness and social justice. In the absence of these mechanisms country has to tackle both social hazard and climate triggered hazards.

The vulnerability of society not only depends on the magnitude but also on the wealth created in the region. Hence the disaster of same magnitude can bring in great losses to developed country if measured in terms of "millions of dollars" where as "number of people dead/affected" will be more in developing countries due to absence of early warning system and other disaster coping mechanisms.

All disasters can't be avoided. But a prompt, well-coordinated and effective response after disasters can minimize loss of life and property. The response requires integrated institutional arrangements, forecasting and early warning systems, good communication system, rapid evacuation of affected communities.

Disaster also affects natural ecosystem on which human well-being depends. The disruption of diverse livelihoods which depends on natural ecosystem also disrupts economy of the region. Ecosystems, which provide natural buffers against hazard events, will be lost. Degraded ecosystems can affect biodiversity through the spread of invasive species.

7.1 Land Movement Disasters

Land movement disasters include wet land movement and dry land movement. Number of occurrence of earthquake and avalanche/land slide is given in Figs. 7.7 and 7.8. The major land movement disasters and relevance is listed in Table 7.7.

About 130 million inhabitants are exposed on average every year to earthquake risk (UNDP 2004). Earthquake will usually occur due to tectonic plate movement but can also be caused due to other phenomena like Reservoir-triggered seismicity (RTS), earthquakes are triggered by the physical processes that accompany the impoundment of large reservoirs. The 1967 Koyna earthquake of magnitude 6.5 is the largest RTS earthquake in the world so far.

Most earthquakes happen along the boundaries of the tectonic plates. Countries around the Pacific Ocean often experience earthquakes as they are located in boundary of the Pacific plate. Earthquakes happen suddenly. About 80% of the world's major earthquakes are experienced along a belt encircling the Pacific Ocean. This belt often refereed as 'Ring of Fire' since it often experiences earthquakes, and other geologic activity.

Earthquakes can lead to slope instability resulting in landslides, a major geological hazard. Soil liquefaction can occur due to shaking. Water-saturated soil thus temporarily transforms from a solid state to liquid. Soil liquefaction may result rigid structures to tilt or sink into the liquefied deposits.

Apart from earthquake, avalanche and land slide can be triggered due to other causes as well. Occurrence and magnitude of Avalanches depends on weather,

Number of Occurrences of Earthquake Disasters by Country:
1974-2003

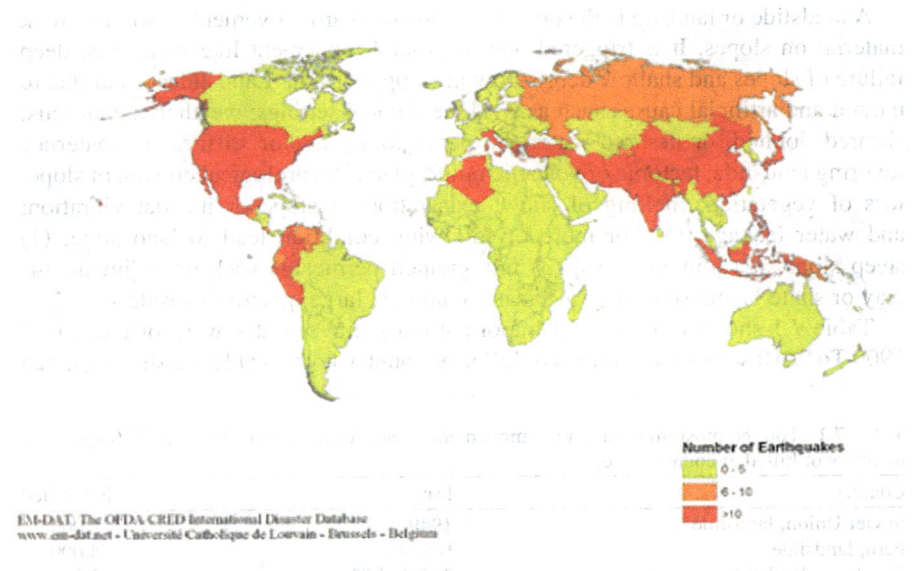

Number of Earthquakes
- 0 - 5
- 6 - 10
- >10

EM-DAT: The OFDA-CRED International Disaster Database
www.em-dat.net - Université Catholique de Louvain - Brussels - Belgium

Fig. 7.7 Number of occurrence of earthquake disasters by country from 1974 to 2003

Number of Occurrences of Avalanche/Landslide Disasters by Country:
1974-2003

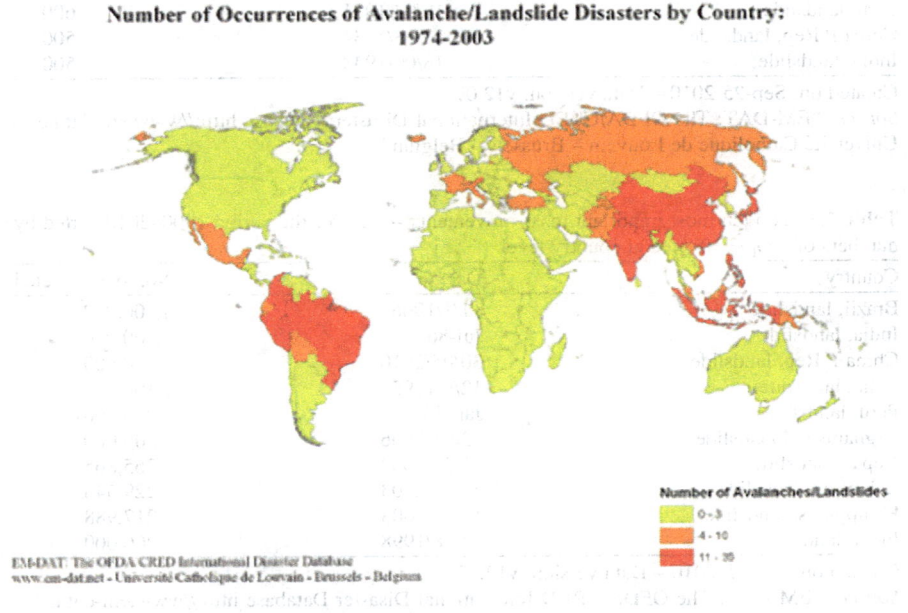

Number of Avalanches/Landslides
- 0 - 3
- 4 - 10
- 11 - 39

EM-DAT: The OFDA-CRED International Disaster Database
www.em-dat.net - Université Catholique de Louvain - Brussels - Belgium

Fig. 7.8 Number of occurrence of avalanche/landslide disasters by country from 1974 to 2003

slope angle, slope orientation, snowfall, snow pack conditions, terrain, and vegetation. Avalanches can be triggered by both natural and anthropogenic causes.

A landslide or landslip is the downward and outward movement of soil and rock material on slopes. It is triggered due to ground movement like rock falls, deep failure of slopes and shallow debris flow in sloppy terrain. Landslides occur due to natural and artificial causes such as weak geological settings; weathered materials; sheared, jointed, or fissured materials; permeability and/or stiffness of materials favoring landslide; tectonic or volcanic uplift; glacial/hydrological erosion of slope; loss of vegetation; melting of snow; excavation of slope or its toe; vibration; and water leakage. One or more of following condition lead to land slide: (1) steep slope; (2) jointed rocks; (3) fine-grained permeable rock or sediment; (4) clay or shale layers subject to lubrication; and (5) large quantity of water.

Table 7.1 shows top ten most important mass movement – wet, for the period 1900 To 2010 sorted by numbers of killed at country level. Table 7.2 shows top ten

Table 7.1 Top ten most important mass movement – wet, for the period 1900 to 2010 sorted by numbers of killed at country level

Country	Date	No. killed
Soviet Union, landslide	1949	12,000
Peru, landslide	Dec-41	5,000
Honduras, landslide	20/09/1973	2,800
Italy, landslide	9/10/1963	1,917
Philippines, landslide	17/02/2006	1,126
India, landslide	1/10/1968	1,000
Colombia, landslide	27/09/1987	640
Peru, landslide	18/03/1971	600
China P Rep, landslide	23/03/1934	500
India, landslide	18/09/1948	500

Created on: Sep-25-2010 – Data version: v12.07
Source: "EM-DAT: The OFDA/CRED International Disaster Database http://www.em-dat.net – Université Catholique de Louvain – Brussels – Belgium"

Table 7.2 Top ten most important mass movements – wet, for the period 1900–2010 sorted by numbers of people affected at country level

Country	Date	No. total affected
Brazil, landslide	11/1/1966	4,000,000
India, landslide	Jul-86	2,500,000
China P Rep, landslide	30/05/2010	2,100,000
India, landslide	12/9/1995	1,100,000
Peru, landslide	Jan-83	700,000
Afghanistan, landslide	13/01/2006	300,000
Nepal, landslide	15/07/2002	265,865
Indonesia, landslide	31/03/2003	229,548
Philippines, landslide	19/12/2003	217,988
India, landslide	17/08/1998	200,000

Created on: Sep-25-2010 – Data version: v12.07
Source: "EM-DAT: The OFDA/CRED International Disaster Database http://www.em-dat.net – Université Catholique de Louvain – Brussels – Belgium"

most important mass movements – wet, for the period 1900–2010 sorted by numbers of people affected at country level. Even though landslides predominantly occur in mountainous regions, they can also occur in areas of low relief. In low-relief areas landslides occur in roadway and building excavations, river buff failures, lateral spreading landslides, collapse of mine-waste piles, and collapse in quarries as well as open-pit mines.

Table 7.3 shows summarized table of mass movement – wet, sorted by continents for the period 1900–2010. Table 7.4 shows top ten most important mass movements – dry, for the period 1900–2010 sorted by numbers of people killed at country level. Most landslides accompany rains. The monsoon of 1998 caused widespread damage in the Central Himalaya, Mandakini river valley in Garhwal (Sah and Bist 1998; Rautela and Thakur 1999 and Kali river valley in Kumaun Himalaya (Valdiya 1998; Paul and Mahajan 1999) with the death toll exceeding 300 (Paul et al. 2000).

Landslides of different types occur frequently in Himalaya, and other mountainous regions of Asia. Landslides occur due to chain of events which could be natural or/and man-made. Numerous coastal areas of South Asia have been affected by underwater landslide and coastal instability. Climate change and associated glacial

Table 7.3 Summarized table of mass movement – wet, sorted by continents for the period 1900–2010

		No. of events	Killed	Total affected	Damage (000 US$)
Africa	Landslide	28	1,154	54,537	–
	Ave. per event		41.2	1,947.8	–
Americas	Avalanche	4	95	154	–
	Ave. per event		23.8	38.5	–
	Landslide	145	18,987	5,454,254	2,021,727
	Ave. per event		130.9	37,615.5	13,942.9
	Rockfall	1	33	–	–
	Ave. per event		33	–	–
Asia	Avalanche	43	2,357	57,697	50,000
	Ave. per event		54.8	1,341.8	1,162.8
	Debris flow	1	106	–	–
	Ave. per event		106	–	–
	Landslide	257	16,496	7,753,322	1,989,916
	Ave. per event		64.2	30,168.6	7,742.9
	Subsidence	1	287	2,838	–
	Ave. per event		287	2,838	–
Europe	Avalanche	34	1,247	13,145	774,889
	Ave. per event		36.7	386.6	22,790.9
	Landslide	33	15,282	25,352	2,334,000
	Ave. per event		463.1	768.2	70,727.3
Oceania	Landslide	17	486	20,315	2,466
	Ave. per event		28.6	1,195	145.1

Created on: Sep-25-2010 – Data version: v12.07
Source: "EM-DAT: The OFDA/CRED International Disaster Database http://www.em-dat.net – Université Catholique de Louvain – Brussels – Belgium"

Table 7.4 Top ten most important mass movements – dry, for the period 1900–2010 sorted by numbers of people killed at country level

Country	Date	No. killed
Peru, landslide	10/1/1962	2,000
Philippines, landslide	21/10/1985	300
China P Rep, landslide	7/3/1983	277
Turkey, avalanche	1/1/1992	261
Colombia, rockfall	28/07/1983	160
Nepal, landslide	10/8/1963	150
Indonesia, landslide	4/5/1987	131
Afghanistan, avalanche	Mar-93	100
Egypt, rockfall	6/9/2008	98
Canada, rockfall	29/04/1903	76

Created on: Sep-25-2010 – Data version: v12.07
Source: "EM-DAT: The OFDA/CRED International Disaster Database http://www.em-dat.net – Université Catholique de Louvain – Brussels – Belgium"

Table 7.5 Top ten most important mass movements – dry, for the period 1900–2010 sorted by numbers of people affected at country level

Country	Date	No. total affected
Soviet Union, landslide	Mar-89	8,000
China P Rep, landslide	6/10/1990	5,105
Guatemala, landslide	4/1/2009	3,028
Colombia, landslide	18/12/1993	2,411
Canada, landslide	12/11/1955	2,006
Russia, avalanche	26/11/1992	1,750
Canada, landslide	4/5/1971	1,500
Turkey, avalanche	1/1/1992	1,069
Papua New Guinea, landslide	6/9/1988	1,000
Indonesia, landslide	4/5/1987	701

Created on: Sep-25-2010 – Data version: v12.07
Source: "EM-DAT: The OFDA/CRED International Disaster Database http://www.em-dat.net – Université Catholique de Louvain – Brussels – Belgium"

melts, glacial lake outburst floods (GLOFs) and sea level rise have contributed to risks of landslides.

Slope saturation by water is one of major cause of landslides. Changing climate resulting in severe rainfall, snowmelt, flooding, variation in ground-water levels, and sea level rise could favor landslides. Landslides can block valleys and stream channels, allowing large amounts of water stagnate leading to upstream flooding. Failure of the blocked material leads to downstream flooding.

Table 7.4 shows top ten most important mass movements – dry, for the period 1900–2010 sorted by numbers of people killed at country level. Table 7.5 shows top ten most important mass movements – dry, for the period 1900–2010 sorted by numbers of people affected at country level.

Table 7.6 shows summarized table of mass movement – dry, sorted by continents for the period 1900–2010. Table 7.7 shows land movement disasters and relevance to climate change. The occurrence of earthquakes in greatly increases the likelihood

Table 7.6 Summarized table of mass movement – dry, sorted by continents for the period 1900–2010

		No. of events	Killed	Total affected	Damage (000 US$)
Africa	Landslide	2	59	200	–
	Ave. per event		29.5	100	–
	Rockfall	2	129	697	–
	Ave. per event		64.5	348.5	–
	Subsidence	1	34	300	–
	Ave. per event		34	300	–
Americas	Avalanche	3	144	44	–
	Ave. per event		48	14.7	–
	Debris flow	1	10	–	–
	Ave. per event		10	–	–
	Landslide	10	2,290	8,945	200,000
	Ave. per event		229	894.5	20,000
	Rockfall	4	277	41	–
	Ave. per event		69.3	10.3	–
Asia	Avalanche	4	423	1,069	–
	Ave. per event		105.8	267.3	–
	Landslide	14	1,184	6,574	1,000
	Ave. per event		84.6	469.6	71.4
	Rockfall	1	50	–	–
	Ave. per event		50	–	–
Europe	Avalanche	6	167	1,802	2,600
	Ave. per event		27.8	300.3	433.3
	Landslide	3	102	8,506	–
	Ave. per event		34	2,835.3	–
Oceania	Unspecified	1	10	–	–
	Ave. per event		10	–	–
	Landslide	1	76	1,000	–
	Ave. per event		76	1,000	–

Created on: Sep-25-2010 – Data version: v12.07
Source: "EM-DAT: The OFDA/CRED International Disaster Database http://www.em-dat.net – Université Catholique de Louvain – Brussels – Belgium"

that landslides due to ground shaking and rapid infiltration of water. Volcanic lava can also lead to land slide by melting snow at a rapid rate leading to rapid flow of rock, soil, ash, and water.

7.2 Water related Disasters

Among all adversities, water-related disasters are the most regular and pose major obstruction to the achievement of human development. The numbers of water-related disasters recorded was highest in Asia. The biggest challenge for water disaster planning is the underlying capacities of the provincial and district authorities. In majority of cases, local resources and capacities are often overlooked, thus depending too much upon external assistance.

Table 7.7 Land movement disasters and relevance to climate change

Land movement disaster	Description	Relevance to climate change
Earthquakes	An earthquake is a sudden shaking of the Earth's crust	Earthquake may or may not happen due to climate change, but may destroy forests which are ultimate sinks for carbon dioxide
Lahars	A lahar is a type of mudflow or landslide that flows down from a volcano	Likely to contribute to GHG emission and aerosols
Rock fall	Rock fall is the natural motion of a detached blocks involving free falling, bouncing, rolling, and sliding	Geology of rocks and climatic factors (such as freeze-thaw; avalanche, temperature changes and precipitation) are responsible for magnitude of the damage
Land slide	A landslide or landslip is a geological phenomenon which includes a wide range of ground movement like – (1) debris flow (slide characterised by movement of rocks/soil and debris mixed with water/ice); (2) Earth flow (viscous flows of saturated, fine-grained materials); (3) debris avalanche (chaotic movement of rocks/soil and debris mixed with water/ice); (4) Sturzstrom (slides are unusually mobile, flowing very far over a low angle, flat, or even slightly uphill terrain); (5) shallow landslide (sliding surface is located within few decimetres to some metres); (6) deep-seated landslide (sliding surface is mostly deeply located below the maximum rooting depth of trees)	Climate plays major role in triggering landslides by changing temperature, precipitation and wind pattern which loosen the rocks and soil leading to landslides
Avalanche	An avalanche is a fast flow of snow down a slope, due to natural/anthropogenic triggers	Changing weather can trigger the avalanches

For a successful water disaster preparedness planning, it is essential to learn from the experiences and best practices for greater relationship and information sharing to enhance the synergy and to expand the resource base for more effective execution of flood preparedness programs.

Floods are among the most disparaging natural hazards causing widespread damage to infrastructure; public and private services; environment; economy; and human settlements. The water disasters and its relevance are listed in Table 7.8. Number of occurrence of flood is given in Fig. 7.9. The Asian region, experiences flood disasters almost every year. While the number of deaths caused by flooding decreased over the last decade, economic losses have increased significantly. This is mainly due to increase in cost of property. These trends demand better

Table 7.8 Water disasters and relevance to climate change

Water disaster	Description	Relevance to climate change
Flood	Covering by water of land not normally covered by water temporarily	Floods are direct consequences of heavy rainfall. The changing precipitation pattern could be one of the main reason
Storm	A storm is disturbed state of an atmosphere, strongly implying severe weather. It may be marked by strong wind, thunder and lightning; and heavy precipitation	The changing precipitation pattern could be one of the main reason
Limnic eruption	A limnic eruption is natural disaster in which carbon dioxide suddenly erupts from deep water bodies, suffocating fauna	Released carbon dioxide may be due to volcanic gas emitted under the water bodies, or from degradation of organic material. In either the case the phenomena would contribute to GHG emission to atmosphere
Tsunami	Series of water waves caused by the displacement of a large volume of water in water body	Tsunami does not happen due to climate change, but may destroy forests which are ultimate sinks for carbon dioxide

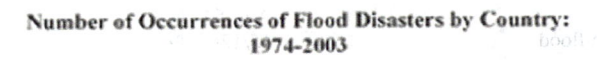

Number of Occurrences of Flood Disasters by Country: 1974-2003

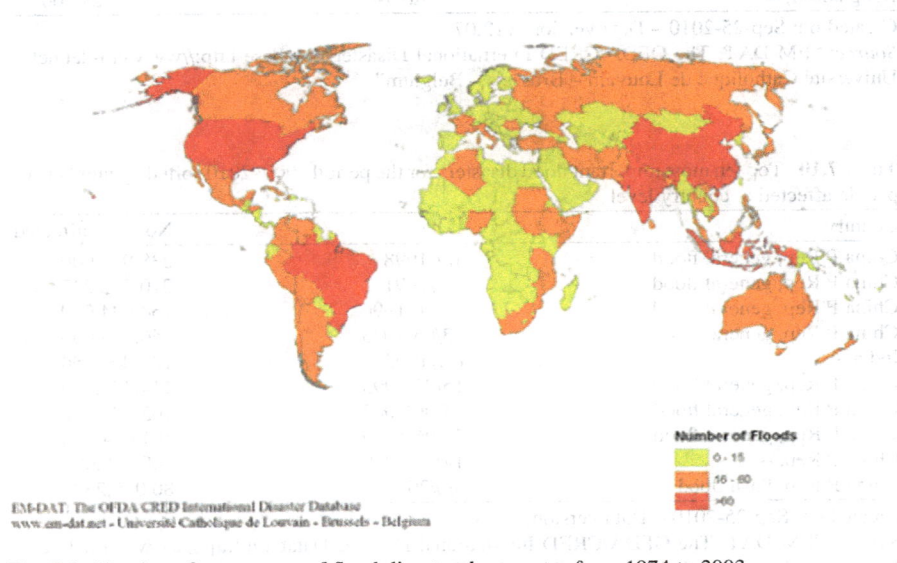

Number of Floods
- 0 - 15
- 16 - 60
- >60

EM-DAT: The OFDA CRED International Disaster Database
www.em-dat.net - Université Catholique de Louvain - Brussels - Belgium

Fig. 7.9 Number of occurrence of flood disasters by country from 1974 to 2003

preparedness at international, national, provincial and local levels to make effective response measures.

Recent years have seen thriving population and hapazard urbanization create tremendously dangerous situations. Flood plain areas that are populated are not safe during prolonged periods of flooding (Todini 1999).

Table 7.9 shows top ten most important flood disasters for the period 1900–2010 sorted by numbers of killed at country level. Table 7.10 shows top ten most important flood disasters for the period 1900–2010 sorted by numbers of people affected at country level. Table 7.11 shows top ten most important flood disasters for the period 1900–2010 sorted by economic damage at country level. Table 7.12 shows summarized table of flood sorted by continents for the period 1900–2010.

Floods usually occur locally and are short-lived events that can happen suddenly with little or no warning. They usually caused by runoff that are greater than normal runoff stream can carry. Rivers can also get flooded due to failure of dams and

Table 7.9 Top ten most important flood disasters for the period 1900–2010 sorted by numbers of killed at country level

Country	Date	No. killed
China P Rep, general flood	Jul-31	3,700,000
China P Rep, –	Jul-59	2,000,000
China P Rep, general flood	Jul-39	500,000
China P Rep, –	1935	142,000
China P Rep, general flood	1911	100,000
China P Rep, –	Jul-49	57,000
Guatemala, –	Oct-49	40,000
China P Rep, –	Aug-54	30,000
Venezuela, flash flood	15/12/1999	30,000
Bangladesh, –	Jul-74	28,700

Created on: Sep-25-2010 – Data version: v12.07
Source: "EM-DAT: The OFDA/CRED International Disaster Database http://www.em-dat.net – Université Catholique de Louvain – Brussels – Belgium"

Table 7.10 Top ten most important flood disasters for the period 1900–2010 sorted by numbers of people affected at country level

Country	Date	No. total affected
China P Rep, general flood	1/7/1998	238,973,000
China P Rep, general flood	1/6/1991	210,232,227
China P Rep, general flood	30/06/1996	154,634,000
China P Rep, general flood	23/06/2003	150,146,000
India, –	8/7/1993	128,000,000
China P Rep, general flood	15/05/1995	114,470,249
China P Rep, general flood	15/06/2007	105,004,000
China P Rep, general flood	23/06/1999	101,024,000
China P Rep, –	14/07/1989	100,010,000
China P Rep, flash flood	8/6/2002	80,035,257

Created on: Sep-25-2010 – Data version: v12.07
Source: "EM-DAT: The OFDA/CRED International Disaster Database http://www.em-dat.net – Université Catholique de Louvain – Brussels – Belgium"

Table 7.11 Top ten most important flood disasters for the period 1900–2010 sorted by economic damage at country level

Country	Date	Damage (000 US$)
China P Rep, general flood	1/7/1998	30,000,000
Korea Dem P Rep, general flood	1/8/1995	15,000,000
China P Rep, general flood	30/06/1996	12,600,000
United States, general flood	24/06/1993	12,000,000
Germany, –	11/8/2002	11,600,000
United States, general flood	9/6/2008	10,000,000
Italy, general flood	1/11/1994	9,300,000
China P Rep, general flood	23/06/1999	8,100,000
Italy, flash flood	14/10/2000	8,000,000
China P Rep, general flood	23/06/2003	7,890,000

Created on: Sep-25-2010 – Data version: v12.07
Source: "EM-DAT: The OFDA/CRED International Disaster Database http://www.em-dat.net – Université Catholique de Louvain – Brussels – Belgium"

blockage of streams. Mountains are most vulnerable to natural hazards and they also harbour rich repositories of biodiversity. Change in climate has resulted in floods, landslides, rock fall, and avalanches threatening endemic species.

September 2010 was the wettest monsoon since 1978 when record rainfall 949 mm (upto September 19, 2010) was noticed as compared to the rainfall 965.1 mm of 1978.

Hathinikund barrage released 6.92 cusec water upto September 19, 2010 which threatened Delhi of a flood. Capital region particularly Common Wealth Games (CWG) Village was under low lying areas had highest risk of flood just before the starting of games.

On September 23, 2010 River Yamuna witnessed 32 years high 207.11 m against danger mark of 204.83 m. This happened after 1978 and the level remained for 8 h. After rising of water level, Old Yamuna Bridge was shut for rail and road traffic. This bridge connects East Delhi with the heart of the national capital. The flood water entered into Batla house, Kashmere Gate, Majnu Ka Tila and Jaitpur. Thousands of flood victims were shifted to tents and others were warned to opt safer places. Outside of the city, flood destroyed crops worth Millions of rupees.

Floods do not always result in destructions. Floods deposit nutrients on the flood plain nourishing flora in the locality. The moist condition due to flood in a peat land promotes plant growth and also slows the decay of dead plant matter.

Detrimental action of flood includes physical damage to infrastructure; casualties and public health; disruption of water supplies and transportation; damage to crops and food supplies. An early warning system and preparedness plan are essential tools in reducing loss of life and socioeconomic impacts of floods.

Changing climate has accelerated glacial melt; risk of avalanches and floods; and expansion of lakes formed from melting glaciers in mountainous region. Many fast growing lakes are increasing the threat of glacial lake outburst floods (GLOFs) leading to disastrous consequences for people, agriculture and infrastructure in the locality.

Table 7.12 Summarized table of flood sorted by continents for the period 1900–2010

		No. of events	Killed	Total affected	Damage (000 US$)
Africa	Unspecified	221	6,909	13,592,362	965,007
	Ave. per event		31.3	61,503.9	4,366.5
	Flash flood	82	2,717	2,161,533	457,086
	Ave. per event		33.1	26,360.2	5,574.2
	General flood	411	13,113	33,903,697	3,907,430
	Ave. per event		31.9	82,490.7	9,507.1
	Storm surge/coastal flood	7	169	1,202,829	42,750
	Ave. per event		24.1	171,832.7	6,107.1
Americas	Unspecified	357	56,972	28,883,356	22,187,497
	Ave. per event		159.6	80,905.8	62,149.9
	Flash flood	61	32,398	2,756,310	4,839,870
	Ave. per event		531.1	45,185.4	79,342.1
	General flood	449	10,603	42,808,154	53,216,840
	Ave. per event		23.6	95,341.1	118,523
	General flood/mudslide	1	11	9,950	–
	Ave. per event		11	9,950	–
	Storm surge/coastal flood	16	1,070	1,053,098	1,212,720
	Ave. per event		66.9	65,818.6	75,795
Asia	Unspecified	534	2,368,542	863,332,073	41,796,758
	Ave. per event		4,435.5	1,616,726.7	78,271.1
	Flash flood	245	23,493	139,032,777	16,531,520
	Ave. per event		95.9	567,480.7	67,475.6
	General flood	719	4,388,988	2,003,394,635	188,843,808
	Ave. per event		6,104.3	2,786,362.5	262,647.9
	Storm surge/coastal flood	40	2,060	18,174,201	8,472,384
	Ave. per event		51.5	454,355	211,809.6
Europe	Unspecified	133	3,289	4,265,569	24,260,105
	Ave. per event		24.7	32,071.9	182,406.8
	Flash flood	40	1,277	534,759	12,799,150
	Ave. per event		31.9	13,369	319,978.8
	General flood	289	1,836	8,262,172	65,947,253
	Ave. per event		6.4	28,588.8	228,191.2
	Storm surge/coastal flood	7	2,028	615,531	342,622
	Ave. per event		289.7	87,933	48,946
Oceania	Unspecified	46	219	432,393	580,021
	Ave. per event		4.8	9,399.8	12,609.2
	Flash flood	18	90	36,939	1,892,100
	Ave. per event		5	2,052.2	105,116.7
	General flood	43	146	161,876	3,639,754
	Ave. per event		3.4	3,764.6	84,645.4
	Storm surge/coastal flood	10	14	78,030	252,500
	Ave. per event		1.4	7,803	25,250

Created on: Sep-25-2010 – Data version: v12.07

Source: "EM-DAT: The OFDA/CRED International Disaster Database http://www.em-dat.net – Université Catholique de Louvain – Brussels – Belgium"

7.3 Weather Disasters

Weather disaster includes but not restricted to blizzards, cyclonic storms, droughts, hailstorms, tornadoes, cloud burst. The weather disasters and relevance is listed in Table 7.17. Number of occurrence of drought/famine throughout the world is given in Fig. 7.10.

Weather disasters are affecting millions of people, especially poor and under-privileged. Their vulnerability is enhanced by ongoing environmental degradation, political system and efficiency of governance. While property losses is measured in dollars or local currency it would be very difficult to measure value of loss of ecosystems such as wetlands, forests, mangroves and sand dunes.

Drought is an extreme climatic event which itself does not trigger an emergency. Whether it becomes an emergency depends on its effect on local people. The same temperature in Arabian Asia and south Asia will have different impacts in the regions. The ability to adapt with drought varies from region to region.

Drought can be classified in to three types: meteorological, agricultural, and hydrological.

Meteorological drought is deficiency of precipitation from "normal rainfall". Agricultural drought is characterized by a shortage in water to crop. Hydrological drought is shortage in surface and subsurface water supplies, resulting in lack of water availability to meet normal water requirements.

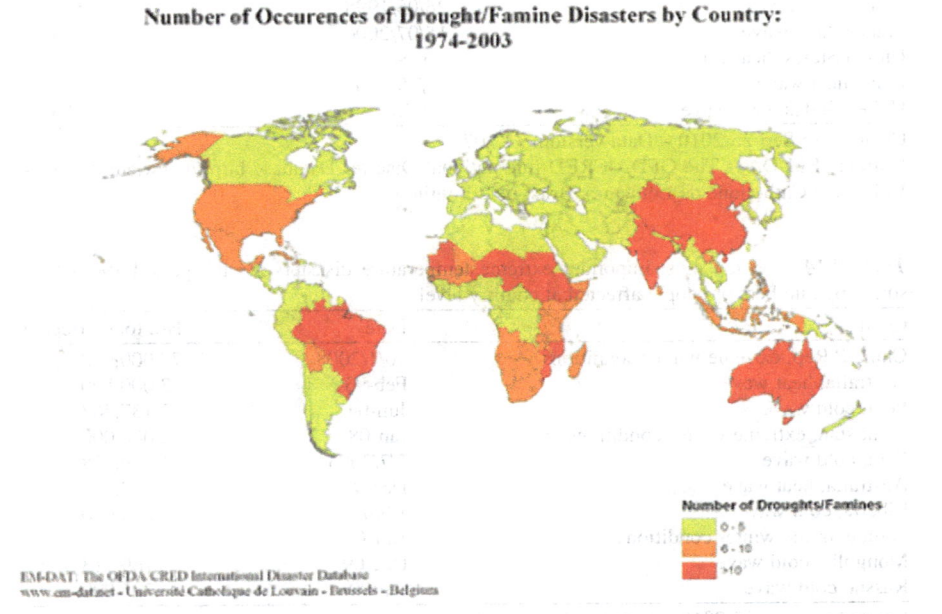

EM-DAT: The OFDA-CRED International Disaster Database
www.em-dat.net - Université Catholique de Louvain - Brussels - Belgium

Fig. 7.10 Number of occurrence of drought/famine disasters by country from 1974 to 2003

The 1987 drought in India was one of the worst droughts of the century which affected 59–60% of the crop area and a population of 285 million (PACS 2008).

Table 7.13 shows top ten most important extreme temperature disasters for the period 1900–2010 sorted by numbers of people killed at country level. Table 7.14 shows top ten most important extreme temperature disasters for the period 1900–2010 sorted by numbers of people affected at country level. Table 7.15 shows top ten most important extreme temperature disasters for the period 1900–2010 sorted by economic damage at country level. Table 7.16 shows summarized table extreme temperature disasters sorted by continents for the period 1900–2010. Table 7.17 shows water disasters and relevance to climate change.

Droughts are slow phenomenon but difficult to forecast its consequences. Its spatial extent is usually spread over large areas. Drought in India has resulted in millions of deaths till date. The main reason for such loss of life is due to dependency of agriculture on monsoon. Over 68% of India is vulnerable to drought (PACS 2008).

Table 7.13 Top ten most important extreme temperature disasters for the period 1900–2010 sorted by numbers of people killed at country level

Country	Date	No. killed
Italy, heat wave	16/07/2003	20,089
France, heat wave	1/8/2003	19,490
Spain, heat wave	1/8/2003	15,090
Germany, heat wave	Aug-03	9,355
Portugal, heat wave	Aug-03	2,696
India, heat wave	26/05/1998	2,541
France, heat wave	15/07/2006	1,388
United States, heat wave	Jun-80	1,260
India, heat wave	14/05/2003	1,210
United States, heat wave	Jul-36	1,193

Created on: Sep-25-2010 – Data version: v12.07
Source: "EM-DAT: The OFDA/CRED International Disaster Database http://www.em-dat.net – Université Catholique de Louvain – Brussels – Belgium"

Table 7.14 Top ten most important extreme temperature disasters for the period 1900–2010 sorted by numbers of people affected at country level

Country	Date	No. total affected
China P Rep, extreme winter conditions	10/1/2008	77,000,000
Australia, heat wave	Feb-93	3,000,500
Peru, cold wave	Jun-04	2,137,467
Tajikistan, extreme winter conditions	Jan-08	2,000,000
Peru, cold wave	7/7/2003	1,839,888
Australia, heat wave	Dec-94	1,000,034
Liberia, cold wave	1990	1,000,000
Peru, extreme winter conditions	Apr-07	884,572
Mongolia, cold wave	Dec-09	769,113
Russia, cold wave	Jan-99	725,000

Created on: Sep-25-2010 – Data version: v12.07
Source: "EM-DAT: The OFDA/CRED International Disaster Database http://www.em-dat.net – Université Catholique de Louvain – Brussels – Belgium"

Table 7.15 Top ten most important extreme temperature disasters for the period 1900–2010 sorted by economic damage at country level

Country	Date	Damage (000 US$)
China P Rep, extreme winter conditions	10/1/2008	21,100,000
France, heat wave	1/8/2003	4,400,000
Italy, heat wave	16/07/2003	4,400,000
United States, heat wave	1/5/1998	4,275,000
United States, cold wave	1977	2,800,000
United States, heat wave	Jun-80	2,000,000
Canada, cold wave	Dec-92	2,000,000
United States, heat wave	Jul-86	1,750,000
Germany, heat wave	Aug-03	1,650,000
United States, cold wave	26/01/2009	1,100,000

Created on: Sep-25-2010 – Data version: v12.07
Source: "EM-DAT: The OFDA/CRED International Disaster Database http://www.em-dat.net – Université Catholique de Louvain – Brussels – Belgium"

Table 7.16 Summarized table of extreme temperature disasters continents for the period 1900–2010

		No. of events	Killed	Total affected	Damage (000 US$)
Africa	Cold wave	5	73	1,000,105	47,000
	Ave. per event		14.6	200,021	9,400
	Heat wave	5	154	–	809
	Ave. per event		30.8	–	161.8
Americas	Cold wave	52	2,030	4,111,449	7,739,850
	Ave. per event		39	79,066.3	148,843.3
	Extreme winter conditions	6	82	884,572	–
	Ave. per event		13.7	147,428.7	–
	Heat wave	31	5,944	2,731	9,025,000
	Ave. per event		191.7	88.1	291,129
Asia	Cold wave	62	7,621	1,837,757	1,185,010
	Ave. per event		122.9	29,641.2	19,113.1
	Extreme winter conditions	8	572	79,109,150	21,940,000
	Ave. per event		71.5	9,888,643.8	2,742,500
	Heat wave	54	10,830	8,801	401,000
	Ave. per event		200.6	163	7,425.9
Europe	Cold wave	75	3,642	770,334	2,291,700
	Ave. per event		48.6	10,271.1	30,556
	Extreme winter conditions	30	1,429	79,239	1,000,000
	Ave. per event		47.6	2,641.3	33,333.3
	Heat wave	57	77,891	2,120	12,363,050
	Ave. per event		1,366.5	37.2	216,895.6
Oceania	Heat wave	6	370	4,602,784	200,000
	Ave. per event		61.7	767,130.7	33,333.3

Created on: Sep-25-2010 – Data version: v12.07
Source: "EM-DAT: The OFDA/CRED International Disaster Database http://www.em-dat.net – Université Catholique de Louvain – Brussels – Belgium"

Figure 7.11 shows number of occurrence of windstorm disasters by country from 1974 to 2003. Storms particularly affect countries located on Indian Ocean even though some countries like Singapore are not affected. The most vulnerable countries are India, China, Bangladesh, Japan, and Vietnam.

Table 7.17 Water disasters and relevance to climate change

Weather disaster	Description	Relevance to climate change
Blizzards	A blizzard is a severe storm condition accompanied by low temperatures, strong winds, and heavy snow. They can bring near-whiteout conditions, which restrict visibility to near zero	The changing precipitation and wind pattern are the reason for blizzards
Cyclonic storms	A cyclone is closed, circular spiralling wind rotating in the same direction as the Earth. Cyclone is also called typhoons or Hurricanes or tropical cyclonic storm or cyclonic storms	The changing precipitation and wind pattern are the main reasons for cyclonic storms
Droughts	A drought is an extended period of deficiency in water supply	The change in precipitation pattern is the main reason for drought
Hailstorms	Hailstorms are storms that produce hailstones that fall on the ground	The changing precipitation and wind pattern is the main reason
Tornadoes	A tornado is a fiercely rotating column of air that is in contact with surface of the earth and cloud	The changing precipitation, oceanic pressure and wind pattern are the main reason
Cloud burst	A cloudburst is an extreme rainfall, which normally lasts no longer than a few minutes and capable of creating flood condition	The accumulation of clouds in a small area is the main reason for cloud burst

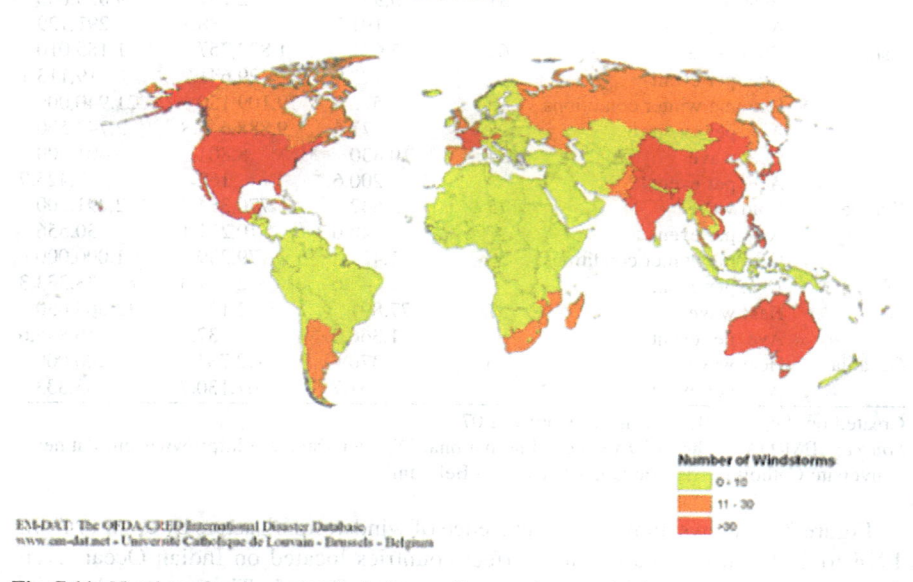

Number of Occurrences of Windstorm Disasters by Country: 1974-2003

Number of Windstorms
- 0 - 10
- 11 - 30
- >30

EM-DAT: The OFDA/CRED International Disaster Database
www.em-dat.net - Université Catholique de Louvain - Brussels - Belgium

Fig. 7.11 Number of occurrence of windstorm disasters by country from 1974 to 2003

The drought affect secondary and tertiary sectors by shortage of raw material to agro-based industries; reduced demand for industrial/consumer products; shift in government resource allocation to drought relief measures instead of investing on infrastructure. Inflation of food commodity is also cause for concern in many Asian countries after drought.

7.4 Fire

Every year, millions of hectares of the world's forests are destroyed by wild fires, resulting in huge damage, health problems, and deaths to humans and other species. Fire could be caused due to natural or anthropogenic causes. It could occur within the natural setup as well as urban ecosystem. Annually fires burn approximately 500 million hectares of woodland, open forests, tropical and sub-tropical savannahs; 10–15 million hectares of boreal and temperate forest; and 20–40 million hectares of tropical forests (Goldammer 1995).

The major natural reasons of wildfire are spontaneous combustion, lightning, volcanic eruption, sparks from rock falls, and coal seam fires. Anthropogenic reasons include cooking forests, shooting by poachers, discarded cigarettes, sparks from equipment, and preparation of forest land for agriculture by intentional firing, and power line arcs. In numerous places fire is a tool for changing forests to agricultural lands (Stolle et al. 2003).

The forest fires of 1997/1998 created massive ecological damage and human suffering. The fires damaged South-east Asia – from Papua New Guinea to Malaysia with fires in Java, Borneo, Sulawesi, Irian Jaya and Sumatra, but Indonesia burned the most, The fires of 1997/1998 in Indonesia are due to fires lit to clear and prepare land and accidental fires in forest and peat swamps (Daniel and Louis 2007). Moscow was covered with dense smog in August 2010 due to peat fires raging around the city. The July heat wave in 2010 led to intense fires across the Russia.

Extended periods of minimal rainfall was the cause for forest fire in Southeast Asia during Ice Age (ADB 2001). Forest fires will increase due to climate change. The El Niño, which caused the drought, affected much of the forests which caught fire in 1997 and 1998. The occurrence and strength of El Niño could be escalating (Trenberth and Hoar 1996, 1997), which may result in more forest fires.

Climate is a critical control factor in fire occurrence and frequency. Figure 7.12 gives interrelation between wild fire, GHG emission and Global Warming. The spread of wildfires depends on the flammable material and arrangement. Burning logs from higher elevation can roll downhill igniting dried patched in lower elevation. Fuel arrangement and distribution is governed by topography, temperature of the area, water bodies, and season. The forest vegetation sheds leaves progressively during dry season and if the region does not receive rain, fuel accumulation will occur for weeks together. During such warm period with lower rain fall, the moisture content of the surface fuels is lowered, and the downed woody material along with loosely packed leaf-litter layer contribute to the build-up

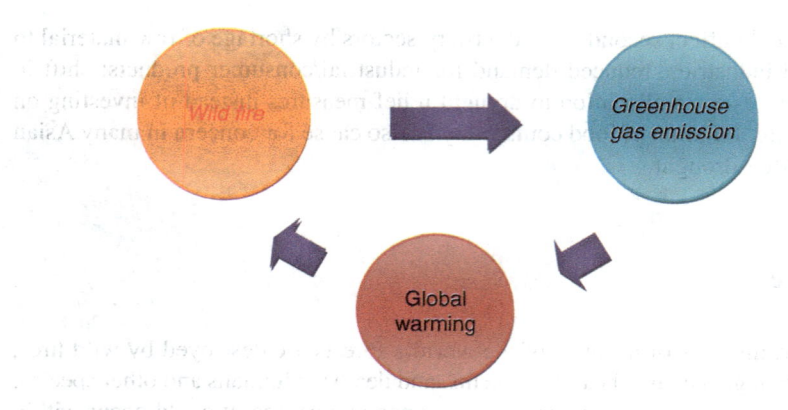

Fig. 7.12 Interrelation with wild fire, GHG emission and global warming

Table 7.18 Fire disasters and relevance to climate change

Fire disaster	Description	Relevance to climate change
Wildfire	A wildfire is uncontrolled fire in combustible vegetation that occurs in wilderness area	Likely to destroy carbon sinks and contribute to GHG
Fire accident	Fire accident is out burst of fire in manmade structures/materials	Likely contribute to GHG

and spread of surface fires. Fuels such as dry climbers become fire ladders resulting in crown fires or "torching" of single trees.

Climate also influences soil microorganism activity. Thus litter decomposition, which is source of methane, is enhanced during favourable climatic conditions. The influencing factors of wild fire are: (1) dependence on forests for livelihood; (2) increased population pressure on forests (3) high temperature and lower atmospheric humidity; (4) fuel load; (5) vegetation types; (6) fire resistance level of ecosystem; (7) topography; (8) extent of fragmentation of the ecosystem; (9) absence of environmental laws; and (10) inefficient enforcement capability.

Table 7.18 gives fire disasters and relevance to climate change. Release of iron by the 1997 Indonesian wild fires lead to extraordinary red tide, leading to reef death by asphyxiation (Neville et al. 2003). Higher temperature, lower precipitation, drop in humidity can increase spread of fire. Fire intensity increases during daytime due to lower humidity and higher temperatures. At night the land cools and humidity increases leading to lower burning intensity.

Table 7.19 shows top ten most important wildfire disasters for the period 1900–2010 sorted by numbers of killed at country level, Table 7.20 shows top ten most important wildfire disasters for the period 1900–2010 sorted by numbers of people affected at country level, Table 7.21 shows top ten most important wildfire disasters for the period 1900–2010 sorted by economic damage at country level, and Table 7.22 shows summarized table of wild fires sorted by continents for the period 1900–2010.

Table 7.19 Top ten most important wildfire disasters for the period 1900–2010 sorted by numbers of killed at country level

Country	Date	No. killed
United States, forest fire	15/10/1918	1,000
Indonesia, forest fire	Sep-97	240
China P Rep, forest fire	May-87	191
Australia, bush/brush fire	2/2/2009	180
United States, forest fire	20/10/1944	121
France, –	Aug-49	80
Australia, scrub/grassland fire	16/02/1983	75
Canada, forest fire	11/7/1911	73
Australia, scrub/grassland fire	1939	71
Greece, forest fire	24/08/2007	67

Created on: Sep-24-2010 – Data version: v12.07
Source: "EM-DAT: The OFDA/CRED International Disaster Database http://www.em-dat.net – Université Catholique de Louvain – Brussels – Belgium"

Table 7.20 Top ten most important wildfire disasters for the period 1900–2010 sorted by numbers of people affected at country level

Country	Date	No. total affected
Indonesia, forest fire	Oct-94	3,000,000
Macedonia FRY, forest fire	Jul-07	1,000,000
United States, scrub/grassland fire	21/10/2007	640,064
Argentina, forest fire	22/01/1987	152,752
Portugal, forest fire	Jan-03	150,000
Paraguay, forest fire	Sep-07	125,000
Russia, forest fire	20/07/1998	100,683
China P Rep, forest fire	May-87	56,313
United States, –	13/11/2008	55,020
Nepal, forest fire	Mar-92	50,000

Created on: Sep-24-2010 – Data version: v12.07
Source: "EM-DAT: The OFDA/CRED International Disaster Database http://www.em-dat.net – Université Catholique de Louvain – Brussels – Belgium"

Table 7.21 Top ten most important wildfire disasters for the period 1900–2010 sorted by economic damage at country level

Country	Date	Damage (000 US$)
Indonesia, forest fire	Sep-97	8,000,000
Canada, forest fire	Jan-89	4,200,000
United States, forest fire	21/10/2003	3,500,000
United States, forest fire	21/10/1991	2,500,000
United States, scrub/grassland fire	21/10/2007	2,500,000
Spain, forest fire	18/07/2005	2,050,000
United States, –	13/11/2008	2,000,000
Greece, forest fire	24/08/2007	1,750,000
Portugal, forest fire	Jan-03	1,730,000
Mongolia, forest fire	8/4/1996	1,712,800

Created on: Sep-24-2010 – Data version: v12.07
Source: "EM-DAT: The OFDA/CRED International Disaster Database http://www.em-dat.net – Université Catholique de Louvain – Brussels – Belgium"

Table 7.22 Table summarized table of wild fires sorted by continents for the period 1900–2010

		No. of events	Killed	Total affected	Damage (000 US$)
Africa	Unspecified	1	49	3,023	–
	Ave. per event		49	3,023	–
	Bush/Brush fire	2	34	150	430,000
	Ave. per event		17	75	215,000
	Forest fire	9	66	11,140	–
	Ave. per event		7.3	1,237.8	–
	Scrub/grassland fire	13	117	14,532	10,000
	Ave. per event		9	1,117.8	769.2
Americas	Unspecified	4	1	56,623	2,016,000
	Ave. per event		0.3	14,155.8	504,000
	Forest fire	94	1,410	464,155	16,301,800
	Ave. per event		15	4,937.8	173,423.4
	Scrub/grassland fire	21	106	706,593	3,067,100
	Ave. per event		5	33,647.3	146,052.4
Asia	Forest fire	49	690	3,246,817	11,633,500
	Ave. per event		14.1	66,261.6	237,418.4
	Scrub/grassland fire	31	–	9,003	–
	Ave. per event		–	290.4	–
Europe	Unspecified	2	82	–	–
	Ave. per event		41	–	–
	Forest fire	88	412	1,287,000	10,343,811
	Ave. per event		4.7	14,625	117,543.3
	Scrub/grassland fire	4	14	800	675,000
	Ave. per event		3.5	200	168,750
Oceania	Unspecified	1	–	–	–
	Ave. per event		–	–	–
	Bush/Brush fire	1	180	9,954	1,300,000
	Ave. per event		180	9,954	1,300,000
	Forest fire	5	24	4,011	468,650
	Ave. per event		4.8	802.2	93,730
	Scrub/grassland fire	25	292	83,175	854,194
	Ave. per event		11.7	3,327	34,167.8

Created on: Sep-24-2010 – Data version: v12.07
Source: "EM-DAT: The OFDA/CRED International Disaster Database http://www.em-dat.net – Université Catholique de Louvain – Brussels – Belgium"

Adaptation to fire regime variations resulting from climate change would benefit from multilevel approach (Adger et al. 2005). Chokkalingam et al. (2007) emphasize the importance of alternative livelihood options during drought years.

7.5 Health and Disease

Health is the physical and mental wellbeing of an organism. While creating of the World Health Organization (WHO 1948), health was defined as:

Health is a state of complete physical, mental and social well-being and not merely the absence of disease or infirmity.

According to Prüss-Üstün and Corvalán (2006), 24% of the global disease burden and 23% of all deaths can be prevented through environmental interventions. Poor people often live in lower environmental conditions and have lower resistance to infection and hence are more vulnerable to diseases. Improving environmental health can reduce poverty due to reduction in loss of workdays.

Figure 7.13 shows number of occurrence of epidemics by country from 1974 to 2003. The health impacts of climate change will not be uniform (Cambell-Lendrum and Woodruff 2006) and the distribution of the health burdens is inverse to the global GHG emissions (Patz et al. 2007). The impact of a changing climate depends on vulnerability of the individual or population as well as magnitude of climate change. The predicted health effects are concentrated in poor countries that already suffer the worst health (Patz et al. 2008; Marland et al. 2003; McMichael et al. 2004).

It is difficult to plan for the future, mainly when the exact nature of climate change is not known (Alistair et al. 1998). The measure countries need to adopt should include (1) Increase medical and paramedical professionals; (2) Generation of sufficient drugs; (3) Improvement in waste management; (4) Up gradation of health care facilities; and (5) Improve spatial distribution of Health care facilities; and (6) Improve quality of medical and paramedical staff.

Climate change can influence on human health. The principal reason for such conclusion is: (1) temperature extremes will increase mortality although there is disagreement about which sex, age group, or race will be most affected; (2) low

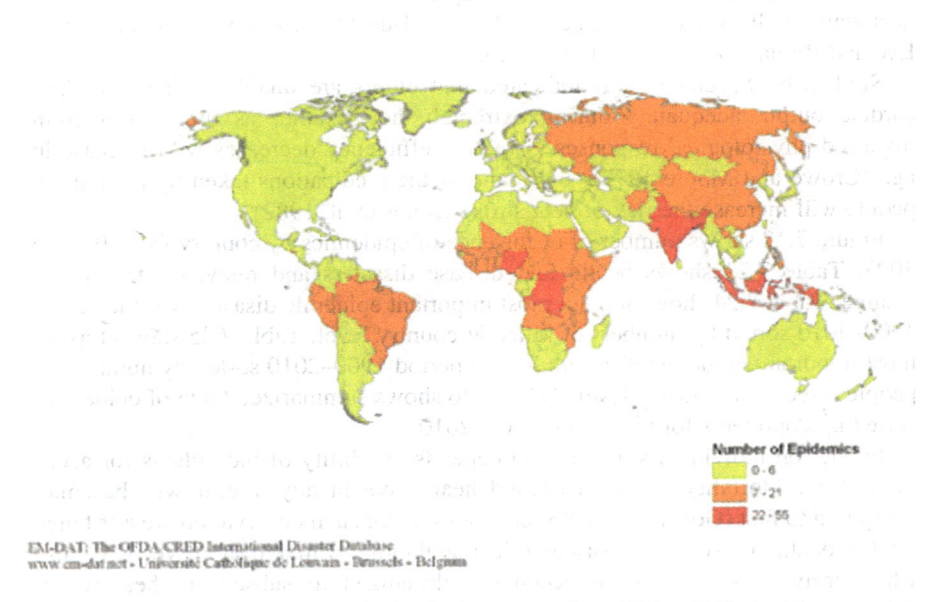

Worldwide epidemics occurrences: 1974-2003

Number of Epidemics
0 - 6
7 - 21
22 - 55

EM-DAT: The OFDA-CRED International Disaster Database
www.em-dat.net - Université Catholique de Louvain - Brussels - Belgium

Fig. 7.13 Number of occurrence of epidemics by country from 1974 to 2003

relative humidity will cause a range of illnesses that may lead to mortality; (3) snowfall can lead to mortality; (4) quick changes in the weather will result in negative physiological response; and (5) Climate change can increase population vectors; and (6) Extreme Climatic disasters can cause injuries.

Health impact due to climate change increases with poor diet, alcoholism, smoking, exposure to pollution, use of tranquilizer. Factors that decrease climate induces health risk are use of well balanced diet, air conditioning, frequent exercising, limited exposure to pollution, consumption of fluids, and living in a well-shaded residence.

Climate change will affect human health directly and indirectly. People are exposed to changing weather patterns (temperature, precipitation, sea-level rise and more frequent extreme events). People will be affected indirectly through changes in the quality of water, air, food, and changes in ecosystems, agriculture, industry, human settlements and the economy.

Seasonal changes in temperature will cause disorders such as bronchitis, glaucoma, goiter, eczema, peptic ulcer, adrenal ulcer, and herpes zoster (Tromp 1963). Studies from Persinger (1980) indicated that changes in ambient temperatures can be correlated to heart failure and cerebrovascular accidents.

The degree of seasonality in the climate of a region also appears to affect mortality rates. Studies by Katayama and Momiyama-Sakamoto (1970) revealed that those countries with smaller seasonal temperature variation will have higher impact compared to countries with higher temperature variations. People in warmer countries will have health imbalance at below normal temperatures as compared to people in cooler countries.

Climate impact on health varies with age, sex, and race. Mortality rates during heat waves will increase with age (Oechsli and Buechley 1970; Bridger et al. 1976; Lye and Ramal 1977; Jones et al. 1982).

Studies by Sprung (1979) indicated that elders are unable to increase their cardiac output adequately during extremely hot weather as they suffer from impaired physiological responses. Sweating efficiency decreases with increase in age (Crowe and Moore 1973), and many of the medications taken by the elderly people will increase the risk of heat stroke (Jones et al. 1982).

Figure 7.12 shows number of occurrence of epidemics by country from 1974 to 2003. Table 7.23 shows health and disease disasters and relevance to climate change. Table 7.24 shows top ten most important epidemic disasters for the period 1900–2010 sorted by numbers of killed at country level. Table 7.25 shows top ten most important epidemic disasters for the period 1900–2010 sorted by numbers of people affected at country level. Table 7.26 shows summarized table of epidemics sorted by continents for the period 1900–2010.

Impact on individual's health also depends on ability of individuals for acclimatization. Mortality during a second heat wave in any a year will be small compared to first (Gover 1938). Possible reasons for such observation are attributed to: (1) death of weak and susceptible population in first heat wave; (2) those who survive first heat wave become acclimatized to subsequent heat waves (Marmor 1975). Cold weather will cause death due to hypothermia, influenza,

Table 7.23 Health and disease disasters and relevance to climate change

Health and disease	Description	Relevance to climate change
Epidemic	Rapid spread of disease in a locality where disease is not prevalent	Spreading is triggered due to epidemic favourable climate
Famine	Wide spread scarcity of food	Usually follows drought but not restricted to drought
Insect infection	An act by which insects spread infection	Climate change is likely to increase vector population and spread of infection through them
Animal stampede	An act of mass impulse among animals	Wild animals can run into human settlements due to wild fire and other climatic disasters in the forest like flood

Table 7.24 Top ten most important epidemic disasters for the period 1900–2010 sorted by numbers of killed at country level

Country	Date	No. killed
Soviet Union, –	1917	2,500,000
India, bacterial infectious diseases	1920	2,000,000
China P Rep, bacterial infectious diseases	1909	1,500,000
India, bacterial infectious diseases	1907	1,300,000
India, bacterial infectious diseases	1920	500,000
India, viral infectious diseases	1926	423,000
Bangladesh, viral infectious diseases	1918	393,000
India, bacterial infectious diseases	1924	300,000
Uganda, –	1901	200,000
Niger, bacterial infectious diseases	1923	100,000

Created on: Sep-25-2010 – Data version: v12.07
Source: "EM-DAT: The OFDA/CRED International Disaster Database http://www.em-dat.net – Université Catholique de Louvain – Brussels – Belgium"

Table 7.25 Top ten most important epidemic disasters for the period 1900–2010 sorted by numbers of people affected at country level

Country	Date	No. total affected
Soviet Union, parasitic infectious diseases	1923	18,000,000
Kenya, parasitic infectious diseases	Jan-94	6,500,000
Canada, viral infectious diseases	Jan-18	2,000,000
Japan, viral infectious diseases	Feb-78	2,000,000
Bangladesh, –	Apr-91	1,500,000
Burundi, parasitic infectious diseases	Oct-00	722,591
Burundi, parasitic infectious diseases	Jan-99	616,034
Bangladesh, –	27/09/1987	600,000
Indonesia, parasitic infectious diseases	1/1/1986	500,000
Zimbabwe, parasitic infectious diseases	23/05/1996	500,000

Created on: Sep-25-2010 – Data version: v12.07
Source: "EM-DAT: The OFDA/CRED International Disaster Database http://www.em-dat.net – Université Catholique de Louvain – Brussels – Belgium"

Table 7.26 Summarized table of epidemics sorted by continents for the period 1900–2010

		No. of events	Killed	Total affected	Damage (000 US$)
Africa	Unspecified	73	207,119	153,199	–
	Ave. per event		2,837.2	2,098.6	–
	Bacterial infectious diseases	482	223,102	2,197,331	–
	Ave. per event		462.9	4,558.8	–
	Parasitic infectious diseases	20	4,723	8,722,500	–
	Ave. per event		236.2	436,125	–
	Viral infectious diseases	141	23,139	1,291,964	–
	Ave. per event		164.1	9,162.9	–
Americas	Unspecified	8	8,693	100,206	–
	Ave. per event		1,086.6	12,525.8	–
	Bacterial infectious diseases	44	6,378	455,365	–
	Ave. per event		145	10,349.2	–
	Parasitic infectious diseases	6	104	518,403	–
	Ave. per event		17.3	86,400.5	–
	Viral infectious diseases	81	52,375	3,379,530	7
	Ave. per event		646.6	41,722.6	0.1
Asia	Unspecified	54	9,939	2,688,181	–
	Ave. per event		184.1	49,781.1	–
	Bacterial infectious diseases	115	5,674,614	773,185	–
	Ave. per event		49,344.5	6,723.3	–
	Parasitic infectious diseases	20	5,284	1,518,005	–
	Ave. per event		264.2	75,900.3	–
	Viral infectious diseases	136	838,340	2,960,906	–
	Ave. per event		6,164.3	21,771.4	–
Europe	Unspecified	6	2,500,000	1,575	–
	Ave. per event		416,666.7	262.5	–
	Bacterial infectious diseases	16	347	167,027	–
	Ave. per event		21.7	10,439.2	–
	Parasitic infectious diseases	3	47	18,000,344	–
	Ave. per event		15.7	6,000,114.7	–
	Viral infectious diseases	24	81	20,888	–
	Ave. per event		3.4	870.3	–
Oceania	Bacterial infectious diseases	5	172	6,475	–
	Ave. per event		34.4	1,295	–
	Viral infectious diseases	13	7,026	11,838	–
	Ave. per event		540.5	910.6	–

Created on: Sep-25-2010 – Data version: v12.07
Source: "EM-DAT: The OFDA/CRED International Disaster Database http://www.em-dat.net –
Université Catholique de Louvain – Brussels – Belgium"

and pneumonia in addition to air pollution related diseases due to increase in concentration air pollutants at ground levels.

Humidity influences the body's ability to cool itself by means. Humidity also affects human comfort as the perceived temperature by humans depends on moisture content in the atmospheric (Persinger 1980). Low moisture content in atmosphere induces stress on the nasal-pharynx and trachea and increases the chance of microbial or viral infection. High moisture content during warm weather lessens the body's ability to evaporate perspiration, possibly leading to heat stress. Cold weather and snow will increase deaths from stroke and heart attack (Rogot and Padgett 1976).

Increase in urban population in Asia has resulted in increase of the poor physical and social conditions for many inhabitants. The health of Asian urban people will be affected by frequent days of extreme heat, air pollution and risk of flooding and storm damage along with diseases due to air pollution, poor water quality, noise pollution, contaminated food, and nosocomial infection. The Indian urban has also seen increase in health care facilities in the past compared to rural population. The trend may be same in other Asian cities as well. The reason for such unequal distribution of health care facilities and doctors are: good salaries and income due to increase in patients as in urban area compared to rural area. But increase in health care establishments also means increase in nosocomial infection. The poor solid waste management especially biomedical waste generated from health care facilities will cause more harm than help to people of cities.

7.6 Cumulative Effect of Man Made and Climatic Hazards

Disasters do not always come individually. Multiple natural disasters can occur at a time or natural disaster may accompany manmade disasters. A 'hazard', is any event, phenomenon, or human activity that is likely to cause loss. Natural and anthropogenic factors may act together to form a hazard. For example, earthquakes which are a natural hazard can also be triggered by mining activities or the construction of large dams. A landslide can happen by a combination of heavy rains, terror strike, earth tremors, explosion for mineral extraction, and deforestation. A 'disaster' is an event that causes serious disruption, beyond the coping capacity of a society. Floods cause loss of life, damage to property, and promote the spread of diseases. While the principal cause of flooding is high rainfall, there are many human activities which enhance the occurrence/magnitude of flooding such as: land degradation; augmented population density along riverbanks; poor land use planning; inadequate drainage in urban area; removal of vegetation in catchment areas; and inadequate management of discharges from river reservoirs. Flooding can also occur due to failure of dams. Drought is aggravated by deforestation. Wildfires may be ignited naturally or anthropogenic activity.

It is important to know possible climatic hazard in a given area and hazard that could occur due to anthropogenic activities. For example an area with oil distillery can prone to fire hazard and earthquake simultaneously. An improperly managed hospital can become source and cause for infection spreading at the time of flooding. Draught may occur along with toxic leak. Flood and earthquake can occur together. Hence planning for disaster preparedness should always consider scenarios due to multiple hazards rather than any one of the disasters at a time.

Globally, there is increase in number of natural disasters and with changing climate Asia is prone to floods, windstorms and landslides, along with other water-related disasters. The measures and precautions taken by governments, international agencies and other organizations have given a promising result indicating action at suitable time will reduce the risk both in terms of loss of life and damage to property.

References

Adger N.W, Arnell N.W and Tompkins E.L. 2005: Successful adaptation to climate change across scales. Global Environmental Change 15, 77–86.
ADB (Asian Development Bank), 2001: Fire, Smoke, and Haze The ASEAN Response Strategy, PP 246
Alistair Woodward, Simon Hales, Philip Weinstein, 1998: Climate change and human health in the Asia Pacific region: who will be most vulnerable?, Clim Res, Vol. 11: 31–38
Bridger, C.A., F.P. Ellis, and H.L. Taylor, 1976: Mortality in St. Louis, Missouri, during heat waves in 1936, 1953, 1954, 1955 and 1966. Environmental Research, 12, 38–48.
Cambell-Lendrum D, Woodruff R, 2006: Comparative risk assessment of the burden of disease from climate change. Environmental Health Perspectives;114:1935–41.
Chokkalingam U, Suyanto S, Permana R.P, Kurniawan I, Mannes J, Darmawan A, Khususyiah N and Susanto R.H 2007: Community fire use, resource change, and livelihood impacts: the downward spiral in the wetlands of Southern Sumatra. Journal of Mitigation and Adaptation Strategies for Global Change 12, 75–100.
Crowe J.P. and Moore R.E, 1973: Physiological and behavioral responses of aged men to passive heating. Journal of Physiology, 236, 43–48.
Daniel Murdiyarso and Louis Lebel, 2007: Southeast Asian forest and land fires: how can vulnerable ecosystems and peoples adapt to changing climate and fire regimes?, iLEAPS Newsletter 4, PP 28–29
Ezio Todini, 1999: An operational decision support system for flood risk mapping, forecasting and management, Urban Water 1, 131–143
Goldammer J.P, 1995: Biomass Burning and the Atmosphere, Paper Presented at Forests and Global Climate Change: Forests and the Global Carbon Cycle; quoted in J. Levine, T. Bobbe, N. Ray, A. Singh, R. G. Witt, 1999: Wild land Fires and the Environment: A Global Synthesis, Division of Environmental Information, Assessment and Early Warning, United Nations Environment Programme, pp4
Gover M, 1938: Mortality during periods of excessive temperature. Public Health Reports, 53, 1112–1143.
Jones T.S, Liang A.P, Rilbourne E.M, Griffin M.R, Patriarca P.A, Wassilak S.G.G, Mullan R.J, Herrick R.F, Donnel H.D, Jr., Choi K; and Thacker S.B; 1982: Morbidity and mortality associated with the July 1980 heat wave in St. Louis and Kansas City, MO. Journal of the American Medical Association, 247, 3327–3330.
Katayama K, and Momiyama-Sakamoto M, 1970: A biometeorological study of mortality from stroke and heart diseases: Its geographical differences in the United States. Meteorology and Geophysics, 21, 127–139.
Lye M. and Ramal A, 1977: The effects of a heat wave on mortality rates in elderly inpatients. The Lancet, 1, 529–531.
Marland G, Boden T.A, Andres R.J, 2003: Global, regional, and national fossil fuel CO_2 emissions. In: Trends: a compendium of data on global change. Oak Ridge TN: Carbon Dioxide Information Analysis Center, Oak Ridge National Laboratory, U.S. Department of Energy
McMichael A, Campbell-Lendrum D, Kovacs S, et al, 2004: Global climate change. [Ezzati M, Lopez A.D, Rodgers A, Murray C.J.L, eds.] Comparative quantifications of health risks: global and regional burden of disease attributable to selected major risk factors. Switzerland: WHO, 2004.

Marmor M, 1975: Heat wave mortality in New York City, 1949 to 1970. Archives of Environmental Health, 30, 131–136.

Neville J. Abram, Michael K. Gagan, Malcolm T. McCulloch, John Chappell, Wahyoe S. Hantoro, 2003: Science, Vol. 301. no. 5635, pp. 952 – 955 DOI: 10.1126/science.1083841

Oechsli F.W and Buechley R.W, 1970: Excess mortality associated with three Los Angeles September hot spells. Environmental Research, 3, 277–284.

Poorest Areas Civil Society (PACS), 2008: Drought In India: Challenges & Initiatives, <www. Corecentre.Co.In/Database/Docs/Docfiles/Drought India.Pdf> Retrieved On 1 December 2010

Persinger M.A, 1980: The Weather Matrix and Human Behavior, New York: Praeger, 327 pp.

Paul S.K, Mahajan A.K, 1999: Malpa rock fall disaster, Kali valley, Kumaun Himalaya. Curr. Sci. 76, 485–487.

Paul S.K, Bartarya S.K, Piyoosh Rautela, Mahajan A.K, 2000: Catastrophic mass movement of 1998 monsoons at Malpa in Kali Valley, Kumaun Himalaya (India), Geomorphology, 35, 169–180

Patz J, Gibbs H, Foley J, Rogers J, Smith K, 2007: Climate change and global health: quantifying a growing ethical crisis. EcoHealth; 4:397–405.

Patz J, Campbell-Lendrum D, Gibbs H, Woodruff R, 2008: Health impacts assessment of global climate change: expanding on comparative risk assessment approaches for policy making. Ann Rev Public Health;29:27–39.

Prüss-Üstün A and Corvalán C, 2006: Preventing disease through healthy environments: Towards an estimate of the environmental burden of disease. Geneva: World Health Organization

Rautela P, Thakur V.C, 1999: Landslide Hazard Zonation in Kaliganga and Madhyamaheshwar valleys of Garhwal Himalaya: a GIS based approach. Himalayan Geol. 20(2), 31–44.

Rogot E, and Padgett S.J, 1976: Associations of coronary and stroke mortality with temperature and snowfall in selected areas of the United States, 1962–1966. American Journal of Epidemiology, 103, 565–575.

Scheidegger A.E, 1994: Hazards: singularities in geomorphic systems. Geomorphology 10, 19–25.

Slaymaker O, 1996: Introduction. In: Slaymaker, O. (Ed.), Geomorphic Hazards. Wiley, Chichester, pp. 1–7.

Sah M.P, Bist K.S, 1998: Catastrophic mass movement of August 1998 in Okhimatu area, Garchival Himalaya. Proc. Int. workshop-cum-training programme on Landslide Hazard, Risk Assessment and Damage Control for Sustainable Development, New Delhi. pp. 259–270.

Sprung C.L, 1979: Hemodynamic alterations of heat stroke in the elderly. Chest, 75, 362–366.

Stolle F, Chomitz K.M, Lambin E.F and Tomich T.P, 2003: Land use and vegetation fires in Jambi Province, Sumatra, Indonesia. Forest Ecology and Management 179, 277–292.

Tromp S.W, 1963: Medical biometeorology, New York: Elsevier.

Trenberth K. E and Hoar T. J, 1996: The 1990–1995 El Niño-Southern Oscillation Event: Longest on Record, Geophysical Research Letters, Vol. 23, No. 1, pp57–60

Trenberth K.E, and Hoar T.J, 1997: El Niño and Climate Change, Geophysics Research Letters, Vol. 24, no.23, pp.3057–3060

UNDP, 2004: A Global Report Reducing Disaster Risk A Challenge For Development

Valdiya K.S, 1998: Catastrophic landslides in Uttaranchal, Central Himalaya. J. Geol. Soc. India 52, 483–486.

WHO, 1948: Preamble to the Constitution of the World Health Organization as adopted by the International Health Conference, New York, 19–22 June, 1946; signed on 22 July 1946 by the representatives of 61 States (Official Records of the World Health Organization, no. 2, p. 100) and entered into force on 7 April 1948. <http://www.who.int/about/definition/en/print.html> retrieved on 29, September 2010.

Chapter 8
Predicting Disaster: Asian Scenario

Prediction Disaster Management has gained appreciable attention in recent years in Asia. The chaos caused by natural disasters such as earthquakes, hurricanes, and floods, have harshly illustrated the inadequacy of pre- and post-disaster planning and action.

The speed of onset of a disaster is important as warning time available depends on it. Earthquakes, landslides, and flash floods happen fast leaving hardly any scope for warning. Tsunamis typically provide warning periods of minutes or hours. Hurricanes and floods provide opportunity for warning several hours or days in advance. Volcanoes which erupt suddenly and surprisingly give indications of eruption weeks or months in advance. Droughts are spread over a period of months or years.

Global climate change will have major effects on both economic and natural resources across the Asia in coming days. Disaster coupled with population growth, unplanned urbanization, and environmental degradation has made people all over the world suffer due to effects of natural disasters. Many of the disasters cannot be avoided. Disaster Risk Reduction (DRR) measures include: (1) hazard assessment, exposure data development, vulnerability mapping and risk assessment; (2) identification of safe areas; (3) adaptation of disaster resilient building codes; (4) disaster prediction and early warning. A system to predict disaster will help to cut down losses drastically. Disaster prediction and early warning system involves data collection, information analysis, and dissipation of prediction. Many of the local communities have knowledge about prediction by observing nature.

Magnitude of destruction with respect to some types of hazards may be altered by taking appropriate measures. But no known technology is available for all disasters. For example, early warning of flood in area to be affected can reduce loss of life by migration, but earthquake will not give chance to such arrangements.

It is now widely recognized that disasters are in existence since millions of years. Past and current experiences hold valuable lessons for reducing vulnerability in future. Globally governments and communities are increasingly focusing their efforts on building resilience and adaptive capacity in order to minimize the impacts of disaster. Many techniques like Environmental Impact Assessment (EIA) and agricultural models have been "borrowed", adjusted and used in vulnerability assessments.

Flooding which occurs every year or every few years will become part of the landscape, and hence development in these areas can be designed and implemented accordingly. Whereas tsunami which may strike any time in the next few decades or centuries makes it difficult to incorporate in planning activities.

The conventional weather and climate prediction practices are based on observation of plants, animals, winds, clouds and lightning patterns. With advance in technology meteorological satellites now monitor the Earth in nearly real time. The satellites perform the following functions: (a) make remote sensing observations which can be converted into meteorological parameters; (b) collect data from sensors placed on platforms; and (c) make broadcasts to provide meteorological information to users.

8.1 Non-climate Induced Risks

Accurate prediction of a hazardous event can save human lives but does ensure reduction in economic losses and social disruption which can only be achieved by measures taken well in advance of several years/months. Mitigation of disasters involves reducing the susceptibility of the area at risk and modifying the hazard vulnerability of the site. Mitigation measures include construction of new infrastructure, modifying of existing facilities. Non-structural mitigation measures include limiting land uses, and risk management through insurance programs, dissipation of information with respect to disaster vulnerable area and proper planning.

Non climate induced risk includes earthquake, technological disasters, etc., which are not triggered by climate change. Many of such disasters can be avoided by proper precautionary measures. Figure 8.1 shows indiscriminate disposal of biomedical waste thrown behind hospital and pigs feeding on the waste. Developing countries often oversee sanitation and hygiene leading to such situation. The recent swine flu and avian flu in the recent past can be attributed to such practices.

While the climate induce risks can be predicted up to certain extent, it may not be possible to predict non-climate induced risks due to the reason that one often

Fig. 8.1 Pigs feeding on biomedical waste from hospital. Such poor sanitation not only produces risk within the country, it can go out of the country by mode of food, air, water. Predicting disasters from such sources will be very difficult

don't know from where such disasters are triggered. In most of non-climate induces risks mathematical models and scientific instruments.

Risk is about exposure to external hazards over which people have limited control; vulnerability is an assessment of capacity to manage hazards without suffering a long-term, potentially irreversible loss of well being (ISSC 2005; UNDP 2007).

8.2 Climate Induced Disasters

Not all disasters can be predicted by global model and in long term. As shown in Table 8.1 some of the disasters occur abruptly giving very short or no notice to the local population. The disorganized nature of weather makes it changeable beyond a few days. This calls for major difference between weather and climate. Weather events are result of random forces and systemic factors.

Like age of individual can't be predicted weather can't be predicted accurately. Projecting changes in climate due to variation in atmospheric composition is much more manageable issue. It will be easy for predicting average age of people in a country or region considering the past living age of population similarly the climate which is average weather conditions can be predicted with much more accuracy than weather. It is also not prudent or possible to depend on global community to predict and solve local problems. Communities have to build the capacity to predict minimize and adapt to local small scale disasters. Past and current experiences help in reducing climate-related adverse impacts. Existing approaches, methods and tools in the fields of agriculture, forestry, ecology and others, will be of great importance in assessing risks posed by climate change.

As depicted in Fig. 8.2, there is change is increase in climate induced disasters from mid of last century, whereas number of people killed was more in the first half of the last century (Fig. 8.3). The reduction in deaths can be contributed to improvement in prediction, early warning and coping mechanism developed in second half of the century.

Even though considerable progress has been made in climate science, challenges still remain. The gaps and needs include resources and local knowledge to improve climate and weather observations, data collection. As on date there are many uncertainties in the forecast of climate variations and extreme events.

Table 8.1 Forecast capability for different types of floods

Flood types	Scale (km)	Duration	Forecast capability
Cloud burst	10 × 10	Few minutes	Small
Flash floods	10 × 10	Few hours	Small
Metropolitan city/urban floods	20 × 20	6–12 h	Small
Riverine floods	1,000 × 1,000	2–7 days	Moderate
Dam burst floods	1,000 × 1,000	Few days	Moderate
Mountain flooding/valley flooding	1,000 × 1,000	Few days	Moderate

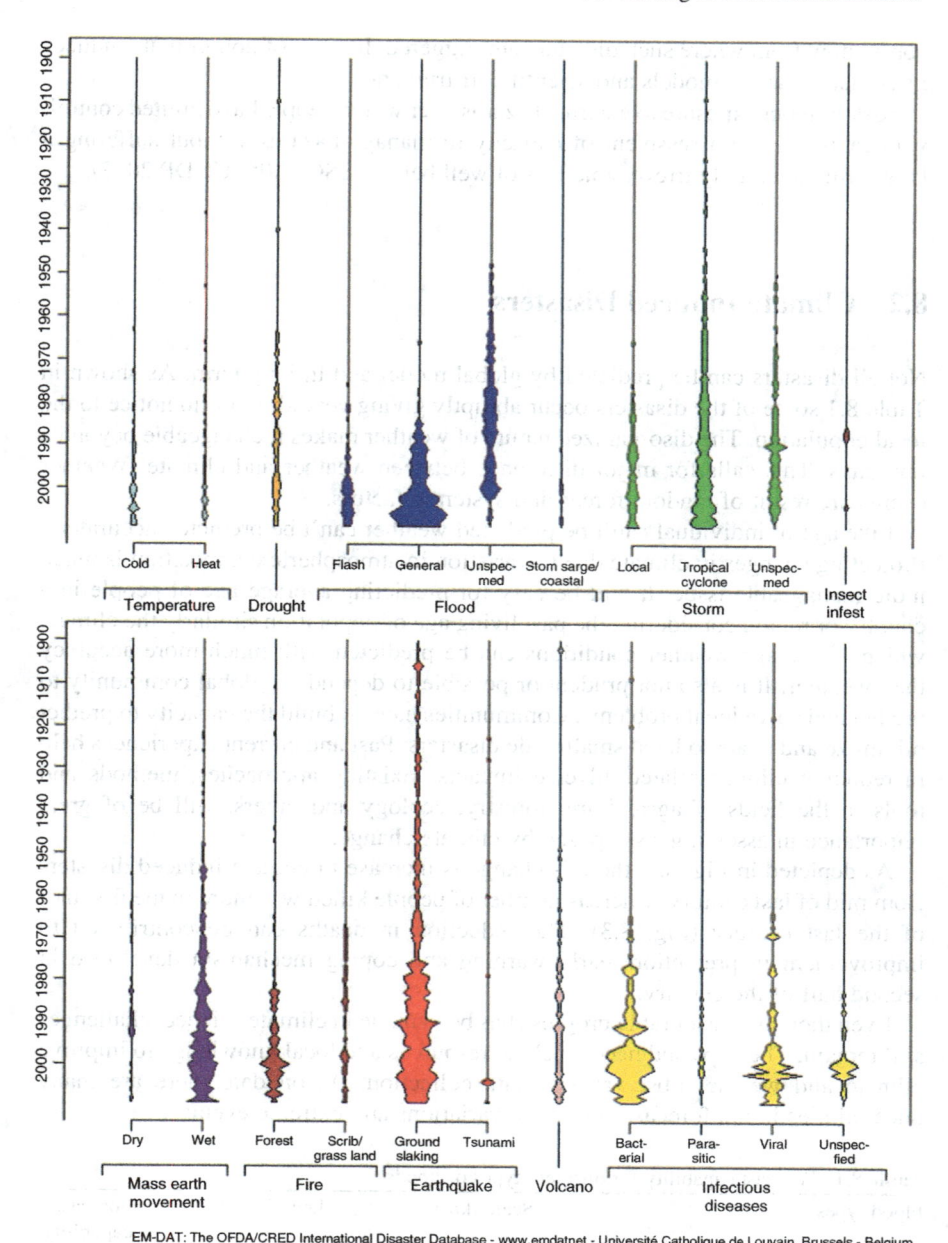

Fig. 8.2 Number of natural disasters reported from 1900 to 2009

Early warning systems as an element to reduce the impact of disasters have been recognised in major international agendas. Some of the noteworthy are UNFCCC, the Barbados Plan of Action for Small Island Developing States, Yokohama

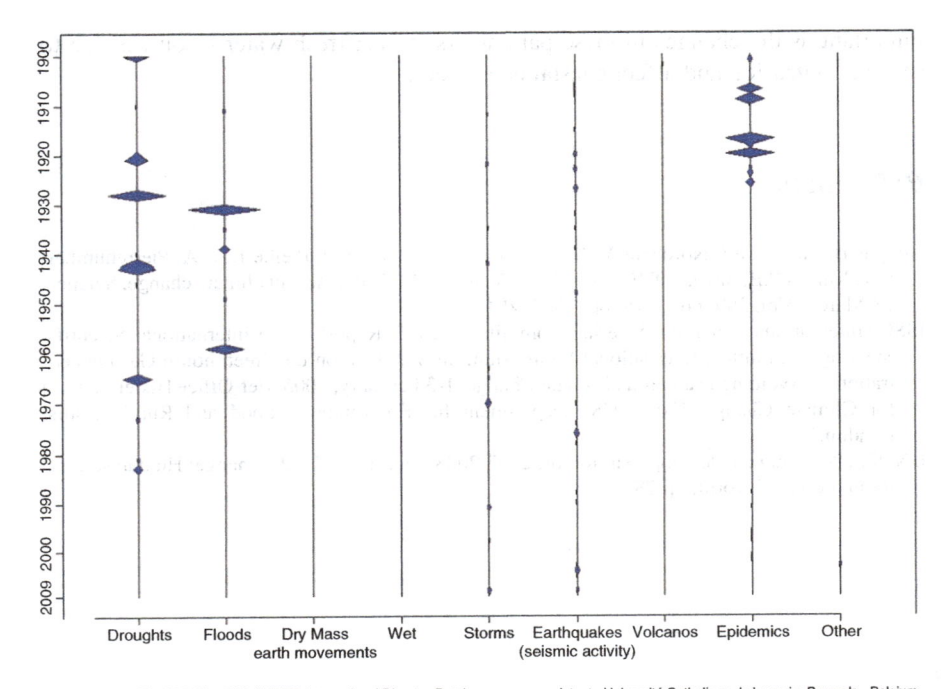

EM-DAT: The OFDA/CRED International Disaster Database - www.emdatnet - Université Catholique de Louvain, Brussels - Belgium

Fig. 8.3 Number of people killed by natural disasters 1900–2009

Strategy, the Johannesburg Plan of Implementation, the Mauritius Strategy and the meeting of G8 ministers in Gleneagles. Many countries in Europe and Asia now have early warning systems for heat waves and floods.

Credible research and monitoring programmes with respect to critical indicators of change are necessary to develop appropriate preparedness strategies and to enhance the adaptability of ecosystems and economies (Alley et al. 2003).

Technology transfer cannot happen by borrowing and installing equipments. It needs long term capacity building at every level right from maintaining repairing the equipment and using the data generated by equipment.

Even short notice of the likely occurrence of a natural phenomenon is of great significance in reducing loss to community. The prediction of a natural event by technologic investigation will help in bringing down possible damage due to disaster. Some hazards, such as droughts and floods, can be predicted accuracy, but other disasters like landslides cannot. Pacific Warning Center, which constantly monitors the oceans, provides advance notice with respect tsunamis that varies from few minutes to a few hours.

Climate monitoring and forecasting should just not concentrate on global temperature and precipitation but shall also include increases in sea level; sea-surface temperature; decreases in sea-ice cover; and ocean circulation. These changes are important as the changes in these parameters impact fresh water resources due to seawater intrusion and affect coastal ecosystem.

References

Alley R.B, Marotzke J, Nordhaus W.D, Overpeck J.T, Peteet D.M, Pielke Jr. R.A, Pierrehumbert R.T, Rhines P.B, Stocker T.F, Talley L.D, Wallace J.M, 2003: Abrupt climate change. Science. 28 March. Vol. 299. no. 5615, pp. 2005–2010

ISSC (International Scientific Steering Committee), 2005: Report of the International Scientific Steering Committee. International Symposium on Stabilization on Green house Gas concentrations – Avoiding Dangerous Climate Change, 1-3 February, 2005 Met-Office Hadely Centre for Climate Change, Exter, UK, Department for Environment, Food and Rural Affairs, London.

UNDP, 2007: Human Development Report 2007/2008, Fighting Climate Change: Human solidarity in a divided world, pp 78

Chapter 9
Early Warning Systems and Their Effectiveness in Asia

Early warning systems allow communities to organize for and tackle the natural hazards. The efficiency of such systems depends on technology available, training of community, disaster preparedness and how corrupt a society is. The efficiency is measured in terms of lives saved and reduction in losses, which is directly related to the execution of response by the people and institutions. The indispensable components of the forecasting, warning and response system consist of a data source, communications, forecasts, decision support, notification, coordination, and responses. A flood forecast and warning programme should be designed to mitigate disaster. To achieve this, it is essential that all of the components of the system be functional.

Countries throughout the world are concerned about the impacts due to natural disasters have on society. Long history of disasters and losses in terms of life and property in some countries shows these societies are not adapted themselves for disasters.

Asia's vulnerability to natural hazards is growing, because of increase in population and migration of people to disaster prone areas.

Few people in the world are vulnerable to major disasters and large population is vulnerable to minor disasters. Irrespective of magnitude of disasters people will witness greater vulnerability in the presence of corruption or poverty. People will witness greatest vulnerability if the society is corrupt and poor. Figure 9.1 shows illustration of vulnerability considering poverty and corruption and Fig. 9.2 shows risk as function of hazard, vulnerability and capacity.

The society can't build capacity if its people is corrupt thus it puts itself into lower capabilities. Typical examples of corruption in education particularly medical and paramedical will lead to poor treatment to victims of disaster.

There is urgent need for intervention from international agencies to increase capacity of developing countries to combat corruption in the absence of which funds will be misused.

Framework of early warning systems comprises of four phases (Fig. 9.3):

1. Monitoring of originator(precursors)
2. Forecasting of a probable event and

R. Chandrappa et al., *Coping with Climate Change*,
DOI 10.1007/978-3-642-19674-4_9, © Springer-Verlag Berlin Heidelberg 2011

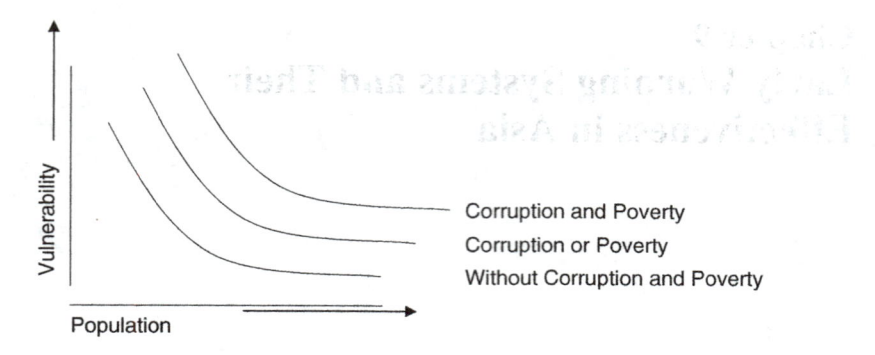

Fig. 9.1 Illustration of vulnerability considering poverty and corruption

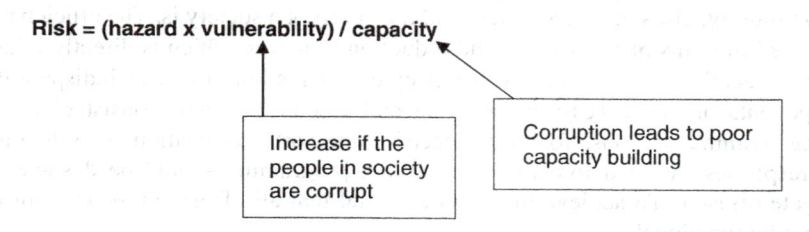

Fig. 9.2 Risk as function of hazard, vulnerability and capacity

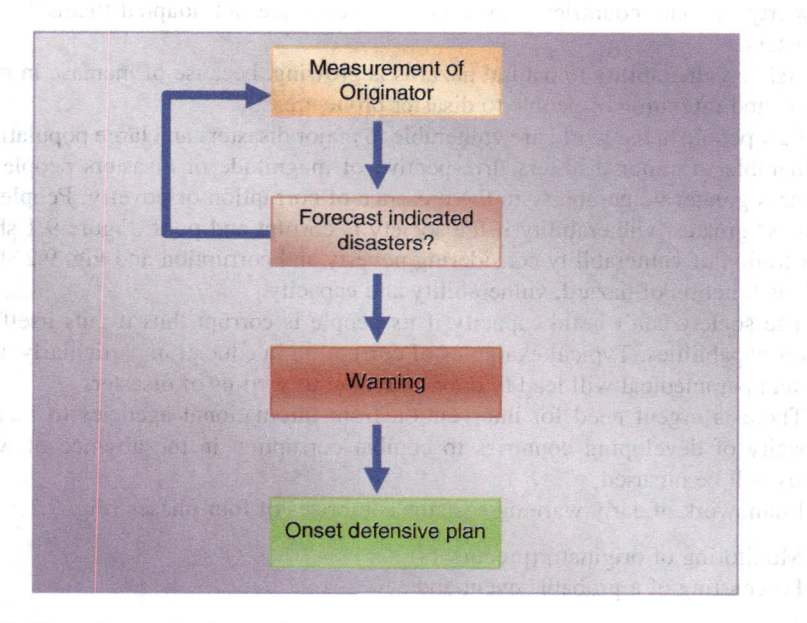

Fig. 9.3 Three phases of early warning system

3. The notification of a warning if catastrophic events are envisaged
4. The onset of emergency response activities

Many citizens in developing Asia are not aware of warning system available within the region. Most often print media apart from radio and television are only means of communicating early warning communication. Small Message Services (SMS) alerts are yet to become popular and need to be incorporated in to country level action plans. Poor access to information and service is reinforced by exploitation by elite people. Livelihood pressure is severe making it difficult for poor to make all arrangements to combat emergency in case of disasters.

Examples of good practice in national early warning systems include the Community Based Flood Early Warning System (CBFEWS) in the Philippines. CBFEWS help local communities prevent losses from floods. The lessons learned from the CBFEWS can be replicated in Asian countries combining indigenous knowledge (UNFCCC 2005).

Advances in science and technology in the recent past have improved the potential of early warning to reduce human loss. Good estimates of the timing and location of cyclones now can be made. Prediction should not aim only to protect human settlements. Climate impacts can affect biodiversity in the locality in addition to fisheries and aquaculture; agriculture; animal husbandry; forestry; tourism; industry; settlements and infrastructure.

The principles of early waning include:

- Early warning practices should aim hazard mitigation and vulnerability reduction
- An agency should be designated by the Government to take up responsibility
- Early warning should address need of local community and provide information about locally predominant disasters and
- The warning process should include demonstrated practices

9.1 Biological Warning Systems

In some places, communities become aware of natural risks through unofficial or informal flood detection and warning processes.

Prediction of many disasters can be done by monitoring data pertaining to

- Climate
- Soil moisture
- Stream flow
- Ground water
- Reservoir and lake levels
- Short, medium and long range forecasts and
- Vegetation health/stress and fire danger

A wealth of traditional knowledge persists for predicting disasters in much community. They observe movements, movements of animals and change in the flora. Though the people do get some warning from government they often come too late and some time do not get at all.

Many animals have senses to grasp upcoming disasters. Birds which fly high and move great distance can particularly predict movement of clouds and hence rain fall approaching its nest. Thus some birds build the nest at grater height to avoid damage to nest. The birds which lay eggs on ground will usually select safe place at higher elevation where water is unlikely to enter. Hence it is essential to gather indigenous knowledge and dissipate the same to enable local population to take notice of it and move themselves to safer place. Even though modern technology can be used to predict disaster with better accuracy and warn people, the biological indication of flora and fauna should not be overseen. The summary of some biological warnings emitted by flora and fauna is given in Table 9.1.

Early warning activities based on disease observation, reporting, and epidemiological examination, supported by information systems that enable integration, analysis, and sharing of animal health data combined with socioeconomic, production, and climatic data help in better understanding of ecological and epidemiological mechanisms responsible for the spread of a given ailment (Martin et al. 2007). 71,332 ill or dead birds were reported and WN virus-positive birds were found more than 3 months before the onset of human cases as part of West Nile (WN) virus surveillance in New York State in 2000. Dead bird surveillance was valuable for early detection of WN virus and for guiding public education and mosquito control efforts (Millicent Eidson et al. 2001).

Several early warning systems based on the response of organisms belonging to different tropic levels have been developed in recent years (Smith and Bailey 1988; Kramer and Botterweg 1991; Baldwin et al. 1994; Borcherding and Volpers 1994; Van Hoof et al. 1994; Sluyts et al. 1996).

Changes in pH in ocean due CO_2 concentration can be studies through plankton demography. Change in population of jelly fish due to large-scale climate influences, is seen in the NorthSea (Lynam et al. 2004).

9.2 Man-Made Warning Systems

Early warning and forecasting are useful for both contingency planning and defining immediate actions in responding to a disaster. Poor people need early warning most. Requirement for early warning vary by livelihood group.

The ultimate goal of disaster forecasting and early warning systems is to protect lives and property. Hence to serve people effectively systems must be integrated and link the technical community, public authorities as well as local communities. Several initiatives have been taken across the world to save lives. The Hazinfo and Community Tsunami Early Warning Centre (CTEC) initiative in Sri Lanka is example which is community-based practices adopted by communities in Sri Lanka.

Table 9.1 Behavior of various flora and fauna prior to for weather change and hydro meteorological disasters

Cyclone	Flood	Rain	Drought
Cattle become restless and stop eating; cattle/dogs wail continuously at night; ants come out of the nest with the eggs and start moving towards safer places; bees move around in cluster; birds move without destination; increase in insect movement; jumping fish in the rivers and ponds; frogs call continuously; crow fly at night; bending trees; water hyacinth in canal; leaves of cotton tree turn upside down; new leaves of the tree fall to the ground	Elderly people experience aches/pains in the joints/bones; persistent croaking of frogs; sighting of unusually large number of cow egret; insect flies in the opposite direction of the river flow	Sparrows playing on the sand; lapwing lays her eggs on the higher portion of the field (these birds never construct a nest but lay their eggs on bare soil); dragon flies swarm over open dry land (swarming over surface water like pond indicate dry season); emergence of centipedes from their holes carrying their eggs in swarms to shift them to safer places; swelling on the lower portion of the camel's legs (probably caused due to higher relative humidity); creepers grow skyward orientation; jacklegs run haphazardly, glow worms approaching water surface; snakes climbing trees; fishes jumps above water, frogs produce there note loudly; buds of small plants wither away, holes appearing on banyan tree barks	Unusual flower blooming in bamboo, withering of crops

Source: Hellen (2008), Philippa (2003), VASAT (2010), Ramanathan (1987), observation by authors

Ministry of Earth Sciences, Government of India has set up an Indian Tsunami Early Warning Center at the Indian National Centre for Ocean Information Services (INCOIS) located in Hyderabad.

Box 9.1 gives principles of effective early warning and Box 9.2 gives activities in recent past which has augmented early warning system and related disaster mitigation activates. 89.1% of global natural disasters victims in 2009 were in Asia (Femke et al. 2009) inferring that there is still scope for lot of improvement in disaster warning, preparedness and response.

Box 9.1 Principles of effective early warning
- Timely detection of the disaster and demarcation of vulnerable areas
- Usage of modern forecasting technology
- Involvement of stakeholders at all levels (local and national)
- Adopt accurate prediction procedures of the disaster phases
- Early warning systems should be based upon risk analysis
- Timely issuance and dissipating of easily understandable warnings
- Vulnerable community need to be aware of the hazards and effects
- Governments need to exercise responsibility to prepare and issue hazard warnings
- Regional institutions shall provide knowledge and advice government
- Timely Action by key disaster response organization
- Timely evacuation of the inhabitants at risk and
- There shall be strong international and regional cooperation

Box 9.2 Activities in recent past which has augmented early warning system and related disaster mitigation activities
The Asian Disaster Reduction Centre (ADRC) was established in Kobe, Hyogo in 1998, to enhance disaster resilience of the member countries, to build safe communities, and to create a society where sustainable development is possible.

Hyogo Framework for **Action 2005–2015: Building the Resilience of Nations and Communities to Disasters** was adopted World Conference held 18–22 January 2005 in Kobe City of Japan's Hyogo Prefecture for guiding framework for the next decade on disaster reduction.

Phuket ministerial declaration on regional cooperation on tsunami early warning arrangements was made in Phuket, Thailand on 29 January 2005 for the purpose of furthering regional cooperation on tsunami early warning arrangements in the Indian Ocean and Southeast Asia.

Indian Ocean Tsunami Warning and Mitigation System (IOTWS) established after December tsunami in 2004 hit the coasts of the Indian Ocean will provide warning to inhabitants of nations bordering the Indian Ocean with respect to tsunamis.

Asian Disaster Preparedness Centre (ADPC) was established in 1986 in response to the recommendation of UN Disaster Relief Organization (UNDRO) – now known as UN Office for the Coordination of Humanitarian Affairs (UN-OCHA), with the aim of strengthening the national disaster risk management systems in the region.

USAID/OFDA and its partners are providing forecasts of rainfall over the Mekong River basin to the Mekong River Commission (MRC) since 2001. USAID/OFDA is also responsible for formation of Asia Flood Network (AFN) in 2003.

Establishing a feasible forecasting and warning system for communities at risk requires the combination of data, forecast tools, and trained people. A forecast system must provide enough lead time for communities to respond. Increasing lead time enhances the potential to lower the level of damages and loss of life and property. Forecasts must be adequately accurate to promote assurance so that communities will respond when warned. If forecasts are not accurate, then trustworthiness of the programme will be questioned and no response actions will occur.

Japan Aerospace Exploration Agency (Jaxa)'s project "Sentinel Asia" shares disaster information in the Asia-Pacific region uses advanced information technologies and satellites.

Figure 9.4 gives natural disasters impacts by disaster sub-group: 2009 versus 2000–2008 annual average; Fig. 9.5 shows top ten countries by damages in 2009 and distributed by disaster type; Fig. 9.6 revels top ten countries by victims in 2009 and distributed by disaster type; Fig. 9.7 depicts top ten countries in terms of disaster mortality in 2009 and distributed by disaster type; and Fig. 9.8 shows top ten countries by number of reported events in 2009.

Table 9.2 gives number of natural disasters in each geological regions of the world; Table 9.3 gives number of victims of natural disasters in different geological regions of the world, Table 9.4 shows damages of natural disasters at various region of the world.

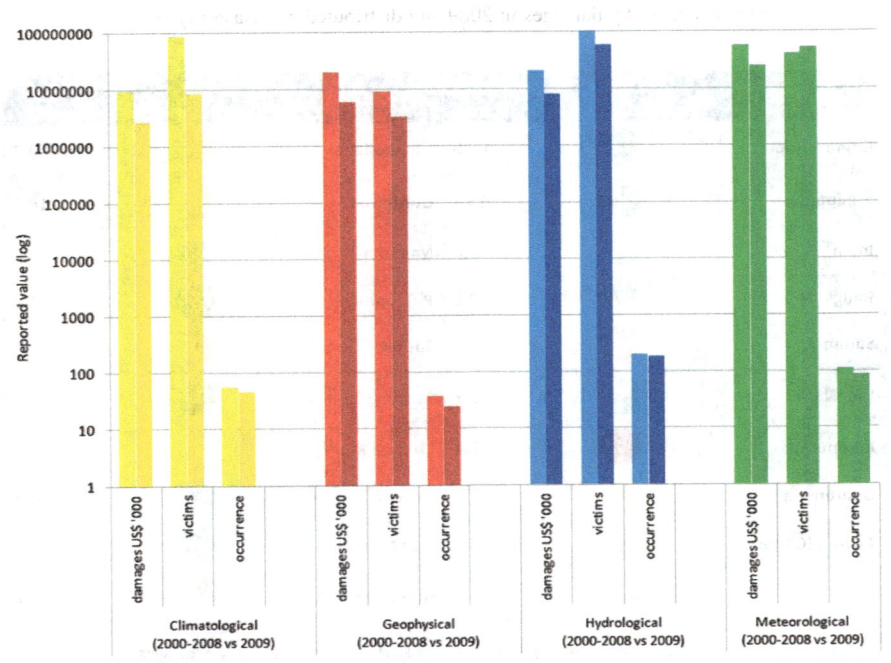

Fig. 9.4 Natural disasters impacts by disaster sub-group: 2009 versus 2000–2008 annual average

Country	Disaster distribution	Damages (US$ Bn.)	Country	Disaster distribution	% of GDP
United States		10.8	Samoa		28.7
China P Rep		5.2	El Salvador		4.4
France		3.2	Tonga		3.6
India		2.7	Lao P Dem Rep		1.9
Italy		2.6	Burkina Faso		1.9
Indonesia		2.4	Fiji		1.6
Spain		1.9	Mongolia		1.2
Australia		1.5	Viet Nam		1.2
Japan		1.4	Honduras		0.7
Viet Nam		1.1	Costa Rica		0.7

■ Climatological   ■ Geophysical   ■ Hydrological   ■ Meteorological

Source: "EM-DAT: The OFDA/CRED International Disaster Database www.emdat.be – Universit? Catholique de Louvain – Brussels – Belgium"

Fig. 9.5 Top ten countries by damages in 2009 and distributed by disaster type

Country	Disaster distribution	No. victims (millions)	Country	Disaster distribution	Victims/ pop. (%)
China P Rep		68.8	Mongolia		19.7
Philippines		13.4	Guatemala		18.4
India		9.0	Namibia		16.6
Bangladesh		4.6	Philippines		14.8
Sudan		4.4	Sudan		10.7
Viet Nam		3.7	Taiwan (China)		10.0
Indonesia		2.9	China P Rep		5.3
Guatemala		2.5	Zambia		4.9
Taiwan (China)		2.3	Viet Nam		4.3
Brazil		1.9	Honduras		4.1

■ Climatological   ■ Geophysical   ■ Hydrological   ■ Meteorological

Source: "EM-DAT: The OFDA/CRED International Disaster Database www.emdat.be – Universit? Catholique de Louvain – Brussels – Belgium"

Fig. 9.6 Top ten countries by victims in 2009 and distributed by disaster type

Country	Disaster distribution	No. of deaths	Country	Disaster distribution	Deaths per 100 000
India		1 806	Samoa		81.5
Indonesia		1 407	American Samoa		51.4
Philippines		1 334	Tonga		8.7
Taiwan		630	El Salvador		4.5
China P Rep		591	Namibia		4.4
Australia		535	Solomon Is		4.1
Peru		419	Bhutan		3.3
Viet Nam		356	Taiwan		2.7
Italy		335	Australia		2.5
El Salvador		275	Sierra Leone		1.9

■ Climatological ■ Geophysical ■ Hydrological ■ Meteorological

Source: "EM-DAT: The OFDA/CRED International Disaster Database www.emdat.be – Universit? Catholique de Louvain – Brussels – Belgium"

Fig. 9.7 Top ten countries in terms of disaster mortality in 2009 and distributed by disaster type

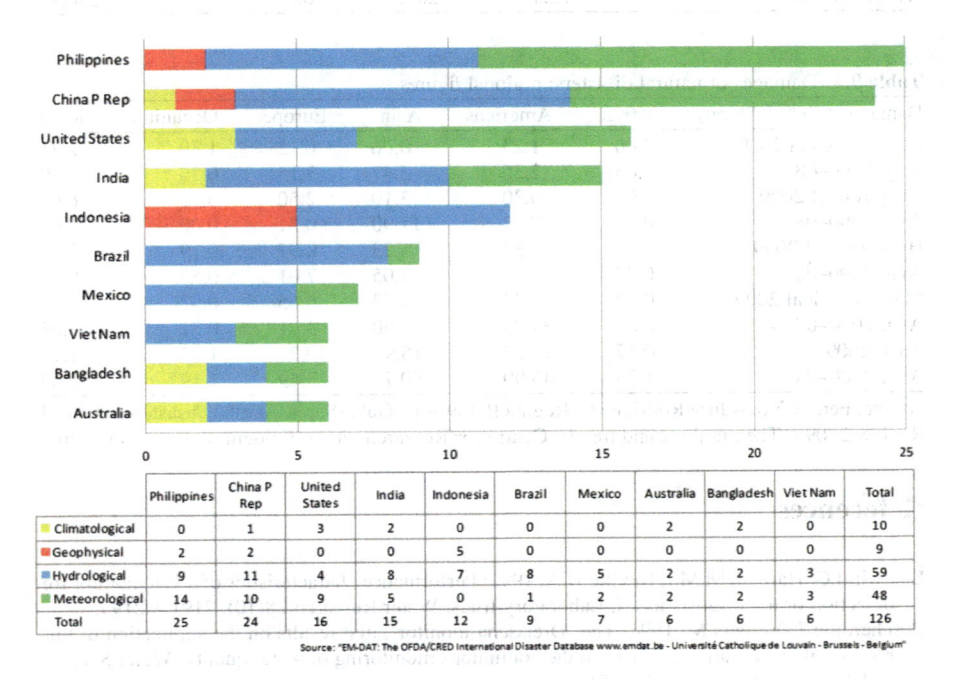

	Philippines	China P Rep	United States	India	Indonesia	Brazil	Mexico	Australia	Bangladesh	Viet Nam	Total
Climatological	0	1	3	2	0	0	0	2	2	0	10
Geophysical	2	2	0	0	5	0	0	0	0	0	9
Hydrological	9	11	4	8	7	8	5	2	2	3	59
Meteorological	14	10	9	5	0	1	2	2	2	3	48
Total	25	24	16	15	12	9	7	6	6	6	126

Source: "EM-DAT: The OFDA/CRED International Disaster Database www.emdat.be - Université Catholique de Louvain - Brussels - Belgium"

Fig. 9.8 Top ten countries by number of reported events in 2009

Table 9.2 Number of natural disasters: regional figures

No. of natural disasters	Africa	Americas	Asia	Europe	Oceania	Global
Climatological 2009	3	11	7	22	2	45
Avg. 2000–08	9	13	13	17	1	54
Geophysical 2009	2	3	14	2	4	25
Avg. 2000–08	3	7	22	3	2	37
Hydrological 2009	46	41	72	15	6	180
Avg. 2000–08	42	39	81	26	5	194
Meteorological 2009	13	18	42	9	3	85
Avg. 2000–08	9	35	42	15	7	108
Total 2009	64	73	135	48	15	335
Avg. 2000–08	64	94	158	61	16	392

Table 9.3 Number of victims of natural disasters: regional figures

No. victims (millions)	Africa	Americas	Asia	Europe	Oceania	Global
Climatological 2009	4.48	3.08	0.78	0.01	0.01	8.37
Avg. 2000–08	11.19	1.02	71.38	0.29	0.00	83.89
Geophysical 2009	0.02	0.18	3.00	0.06	0.01	3.27
Avg. 2000–08	0.09	0.39	8.54	0.01	0.01	9.03
Hydrological 2009	2.50	2.26	52.47	0.04	0.03	57.29
Avg. 2000–08	2.31	2.94	93.51	0.37	0.02	99.15
Meteorological 2009	0.16	0.23	50.18	0.00	0.02	50.59
Avg. 2000–08	0.48	2.88	35.03	0.36	0.04	38.79
Total 2009	7.16	5.75	106.44	0.11	0.07	119.52
Avg. 2000–08	14.07	7.24	208.46	1.03	0.06	230.86

Table 9.4 Damages of natural disasters: regional figures

Damages (2009 US$ bn)	Africa	Americas	Asia	Europe	Oceania	Global
Climatological 2009	0.00	1.23	0.06	0.12	1.30	2.71
Avg. 2000–08	0.05	2.36	3.47	3.15	0.36	9.39
Geophysical 2009	0.00	0.30	3.10	2.50	0.16	6.06
Avg. 2000–08	0.73	0.72	17.90	0.31	0.00	19.67
Hydrological 2009	0.15	1.33	5.23	0.97	0.19	7.88
Avg. 2000–08	0.37	2.99	9.05	7.01	0.52	19.94
Meteorological 2009	0.02	10.37	7.53	6.65	0.07	24.64
Avg. 2000–08	0.08	39.93	10.30	3.01	0.31	53.63
Total 2009	0.17	13.23	15.91	10.24	1.73	41.28
Avg. 2000–08	1.23	45.99	40.72	13.49	1.19	102.63

Source: Femke Vos – Jose Rodriguez – Regina Below – D. Guha-Sapir, Annual Disaster Statistical Review 2009 – The numbers and trends, Centre for Research on the Epidemiology of Disasters

References

Baldwin I.G, Harman M.M.I, Neville D.A, 1994: Performance characteristics of a fish monitor for detection of toxic substances-I. Laboratory trials. Water Research 28(10), 2191 ± 2199.

Borcherding J, Volpers M, 1994: The 'Dreissena-monitor', first results on the application of this biological early warning system in the continuous monitoring of water quality. Water Science and Technology 29(3), 199 ± 201.

Femke Vos - Jose Rodriguez - Regina Below - D. Guha-Sapir, 2009: Annual Disaster Statistical Review 2009 – The numbers and trends, Centre for Research on the Epidemiology of Disasters

Hellen Nyakundi, 2008: 'An Investigation into Community and Response to Flood Risks', ProVention/World Bank Young Researcher report, *Hellen Nyakundi, Dr Isaac Mwanzo, Dr A. Yitambe and Stephen Mogere, Kenyatta University, Kenya,* Community perceptions and response to flood risks in Nyando district, Western Kenya, http://www.proventionconsortium. org/themes/default/pdfs/CRA/Kenya.pdf downloaded on 21 July 2010

Kramer K.J.M, Botterweg J, 1991: Aquatic biological early warning systems: an overview. In: Jeffrey, D.W., Madden, B. (Eds.), Bioindicators and Environmental Management. Academic Press, London, pp. 95 ± 126.

Lynam C.P, Hay S.J, Brierley A.S, 2004: Interannual variability in abundance of North Sea jellyfish and links to the North Atlantic Oscillation. Limnol. Oceanogr. 49, 637–643

Martin V, Von Dobschuetz S, Lemenach A, Rass N, Schoustra W and Desimone L, Supplement 2007: Early Warning, Database, and Information Systems For Avian Influenza Surveillance1, Journal Of Wildlife Diseases, 43(3), Pp. S71–S76

Millicent Eidson, Laura Kramer, Ward Stone, Yoichiro Hagiwara, Kate Schmit, 2001: The New York State West Nile Virus Dead Bird Surveillance as an Early Warning System for West Nile Virus, *July–August 2001, Emerging Infectious Diseases Vol. 7, No. 4, pp 631–635*

Philippa Howell, 2003: Indigenous early warning indicators of cyclones: Potential application in coastal Bangladesh, Disaster Studies Working Paper 6, Benifield hazard Research Centre, Aon Benfield UCL Hazard. www.abuhrc.org/Publications/Working%20Paper%206.pdf downloaded on 21 July2010

Ramanathan A.S, 1987: Indian Journal of History of Science, 22(3): page 191–197

Smith E.H, Bailey H.C, 1988: Development of a system for continuous biomonitoring of a domestic water source for early warning of contaminants. In: Gruber, D.S., Diamond, J.M. (Eds.), Automated Biomonitoring. Ellis Horwood, Chichester, pp. 182 ± 205.

Sluyts H, Van Hoof F, Cornet A, Paulussen J, 1996: A dynamic new alarm system for use in biological early warning systems. Environmental Toxicology and Chemistry 15(8), 1317 ± 1323

UNFCCC, 2005: Report of the Asian Regional Workshop on Adaptation. http://maindb.unfccc.int/ library/view_pdf.pl?url=http://unfccc.int/resource/docs/2007/sbi/eng/13.pdf retrieved on September 30, 2010.

Van Hoof F, Sluyts H, Paulussen J, Berckmans D, Bloemen H, 1994: Evaluation of a biomonitor based on the phototactic behaviour of Daphnia magna using infrared detection and digital image processing. Water Science and Technology 30(10), 79 ± 86.

VASAT (Virtual Academy for Semi Arid Tropics), 2010: http://vasat.icrisat.ac.in/?q=node/304 downloaded on 22 July 2010

Chapter 10
Disaster Risk Reduction and Climate Change Adaptation in Asia

Global climate change is a tangible reality and already, showing devastating effects on humankind and nature in the world. Change in temperature in space and time show evidence of human influence on the climate system (Thomas et al. 2006). Climate change and disasters are serious risk to poverty reduction and erodes wealth built by individuals and nations. Nearly half of the climate related disasters happen in Asia and unless adapted to venerable impacts, people and nations will lose wealth and have to create it again. The international efforts to reduce GHG emissions are already happening to avoid the worst-case scenarios at the end of this century. Climate change is bound to continue regardless of how well those efforts succeed, due to the GHG already emitted, will stay in the atmosphere for a long time. Therefore, there is no choice but to deal with these changes by "adaptation". Adaptation will work best if strategies to reduce climate-related risks are incorporated in ongoing development activity. This comprehensive approach to manage the rising risks is called "climate risk management". The planet's crises – rapid climate variation, degradation of ecosystem, increase in human population, scarcity of resources are affecting the people worldwide. International community seems to be concerned less about climate change than failing financial institutions (CCCD 2009).

Adaptation means getting adjusted to changing scenario and adoption means acceptance of changing scenario.

Different organisms live in specific range of temperature, precipitation, humidity, and sunlight. Anything out the range affects comfortable living and force living organism to adapt or perish. Following are the principles of adaptation

Animals
- Migration to regions of comfortable climate
- Follow migration pattern of prey
- Change in phenology like hibernation period, flowering period, migration period, egg laying period, etc. and
- Change morphology and reproduction

Plants
- Shift their abundance by germinating and flourishing in the climatic regions comfortable to them

R. Chandrappa et al., *Coping with Climate Change*, DOI 10.1007/978-3-642-19674-4_10, © Springer-Verlag Berlin Heidelberg 2011

Human beings
- Travel to comfortable location of earth during the periods of uncomfortable weather at home land
- Control population
- Change food habits, living style, agricultural practices, crop pattern, agriculture practice
- Acquire new skills for lively hood
- Forecasting climate and related disasters
- Form local, regional and national policies and increase institutional capacity
- Cut down corruption and bring in good governance practice along with international cooperation
- Conduct research and share the findings to mitigate changing climate and adapt to changed climate
- Prepare disaster management plan and emergency preparedness plan
- Create infrastructure for coping with climate change and
- Plan for intergenerational and intercontinental adaptation strategy

Adaptations vary from place to place in the system in which they occur and the climatic stimuli. Box 10.2 shows principles of adaptation and implementation difficulties, over the year Asian countries have witnessed drought and people have adapted themselves. While poor people get adjusted to temperature by changing food habit and wetting the cloths during higher temperature, affordable people will use air cooler/conditioners. The people who can afford will purchase more of sun-cream, ice-cream, beer, and cool drinks.

Statistics can't capture severity of disasters and sufferings of living organisms including human beings. There is uncertainty associated with type, location magnitude of disasters. Poor people have fewer choices and beggars will not have any choice. Disasters make children to walk away for schools and may have to choose prostitution and crime for survival. Some time people affected will have to sell the property at lesser price to overcome hunger and to afford other fundamental needs like health.

The impacts of climate change are mixed and overlap with each other. They include food prices, the financial crisis, energy shortage, ecosystem degradation, demographic changes, land use changes, crime and health.

Countries which have formed own governments after long ruling of foreigners had hardly any experience in adaptation and citizens suffered either by hunger or by death. Governments which concentrated their resources on conflicts and military had hardly had any time to save their citizens against natural disasters. Planned preventative and precautionary adaptation has the prospective to reduce vulnerability to climate impact.

Adaptation needs differ across geographical scales (local, national, regional, global) and temporal scales (short term and long term). Box 10.1 gives example of national, regional and international adaptation. Some governments have helped the citizens to adapt by creating construction jobs in drought hit area and have given subsidized food for the poor and marginalized. In the past governments have included relief cess/surcharge for the taxes collected. Another way of adapting is requesting 1 day salaries from government employees.

> **Box 10.1 Example of national, regional and international adaptation**
>
> Bangladesh, which often suffers due to cyclones, storms and floods, has developed action plan consolidating agriculture, health, livelihoods, disaster management, environment, and development. A trust fund has also created.
>
> The International Centre for Integrated Mountain Development (ICIMOD) helps eight countries in the Hindu Kush-Himalayas with respect to local adaptation. This arrangement is an example for regional adaptation as they share same ecosystem and international adaptation as it involves multiple countries.
>
> The World Meteorological Organization (WMO) brings together data from 187 member countries and collects and share climate data. This forms an excellent example for international adaptation in place.

Adaptive capacity differs among countries, regions, nations, cultures and socioeconomic groups. The women whose clothing's were not suitable for running and escaping during flood/tsunami were more vulnerable than men. Apart from the dress the society where women are not trained in swimming due to social and cultural aspect make them more venerable during floods. Vegetarians and people who worship animals will have tough time during drought when only food available will be their dead cattle. A corrupt society where in officers does not spend money for cause or report erroneously will ultimately make society more vulnerable to climate impact.

With more and involvement of more international community to find solution for climate related problems, the number of adaptation methods are increasing.

Climate change often leads to disaster. Disaster risk management is a relatively new area of social concern and practice. It is a very relevant for cooperation during disasters, because natural disasters have increased over the year destroying investments and setback progress in development. Victim of large-scale impacts of earthquakes, tornadoes, typhoons, floods or droughts are barely able to respond, and recovering can take years or decades. Following the United Nations initiative for an International Decade for Natural Disaster Reduction (1990–99), this theme has reached much higher on the international agenda. Growing number of development cooperation is trying to cater for more prevention in their activities.

Climate change is happening faster than the predictions made in the past. Human population should respond to mitigation of climate change either by sacrificing luxury or otherwise by decreasing GHG. Since the past emissions are already causing climate change, we must adapt to climate change. The climate change will affect all but will affect poor people more. It will not spare rich people. If all the poor people are gone who will serve the rich people? Adaptation needs reorientation of global priorities to make sure the availability of sufficient resources (Pachauri 2010)

Developing countries and their poor people are most vulnerable to climate impacts and following are the factors that increase the vulnerability:

- **Poverty**: During disasters some households recover faster than the other. There recovery is often liked to wealth they own. The recovery process is slow among the poorest.
- **Lack of disaster resistance infrastructure/houses**: Due to poverty people will not be able to construct disaster resistance infrastructure/houses pushing themselves to injury, death and loss of wealth.
- **Lack of disaster defence infrastructure**: Most of the developing countries lack disaster defence infrastructure like dykes and levees to control flood.
- **Limited access to insurance**: Many countries do not have high penetration of insurance and most often it is seen as unnecessary expenditure. This often leads to difficulties during recovery from disasters.
- **Corruption**: Corruption will influence the magnitude of vulnerability due to improper decisions without considering consequences. The practice will also affect recovery phase and divert the funds a few influential people instead of affected people.
- **Institutional failure:** Institutional failures at all levels irrespective of the reason will increase the vulnerability of community both pre-disaster and post disaster phases. The lower capacity of institution to predict and cope with disasters is directly proportional to the service rendered by them.
- **Conflicts within the regions**: Conflicts within the regions like terrorism, war and riots demands more expenditure towards such conflict resulting in shortage of resource to cope with disasters.
- **Poor international relations**: International relations especially with neighbouring countries are essential to tackle the large scale disasters. The recovery phase often needs speed to save lives and property. In such instances the help from neighbouring countries shall be boon for the affected community. Disaster vulnerability in case large scale disasters will be higher if the international relation is poor.

Adaptation should happen at different level considering the principle of common goal but differential responsibility. At individual level each on should take responsibility to adopt. The next level is at family, which means the families should know the fore coming disasters in the area they live in. At local level the local institution should anticipate climate change in the region and be ready for it. At national level nations should form strategy to protect its tax payers and spend for the cause without siphoning funds for personal cause. Regional level which share common river basin or ecosystem should come together and act to cope up with changes. The last level is international level which is already acting with the principle of international cooperation should act further with each country contributing a bit instead of leaving it to some players.

Adaptation can be of different types:

Anticipatory adaptation (also referred as proactive adaptation) – It is adaptation that takes place before *climate impacts* are observed.

Autonomous adaptation (also referred as spontaneous adaptation) – It is adaptation that triggers *welfare* changes in *human systems*.

Planned adaptation – It is the adaptation that is the result of a deliberate policy decision.

For human systems, the success of adaptation depends on knowledge and resources. For other organism adaptation depends on its ability to withstand adverse effect of climate or migrate to new location. Adaptation to climate change by human and other organism is highly complex and dynamic. It will be difficult for many species to modify behaviour or migrate to new location for adapting to climate change. Many species are at present stressed by a variety of factors like urban development, pollution, invasive species, and isolated habitats. Such conditions along with the rapid climate change are likely to bring difficulty for leading smooth life. Climate change has the potential to harm societies and ecosystems. It is likely to affect agriculture, forestry, water resources, human health, coastal settlements, and natural ecosystems. While mitigation is effort to reduce the probability of adverse conditions, adaptation is the effort to reduce severity of impacts if adverse conditions are happening. Throughout the ages, human societies and ecosystem have adapted to differing climates and environmental changes, although not always successfully. Human settlements in climatically diverse location have learned how to live in wide variety of climatic condition. Adaptation began by migration to places, modification of diet and life style. Now adaptation to the climate changes is challenge to human community.

The adaptive capacity of natural ecosystems is limited compared to human systems wherein changes are often anticipated in near future and generations to come. The adaptation can be proactive or reactive depending on the intelligence and available resource of a society. Proactive adaptation include improvement in building technology, better regulation, early warning system, good policy and governance. Reactive adaption includes effort a system makes to overcome the impact due to changes already felt like shortage of water resources or rise in sea level.

10.1 Adaptation and Adoption: Human Strategy

The progress of adaptation to climate change requires: improved governance; mainstreaming climate issues into all national, sub-national, and sectoral planning processes; encouraging a ministry with a broad mandate, to be fully involved; combining approaches, empowerment of communities, vulnerability assessments, access to good quality information, Integration of impacts, Increasing the resilience of livelihoods and infrastructure (African Development Bank et al. 2002). Agriculture would expand and increase in productivity in northern areas where temperature was relatively cold. Reduction in moisture in the summer may increase land degradation and desertification. People in low-lying coastal areas of temperate and tropical Asia are likely to displaced due to Sea level rise and an increase in intensity of tropical cyclones

Adaptation must be fast, scaled, focused and integrate action across different sectors (CCCD 2009).

Asia had witnessed increase in temperate, floods, droughts, forest fires, tropical cyclones, water stress, flood, drought and sea level rise which had diminished food security in countries of arid, tropical, and temperate Asia.

Low livelihood diversification and low social capital is the main reasons for poverty in many Asian countries. Principles that should guide the monitoring and evaluation of adaptation efforts are impartiality/independence, credibility, and usefulness (OECD-DAC 2008).

Past adaptation practices by households in many countries vary according to hazard type and asset base holdings. The most common form of adaptation is temporary migration. Study by World Bank (2010) revealed that 37% of surveyed households indicated migration for day labour work by adult men (World Bank 2010). Storage of food and drinking water before extreme events is also common adaptation measures both in developed and developing countries. It is also common practice to build livestock platforms to guard animals in disaster vulnerable area.

Bangladesh has put in place a wide set of risk reduction measures – both structural and non-structural, to enhance its disaster preparedness system. Farmers have adapted floods by switching from low-yielding rice to high yielding rice crops resulting in rise in agricultural production.

The impacts of tropical cyclones, storm surges, floods, and other climatic hazards are concentrated in specific regions of some countries. Poor people in such area are most vulnerable and have the lowest capacity to address the impacts, hence are also affected unreasonably. Different countries have different adaptation plans. Major adaptation measures taken by Bangladesh are – Coastal embankments,

Box 10.2 Principles of adaptation and implementing difficulties

Principles
Maintain ecological resilience
Accommodate change
Develop knowledge and dissipate
Plan strategy and implement
Integrate action across cross sectors

Implementation Difficulties
Regional priorities
Corruption
Local conflicts
Disasters disrupting economy
Knowledge deficiency
Fund crunch
Political priorities
Lively hood pressure
Degraded ecosystem
Prioritizing profit

Foreshore afforestation to protect sea-facing dykes, Cyclone shelters, Early Warning Systems, and Decentralization of Relief Operations.

At present, about 50% of the world's population live in urban areas and is expected to rise to more than 60% in the next 30 years. The vulnerable people of many developing countries are already exposed to scarcity of clean drinking water and poor sanitation, and often live in high-risk areas such as flood plains and coastal areas (Haines et al. 2006). Built areas will result in urban heat islands (UHI) where in temperature will be 5–6°C warmer than neighboring countryside (Oke 1982). Builtup area retains more solar energy during the day, and at the same time have lower rates of radiant cooling during the night. Urban areas have lesser wind speeds, a lesser amount of convective heat losses and evapotranspiration. Urban activity like space heating, air conditioning, transportation, cooking and industrial activity add heat into the urban environment resulting in distinct weekly cycles in UHI intensity (Wilby 2003). The physical constituents and human activities in urban centers interact with other climate drivers like – runoff from impervious surfaces can have impact on downstream risks of flooding and erosion (Hollis 1988); altering river temperatures and water quality by discharges of storm water (Paul and Meyer 2001). Urban air pollution can increase during heat waves leading to mortality (Stedman 2004).

Urban area requires more preparedness and planning due to reason explained in above paragraph. It is common practice in developing to legalize illegal building by levying small fine or bribe or both. The fastest growing city in Asia like Bangalore has come on natural drains and altering natural contours to greatest extent making it difficult to adopt during rains/floods and other disasters. Such bloom has occurred due to mushrooming of educational institutions and industries to capture market or talent pool. The cutting of roots and loosening of soil during trenching has weekend the grip leading to fall of trees/electric poles on the road without any pre-indication causing further traffic congestion and subsequent consequences. Coping with such scenarios either calls for proper planning or being prepared to attend emergency during such accidents. The fast growing cities have also imparted huge water demand and demand for disposal of sewage. The development often occur fast to capture real estate market which is often unpredictable. Such fast development usually does not precede replacement of sewer line, rain water drain and widening of roads.

In a nutshell Fig.10.1 shows impact of climate change on urban area. Coping mechanism in urban area include:

- Providing infrastructure in accordance with speed of development
- Providing proper breathing space and open area for infiltration of rain water
- Formation of new built up area without altering natural drainage pattern
- Removal of week trees and structure
- Formation of disaster management cell and
- Control urban air pollution by curbing traffic and proving reliable mass transport system

Local communities react to weather events without understanding future risks. Traditional coping practices should be fully explored and adjusted for enhancing

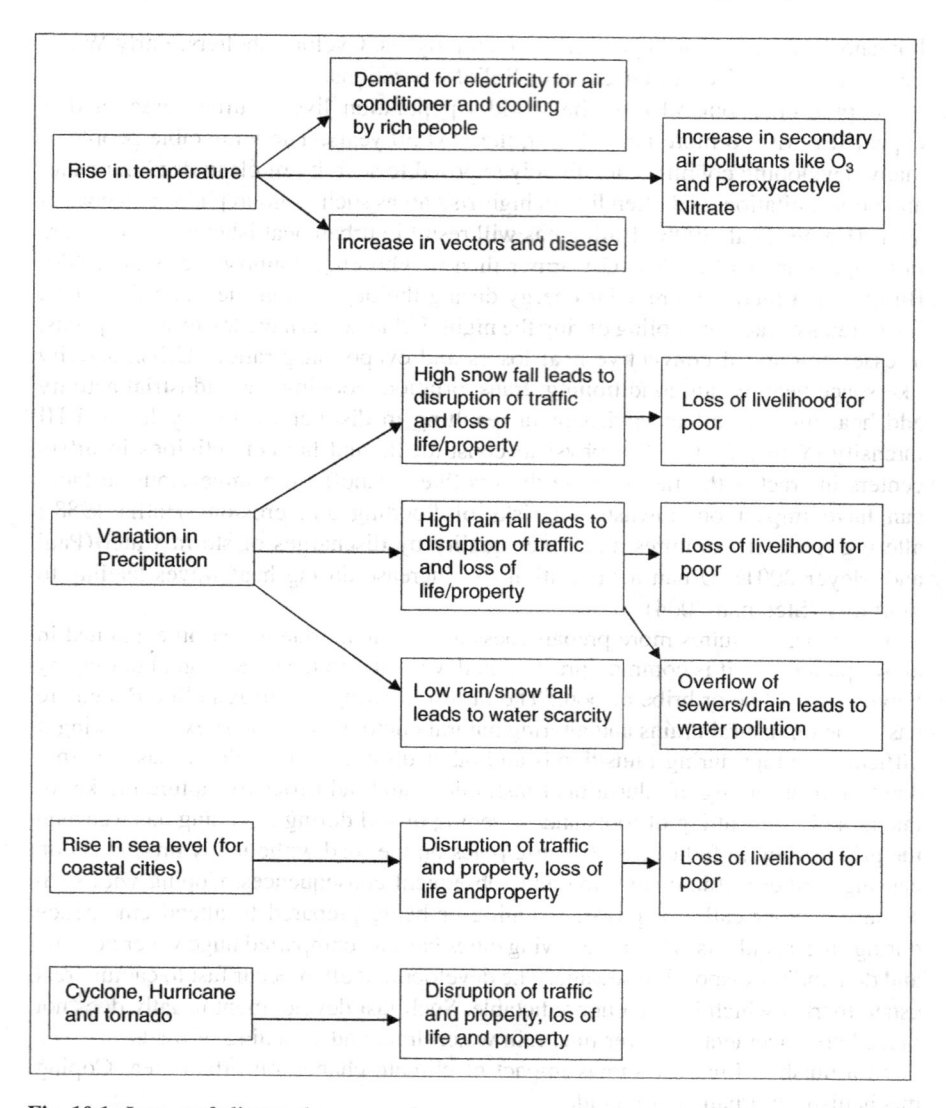

Fig. 10.1 Impact of climate change on urban area

adaptive capacities. The National Adaptation Action Plan is useful for identifying adaptation strategies and the immediate needs of countries (UNFCCC 2005).

Most of the developing countries typically have following characteristics:

- A high proportion of the population lives in poor-quality housing
- Many informal settlements are at high risk like leakage during rains and can catch fire easily

- Many houses are located on hazardous sites which will easily affect inhabitants during floods, landslides or earthquakes
- Some time it is also common practice for poor people to settle in government land next to railway track or open sewers and
- Risk levels are increased by a lack of infrastructure and escape routes in many areas

Climate change has both direct and indirect impacts on biodiversity. The direct impacts of climate change on biodiversity include:

- **Change in metabolism:** Increase in temperature is likely result in increased photosynthesis up to certain extent, but change in water availability and humidity will change metabolic rate.
- **Changes in phenology:** Animal and plant life cycle event will vary due to change in rhythm with respect to raining pattern. Change in rains will affect sprouting of seeds, change in weather will affect migration pattern of migratory birds and reproduction.
- **Changes in species demography:** Species requiring cold weather likely to move towards cooler region. Changes in prey distribution likely to affect distribution of predators.
- **Changes in ecosystem processes (like food chain, energy flow):** Population of species like jelly fish, insects will increase while temperature sensitive species will migrate or vanish creating imbalance in human-animal conflict.
- **Change in space:** Rise in sea level will increase space for marine species but decrease space for terrestrial species.
- **Conflict between Humans and Other Species:** Decrease in land availability due to sea level rise may force humans to migrate towards new regions in search for shelter. Change in animal demography may result in arrival of human unfriendly species including predators near human settlement creating conflict.

The direct impacts of climate change on human settlement include:

- **Loss of life and property due to increase in extreme events:** Increase in temperature likely result in increased extreme events and consequent damage to life and property.
- **Change in health:** The climate changes will in turn bring changes in health by increasing epidemics, injury and other health disorders.
- **Changes in agricultural production:** Agricultural production does vary. Warming may have positive impacts on crop yields if moisture is not a restriction, but rise in the frequency of extreme events or pests may offset any potential benefits.
- **Drop in fish catch:** Overfishing as well as well as climate change has resulted drop in fish reserves and is likely drop further.

The possible strategies for conserve existing biodiversity and human settlements are:

Food diversification: Currently Asians consume lot of rice which demand high water and drop in its production demands changing food habits and food diversification.

Identify food sources of other species and avoid human intervention: Major food sources of Whales and sharks have been captured by humans. Other food sources like fruits for herbivore animals are often consumed by humans. Poaching of wild life is also reducing prey for predators. These interventions with climate change will increase stress on ecology which can be avoided by avoiding human intervention by formation of national and international legislations and implementation.

Identify and conserve high quality habitats: Change in land use due to human activates in forest area is creating barrier for free movement of animals and removal of trees. Human intervention in coastal area and aquatic ecosystem has also increased stress on ecosystem. Creating protected area will help species to live in their habitat.

Promote seed bank collection: Conservation of selected varieties seeds through long-term seed storage techniques, such as cryo-preservation will be of most useful to overcome the potential loss of forest species due to climate change.

Create and implement national legislation: Create national Legislation to conserve biodiversity and implement by providing adequate resources.

Discourage use of timber: The current level of usage of timber has lead to destruction of forest throughout the world. Considering its importance as climate indicator and carbon sink it is essential that use of other eco-friendly products like boards made from agro-waste instead of timber for furniture and construction activities.

Implement international agreements: Implement international agreement for conservation of biodiversity by providing adequate resources.

Preserve and make space for the natural development of rivers, coasts and other water bodies: Increase in back water and swelling of water in deltas demand more space. Allow sufficient space so that species in the banks of water bodies can move to natural setup not submerged.

Reduction of wastage in forest harvesting: Reduction in wastage during harvesting is highly essential to reduce the magnitude of wastage under current forestry practices.

Increase efficiency in wood processing: Increasing efficiency levels in wood processing industries would help to preserve diminishing forest.

Avoid fragmentation of natural setup and connect fragmented area: Human settlements and developments and have fragmented inter linkage between natural setup. International and interstate fences restrict movement of animals into new area due to climate change. There in need for connecting fragmented ecosystems to allow species movement.

Ex-situ conservation: Translocation of species and conservation may look attractive but looks very artificial and leads to problem which is not encountered or predictable as on date. Tran located species also carry pathogens and other parasites which might impact local species.

Capacity building of institutions: The institutions – government, private and nongovernmental needs high capacity building to act themselves with least external guidance in such a way that there acts are best solution for the problem faced in the locality.

Raise awareness: Awareness shall reach the grass root level so that people in the community understands importance of species without which the conservation is not possible.

Continuous and quality research: There is high need for continuous and quality research to build on the existing understanding of the subject for changing climate and impact on biodiversity.

Stakeholder participation: Without stake holder participation conservation does not happen by only government and international agency. There shall be acceptability from stakeholders for the conservation efforts from government and international agency without which the efforts will be nullified.

The possible strategy for lively hood of poor and exposure to disasters are:

Training in new skills: People in village will have limited access to education, especially new skills which have tremendous market. Some of new skills imparted to coastal areas of India fundamentals of information technology so that they can form groups and use the computer for searching desired information. Such practice will not only improve life quality but also dependence on single activity for livelihood.

Help in migration to new area: Migration is last priority for both people and government. Proving sufficient information of possible demand for jobs in non disaster prone area will help people in migration. Evacuation in cyclone and flood hit areas shall happen with the external help as poor community can't move greater distance in small time.

Avoid development of slums and unauthorized constructions: The cities in developing countries are often scattered with slums which come due to lack of proper monitoring and protection of government lands. Such development will hinder natural drainage of water enhancing chances of flood and encourage further illegal encroachments of government lands.

Do not encourage development in locations exposed to severe weather risks: It is common practice in most of the developing countries to encourage investors by giving incentives. But precaution should be taken to avoid development in area affected by weather risk.

Educate people to build shelter which can withstand changing climate: People in disaster prone area often spend their earning in building shelter. But disaster destroys the shelters and forces to rebuild again. Teaching construction of disaster resilient shelter will help the poor in long way.

Reconsider zoning, planning and building regulations: The development in the past has been governed by necessities of the past. The changing time often demands change in land uses like changing residential zones to commercial and moving industries to free space in the outskirts of cities. Scientific rezoning is often necessary for minimization of disaster risks.

Educate people to adopt for new crops: Congenitally Asian would preserve seeds by themselves and use them for agriculture. The past decades have seen surge in seed companies and people have used them for great extent due to benefits they have experienced (like disease resistance, long shelf life, resistance to weather fluctuations). Educating the people to adopt new crops for changing climate would help them in log way.

Support livestock with fodders from other areas: Livestock are part and parcel of rural community. Drought and flood often destroys the live stocks due to lack of food. Supply the fodder and make separate evacuation plan for live stock during disasters.

Adapt water conservation technique: Use of drip irrigation, rain water harvesting, mulching, and other techniques has not been spread in the way it need to in Asia. As a result people tend to drill bore wells every time they need water and resulting in water stress in many area.

Foresee health impact and store sufficient medicine and food: Often doctors and paramedical professionals prefer to live in urban area. Some of the rural area usually will not have proper supply and stock. Most of food grains owned by national food corporations are also not spread itself uniformly. Spreading of storage area of food corporations in more number of places will help people to get food in time during disasters.

Form community groups to help each other during disaster: Local community groups equipped properly are best ways to cope with disasters until the external help arrives to site.

Form local disaster preparedness plan and disaster preparedness team: Many countries have legislation and local disaster preparedness team. But coordination and meeting is often missing as some of the organizations are suffering from staff deficiency and overburdened with other tasks. Making legislation just for statistic/record purpose or to impress international community will not deliver required results.

Provide early warning to enable people to migrate to safer place: Early warning in many place spread at slow speed. Such practice will not be effective and hence message should be spread fast.

Water and sanitation: Water and sanitation during disasters is still an issue in many Asian countries. Proper forecasting and preparedness will certainly bring down impact. It is highly recommended that disaster shelters should have sufficient ration of water, food grain and sanitation facility.

Transportation network: Many Asian villages and settlements still lack connectivity to other places making it difficult to mobilise resources and mobilise affected people. Without proper connectivity coping with climate change will not be very difficult.

Improve governance: Improper governance with corruption, inefficiency, lethargy will only increase the magnitude of climate impact. Box 10.3 gives impact of corruption on disaster management.

Make policy suitable to situation and time: Most often policy in developing countries are copy of policy in developed countries, which may not fit to the geological setup and culture.

Combat corruption: Very less work has been done and published with respect to corruption cycle. Even though press and internet through light on movement of corrupt money among different organizations and community, it is very difficult to track the movement of black money which has entered the system through corrupt way.

Box 10.3 Impact of corruption on disaster management

The world has always witnessed huge flow of resources during disasters. The corruption will lead to erroneous accounts and help the corrupt people to siphon out the funds allocated to these relief works. The governments and international community will usually forget to study such misappropriation and academicians does not take research in such subjects as well. Many times it can be observed that infrastructures built during the relief work are of poor quality and fail as soon as the it is inaugurated. Community will not have any resource left either to repair this inferior infrastructure or remove it. In some cases such infrastructures may crumble down leading to death and injury of the people. It can also be seen that some of the infrastructure created will be occupied by influential people for self benefit and vulnerable people will continue to vulnerable to disasters.

Study of livelihoods and governance that led to high losses in the Orissa cyclone in 1999 has pointed to corruption at all levels, unnecessary bureaucracy and political rivalry (Scott 1985).

Problem caused by inefficient governance is the opportunity it allows for corruption in both the state and non-governmental sectors. Several political actors in disaster relief have been observed following discriminatory policies in distributing relief and recovery assistance, favouring one section of population over others. Such action leads to the marginalization the most vulnerable and also undermines the legitimacy of responsible organizations (UNDP 2004).

In many countries environmental legislations are not implemented seriously due to the presence of corruption making way to discharge of pollutants. This clearly makes way for corruption to dominate in the scene and contribute to emission of GHGs without knowledge of international community thus leading to climate change. Irrespective of magnitude of disasters, people will witness greater vulnerability in the presence of corruption or poverty. People will witness greatest vulnerability if the society is corrupt and poor. The society can't build capacity if it is corrupt thus it puts itself into lower capabilities. There is urgent need for intervention from international agencies to increase capacity of developing countries to combat corruption. In the absence of capacity to tackle corruption funds will be misused. The majority of corruption happens through passing on currency (cash) between intra-organization and inter-organisation. Eliminating the current currency and introducing e-currency (like demat form of equity) will track movement of money and reduce corruption (Ramesha 2008).

The corruption can be classified into 'cash and kind'. The majority of corruption happens through passing on currency (cash) between intra-organization and inter-organisation.

Anti corruption laws have failed miserably in many countries as government is monopoly manufacturer and distributor of laws in all the countries. Many of the corruption interrogation do not involve brain mapping and lie-detection.

Figure 10.2 shows movement of corrupt money among corrupt people (represented by black sheep). Figure 10.3 shows flow of black money among public, private and political parties. It is very difficult to quantify and track movement of black money as on date.

Promote de-urbanisation: There is high migration of rural population to cities especially cities which are already highly populated and highly vulnerable to unforeseen disasters. It is not be possible to efficiently put disaster management

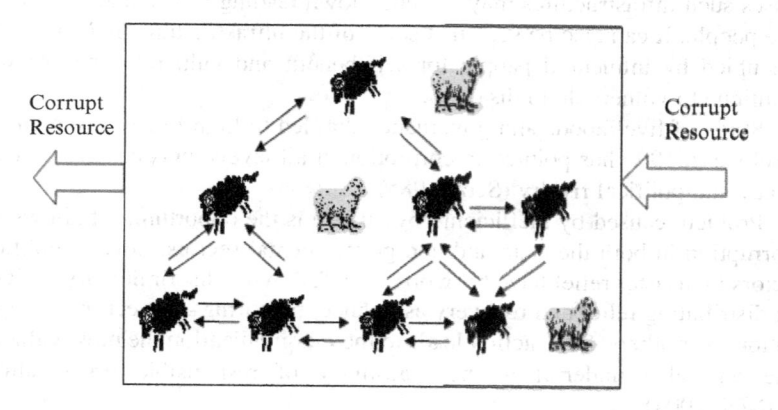

Fig. 10.2 Flow of corrupt resource within the organization. *Black* sheep representing corrupt people and *white* representing honest people

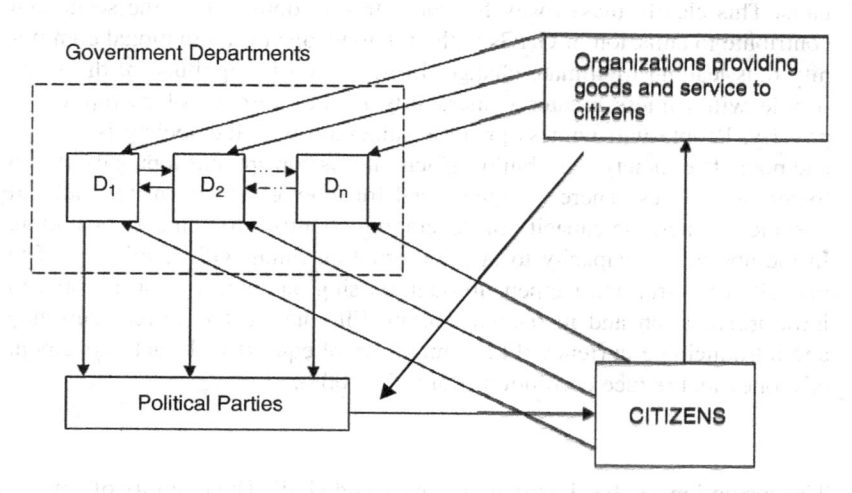

Fig. 10.3 Flow of corrupt money among public, private and political parties (d_1, d_2,...d_n are different departments within the government)

practice in overpopulated area. There should be proper planning so that distribution of population will be uniform throughout the country.

Construct Infrastructure to protect life and property: Construct infrastructure like embankments and cyclone shelters to safeguard citizens.

10.2 Adaptation and Adoption: Strategy of Other Species

Over the years animals and plants have shown different strategy for survival. Shifts in timing of life cycle events, and shifts in ranges, are two major ways that animals adapt to and their ecological communities. Bird species that depends on fish and crustacean population for food are vulnerable to changing population their prey. Many birds now arrive in spring breeding grounds earlier, and lay eggs earlier due to change in climate. Many long-distance migratory birds, are also vulnerable due to non availability of food on their migratory path due to changes that has occurred in migratory path. Studies by Parmesan and Yohe (2003) reveal birds are shifting their ranges pole-ward or to higher altitudes in tropical mountains. Studies by Derek et al. (2010) indicate that changes in ocean temperature, along with other human impacts, may rearrange the distribution of life in the ocean.

Species are responding to changes in climate, with altered population sizes and breeding patterns. Change in flowering dates are likely due to changing season pattern. Another way of animal adaptation is to migrate to a region which is comfortable for given species. The notable species which have shown migration are turtles, fish and birds. Species which requires cold climate move towards Polar Regions changing the species demography. Plants cannot migrate, but all plants have seed dispersal mechanisms. Seedlings thrive in a desirable climatic condition, but fail in a less desirable climatic condition.

Acclimatization is not new phenomena with respect to living organism. Ecological system does not respond to change in climate in smooth manner if the climate change is abrupt and is beyond acclimatization capacity of individual species.

Arid and semi arid region experience wide fluctuations in rainfall, and their native plants and animals are adapted to survive in extreme climatic conditions. Desertification may occur in this region where the probable evapotranspiration is greater than 70% of the total precipitation. Changes in land-use practices can affect surface temperatures resulting in differences in albedo and thus to temperature differences. Many of the ecological systems in these areas are unlikely to create conditions considerably outside the present range of variation during the early climatic changes (IPCC 1996); therefore, impacts from climate change may not be evident for several decades.

10.3 Adaptation to Climate Change in the Context of Sustainable Development and Equity

Small events such as droughts, floods, or pest out breaks can trigger ecological changes that are difficult to reverse (CCCD 2009). This has been observed in many cases.

Three main dimensions of sustainable development are economic, social and environmental. Integration takes place at different levels: Local level, Sectoral level, National Level, Regional Level, Global level.

Adaptation requires natural resource management, strengthening food security, increase of social and human capital and amplification of institutional systems (Adger et al. 2003). Vulnerability often highlights the importance of poverty and inequality (Adger and Kelly 1999). The connection between climate change and sustainable development are many and varied (Cohen et al. 1998; Robinson and Herbert 2001; Banuri and Gupta 2000). Adaptation has emerged as a policy priority both within and outside the climate change negotiations (Parry et al. 2005).

Adaptation should not lead to creation of new environmental problems like e-waste generation after life time of solar cells; disposal of nuclear waste; disposal of waste from wind energy; and disposal of air conditioners and gases filled to them.

Productivity of agriculture depends mainly on water availability. Reduction in tillage and mulching to amplify available water and reduce surface erosion can promote augmented soil carbon. Efficient use of manure and commercial fertilizer also can boost productivity and soil carbon. Eliminating or reducing summer fallow through better water management will increase carbon and decrease soil erosion in semi-arid croplands (IPCC 1996). Grazing control is an important option for the maintenance of soil carbon. Decrease in animal numbers can augment carbon storage by enhancing plant cover.

The extent to which sustainable development are vulnerable depends both on changes in climate and on the ability of the impacted system to adapt (Barry et al. 2001). Adaptive management processes and the formation of new institutions should occur in social environment that enables affected people to voice their interests and exercise political as well as economic power (Dreze and Sen 1989; Sen 1999a, b). Adaptation should be fair for people with different income, region, gender, religion, age.

Box 10.4 gives glimpse of gender equality and sensitivity while coping with climate change. Gender inequalities crisscross climate vulnerability. Women are historically have limited access to resources and decision making. Such situation often leads to erroneous decisions and neglects the problem of women while coping with climate change. Women and girl in developing world are allocated work of collecting wood, fetching water and other routine tasks while men engage in decision making. In many instances women are used in fashion shows and glamour shows rather than involvement in decision making. While some women are busy in 'cat walking' to make easy money other members of same gender walk miles to fetch water during dry seasons and drought. In many countries women were given

> **Box 10.4 Coping with climate change gender equality vs. gender sensitivity**
>
> Gender equality means indiscrimination on grounds of a person's sex in the allotment of resources, benefits and access to services. A society with gender equality share equal responsibilities, power and authority. The gender gap is the gap between women and men in terms of how they benefit from opportunities, education, employment, services, etc. Some societies often curb the potential of girls and women. Discrimination usually denies them health care and education. It avoids dissemination of information that they can use to protect themselves from diseases and other threats. Discrimination makes girls and women to earn a living.
>
> Gender equity ensures that development policies and interventions will position weaker sex better off economically and in terms of social responsibility after intervention.
>
> Gender sensitivity is the ability to recognize gender issues, and women's different perceptions and interests due to differing gender roles and culture.
>
> Adoption should be fair and climate change adaptation often needs interventions wherein equal responsibility shall be given to both gender in decision making platforms and an opportunity shall be extended to both the genders before making decisions. On the other hand implementation shall take sensitive needs of women in terms of sanitation, hygiene and mother hood. Intervening agencies and organization have fair idea of intervention acceptable to culture.

voting right much later than men were given. In some countries right to stand in election came after many years of getting voting right. Winning right to vote and stand in election does not mean all the women has been given decision making power. Most of the time wife of powerful politician will stand in election and will make decisions according to wish of her husband. In summary involvement of women in planning does not solve the problem unless there is expression and projection of problems faced by women without external pressure.

In the era of where criminal activities are common practice to eliminate rivals, a fresh and honest people does not have chance to enter the politics at all. Adaptation should not only help influential human community but it should also help others as well as other species. To achieve this there is need for good understanding of impact of each of the alternative adaptation for different community and species. Best way to achieve this is create solutions for adaptations which suit weaker people and species.

Climatic conditions vary very much in a region that ranges from mediterranean climate to extreme deserts. The economies in these regions also vary. The impacts of climate change are likely to be relatively small in counties whose economy depends on oil compared with countries depending on agricultural systems and the environment.

Countries and provinces within in the countries may have to adapt different strategies. Adaptation strategy needs international cooperation, national strategy, local initiative and action from individuals to overcome the vulnerability.

Freshwater resources are used for wide range of uses like drinking, irrigation, industry, agriculture, hydro-power plants, recreation, wild life, etc. It is required for maintaining food security, industrial economy, livelihood and economic growth. As shown in Fig. 10.4 overuse and misuse of surface water resources has resulted in shortage of water. Further the use of lakes, river beds for soil, sand, and clay has altered their natural capacity and shape. The change in land use has reduced capacity of water holding bodies due to silting of topsoil eroded during rains. Unscientific water uses like furrow irrigation and flood irrigation are still practiced (Fig. 10.5) in many part of Asia resulting in water loss due to evapotranspiration. Such practices are adapted due to lack of funds to install efficient irrigation technology like drip irrigation as well as absence of long term implication of water wastage.

Figures 10.6–10.8 shows typical rural setup in Asia where in old building are deteriorating due to poor maintenance and changing weather. Rural areas are often devoid of quality education, infrastructure due to its negligent contribution to

Fig. 10.4 Water stress has affected many places. The tank is altered for collecting clay for brick manufacturing (photo of Nagavalli Tank, Tumkur District, Karnataka State India)

Fig. 10.5 Flood irrigation and furrow irrigation does not need much investment but demands more water due high water loss

Fig. 10.6 A typical rural setup will poor planning and week structures

Fig. 10.7 Efforts to strengthen walls of a temple against climate change by providing support against week walls

Fig. 10.8 Impact of climate change on Kalleshwara temple in Tumkur district, Karnataka state, India

countries economy. While many village roads are devoid of pavements, the walls of houses are constructed with mud or low cost construction material.

References

Adger N, and Kelly M, 1999: "Social Vulnerability to Climate Change and the Architecture of Entitlements", Mitigation and Adaptation Strategies for Global Change, Vol 4, pages: 253–266.

Adger W.N, Khan S.R, and Brooks N, 2003: Measuring and Enhancing Adaptive, Capacity, UNDP Adaptation Policy Framework Technical Paper 7, New York.

African Development Bank, Asian Development Bank, Department for International, Development, United Kingdom, Directorate-General for Development, European Commission, Federal Ministry for Economic Cooperation and Development, Germany, Ministry of Foreign Affairs - Development Cooperation, The Netherlands Organization for Economic Cooperation and Development, United Nations Development Programme United Nations Environment Programme, The World Bank, 2002: Poverty and Climate Change Reducing the Vulnerability of the Poor through Adaptation, <www.undp.org/energy/docs/poverty-and-climate-change-72dpi-part1.pdf> downloaded on 27.7.2010

Banuri T, and Gupta, S, 2000: "The Clean Development Mechanism and Sustainable Development: An Economic Analysis", Manila: Asian Development Bank.

Barry Smit, Olga Pilifosova, Burton I, Challenger B, Huq S, Klein R.J.T, Yohe G, Adger N, Downing T, Harvey E, Kane S, Parry M, Skinner M, Smith J, Wandel J, 2001: Adaptation to Climate Change in the Context of Sustainable Development and Equity, Climate Change 2001, Impacts Adaptation and Vulnerability [A. Patwardhan And J.-F., Soussana (Eds)], Cambridge University Press, Cambridge, United Kingdom and New York, NY, USA.

CCCD (Commission on Climate Change and Development) 2009: Closing the Gaps: Disaster risk reduction and adaption to climate change in developing countries.

Cohen S, Demeritt D, Robinson J, and Rothman D, 1998: Climate Change and sustainable development: towards dialogue, Global Environmental Change, Vol. 8(4), pages: 341–371.

Derek P. Tittensor, Camilo Mora, Walter Jetz, Heike K. Lotze, Daniel Ricard, Edward Vanden Berghe & Boris Worm, 2010: Global patterns and predictors of marine biodiversity across taxa, Nature 466, doi:10.1038/nature09329, PP 1098–1101

Dreze J, Sen A, 1989: Hunger and Public Action, Clarendon Press, Oxford

Haines A, Kovats R.S, Campbell-Lendrum D and Corvalan C, 2006: Climate change and human health: impacts, vulnerability, and public health. Lancet, 367, pp. 2101–2109.

Hollis G.E, 1988: Rain, roads, roofs and runoff: hydrology in cities. Geography, 73, pp. 9–18.

IPCC, 1996: Climate Change 1995: Economic and Social Dimensions of Climate Change. Contribution of Working Group III to the Second Assessment Report of the Intergovernmental Panel on Climate Change [Bruce, J.P., H. Lee, and E.F. Haites (eds.)]. Cambridge University Press, Cambridge, United Kingdom and New York, NY, USA, 448 pp.

OECD-DAC (Organisation for Economic Co-operation and Development - Development Assistance committee) 2008: OECD-DAC Network on Development Evaluating, "Evaluating Development Cooperation: Summary of Key Norma and standards" (Paris : 2008).

Oke T.R, 1982: The energetic basis of the urban heat island. Quarterly Journal of the Royal Meteorological Society, 108, pp. 1–24.

Pachauri Rajendra, 2010: Insights Into The Climate Challenge; Global Sustainability - A Nobel Cause , Edited By Hans Joachim Schellnhuber, Mario Molina, Nicholas Stern, Veronika Huber, Susanne Kadner, 143–154

Parmesan C & Yohe G, 2003: A globally coherent fingerprint of climate change impacts across natural systems. Nature 421: 37.

Parry J.E, Hammill A, and Drexhage J, 2005: Climate Change and Adaptation, IISD,

Paul M.J and Meyer J.L 2001: Streams in the urban landscape. Annual Review of Ecology and Systematics, 32, pp. 333–365.

Ramesha Chandrappa, 2008: Impact of corruption on Disaster Risk Reduction. Individual study project for Partial fulfilment of Certificate of Disaster Risk Reduction, *Ecole Polytechnique Fédérale de Lausanne*

Robinson J, and Herbert D, 2001: "Integrating climate change and sustainable development", International Journal of Global Environmental Issues, Vol 1(2), pages: 130–148.

Sen A, 1999a: Development as Freedom, Alfred Knopf, New York, NY

Sen A, 1999b: Poverty and Famines, Oxford University Press, Delhi

Scott J.C, 1985: Weapons of the weak: everyday forms of peasant resistance, Yale University press, London.

Stedman J.R, 2004: The predicted number of air pollution related deaths in the UK during the August 2003 heatwave. Atmospheric Environment, 38, pp. 1083–1085.

Thomas R. Karl, Susan J. Hassol, Christopher D. Miller, and William L. Murray, editors, 2006: Temperature Trends in the Lower Atmosphere: Steps for Understanding and Reconciling Differences.. A Report by the Climate Change Science Program and the Subcommittee on Global Change Research, Washington, DC.

UNDP (United Nations Development Programme) 2004, Reducing Disaster Risk a Challenge for development,

UNFCCC, 2005. Report of the Asian Regional Workshop on Adaptation. http://maindb.unfccc.int/library/view_pdf.pl?url=http://unfccc.int/resource/docs/2007/sbi/eng/13.pdf retrieved on September 30, 2010.

Wilby R.L, 2003: Weekly warming. Weather, 58, 446–447.

World Bank, 2010: Economics of adoption to climate change, A Synthesis Report - final consultation draft

Chapter 11
Prediction of Regional Climate Change

Progress in the science in recent decades has potentially benefitted society by seasonal-to-interannual climate prediction. It is now possible to have more accurate advance warning. In recent years, use of the climate information has increased. But still there is significant data scarcity in Asia, particularly in mountainous and coastal ecosystems.

Weather events cannot be predicted beyond about 10–14 days because equations that governs the atmosphere is non-linear. However, the climate is made up of the average weather which has deviations. The deviations from the annual cycle are termed as climate anomalies.

Plotting the actual temperature and the normal temperature in the observed Sea-surface Temperature (SST) shows the SST anomalies. These variations in SST force changes in climate extending around the globe. SST anomaly patterns often continue for a few months to seasons, and sometime even longer. This helps prediction to some atmospheric climate anomalies based on assumption that SST anomalies will continue into the period being forecast. There is one major additional factor which makes wind curved when it flows from the point of high surface pressure to the low surface pressure. This factor is called the Coriolis force.

Coriolis Effect is an inertial force explained by the nineteenth-century French engineer-mathematician Gustave-Gaspard Coriolis in 1835. The effect of the Coriolis force is a noticeable deviation of the path of an object that moves within a rotating coordinate system. The object in reality does not deviate from its path, but it appears to do so.

The next factor need to be considered is three-dimensional nature of the atmosphere. Further warmer SST induces rising air. As air rises, it cools and in the process no longer holds moisture as a gaseous form resulting in condensation of moisture into water droplets resulting in clouds. In the process of water condensation heat is released which is important in the tropics since the increased temperature makes the air more buoyant and encourages further rise. At an elevation of around 10 km from ground level, air can no longer rise, even if it is releasing latent heat and hence at this point, air spreads out.

In summary, regional prediction should consider anomaly of SST that induces large atmospheric circulation, location from equator, height of the location, water

R. Chandrappa et al., *Coping with Climate Change*,
DOI 10.1007/978-3-642-19674-4_11, © Springer-Verlag Berlin Heidelberg 2011

bodies in the location, clouds, and vegetation, prevalence of El Niño/La Niña/ monsoon.

The change in Earth's climate system involves scientific principles and hence future climate can be predicted through careful, systematic study using following methodology:

- Current observations and understanding of science are the basis for future use and predicting climate change
- Data should be collected date from the bottom of the ocean to the surface of the sun
- Instruments should be used in all possible locations and parameters by using weather stations, buoys, satellites, and other way and means
- All monitored values should be documented and maintain records be kept for future use
- All scientific evidence like tree rings, ice cores, sedimentary layers, native knowledge shall be used for proper prediction and
- Mathematical models of all types including empirical, stochastic and black box shall be used and validated frequently for better prediction

Climate and weather is not uniform throughout the earth. Greater the location from equator, colder will be the temperatures due to the decreasing angle of incidence. Air masses formed over Polar Regions are cooler and, air masses formed over the tropics are warmer. As the upper regions of the troposphere are extremely cold, the precipitation that falls on mountains usually would be snow. Air masses formed over water would be more humid than those formed over the land. Mountains also affect climate by acting as barrier of movements of air parcels. Mountains also form rain shadows by blocking the path of precipitation resulting in desert.

Clouds act as shields for sunlight making temperatures drop during the daytime and warmer at night by trapping the heat in the layers of air below cloud. Vegetation forms shade and contributing to coolness of the region.

Water vapour in the atmosphere traps infrared radiation making temperature hotter. Dry desert air is not capable of blocking sunlight during the day time or trapping heat during the night time. During day most of the sun's heat reaches the ground, and as soon as the sun sets the desert cools quickly.

Urban areas do contribute to temperature due to combustion of fuel. Urbanisation not only radiates thermal energy it also acts as major source for GHGs which trap heat waves.

Uneven heating between the equator and poles causes global winds. Intense sunlight near equator heats the surface to a high temperature leading to rising of air. Air above the polar region is cooler and denser leading to sinking of air setting global winds in motion.

Figure 11.1 shows the major phenomenon responsible for climate change at global level. Figure 11.2 shows major phenomenon responsible for regional climate change. Some of the components of water cycle like evaporation may occur in one region and precipitate can occur in other region. Similarly carbon emitted from one region can sink in other region.

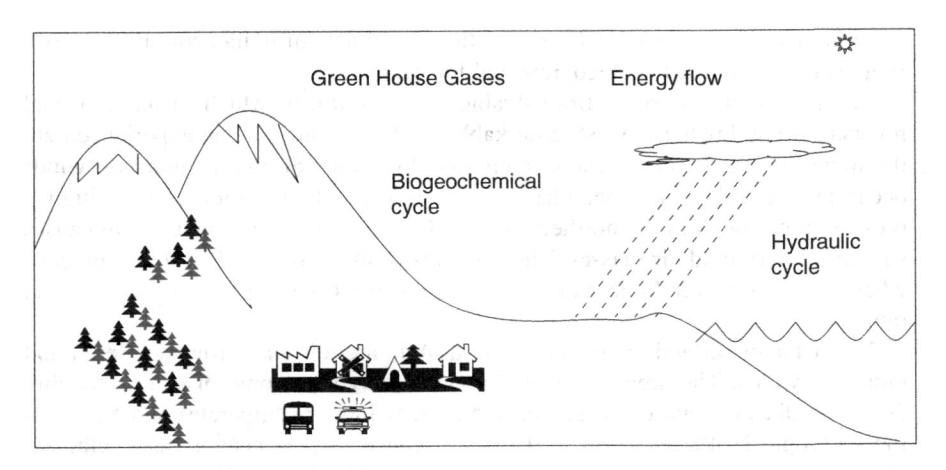

Fig. 11.1 Major phenomenon responsible for climate change

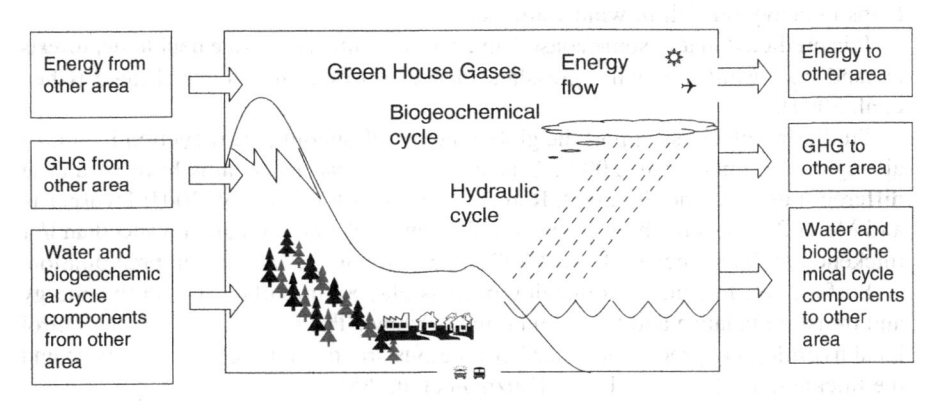

Fig. 11.2 Major phenomenon responsible for regional climate change

11.1 Factors Affecting Asian Regional Climate

Asia is very vast making it largest among the continents on the earth. Except Indonesia and Malaysia, rest of the Asian countries are located in northern hemisphere spreading itself near to northern pole. Russia being most near to pole, is cooler compared to south India, Sri Lanka, Myanmar, Thailand, Philippines, Laos, Vietnam and Cambodia of country. Apart from geological location, the climate of a country is affected by oceans surrounded around the continent. To the north of Asia is Arctic Ocean, and to south is Indian Ocean. To the east of Asia is North Pacific Ocean and west is European land mass.

South Asia is referred as Indian sub continent and area does not have uniform climatic condition. The area is characterised by great difference in rain fall, relative humidity and temperature. Monsoon unifies the climate of Indian continent. Hence India is often regarded as 'Geographical Laboratory',

The monsoon is derived from Arabic word *'Mausim'* which means seasonal reversal of wind pattern. Most remarkable of climate phenomena experienced are the monsoon winds that occur in numerous locations. Strongest monsoon winds occur in India. This weather phenomenon is created in winter when a high – pressure area forms in the northern part of Asia due to which young storm causes sinking of very cold air masses. These storms push strong winds to the southeast, where they pass over India and move out into the ocean dissipating clouds and rains.

Major factors affecting climate in South Asia are: distance from sea, relief and monsoon winds. The area is located in such a way that tropic of cancer roughly divides India into two equal halves – Northern India (Temperate) and Southern India (Tropical). Places north of Tropic of cancer experiences extreme climate. Places south of Tropic of cancer, experiences equable climate. Himalayas prevents cold winds from North, blocks monsoon winds and gives rainfall. Western Ghats helps in heavy rainfall in wind ward side.

It is predicted that in some coastal and central China temperate deciduous forests expand northward, replacing grassland due to warmer and wetter climate (Chen et al. 2004).

Studies on glaciers support the global picture of ongoing ice reduction in almost all regions (Lemke et al. 2007). Glaciers in the Asian mountains have shrunk at different rates (Su and Shi 2002; Ren et al. 2004; Solomina et al. 2004; Dyurgerov and Meier 2005), some high glaciers in the central Karakoram are advanced and/or thickened at their tongues (Hewitt 2005), possibly due to increase in precipitation.

Surface soil freezing and thawing process play an important role in the energy and moisture balance and hence in climate change. The key reasons which control local hydrological processes in northern regions are the presence of permafrost and the thickness of the active layer (Hinzman et al. 2003).

In addition to varying topography Asia is currently witnessing brisk industrial activity in India and China resulting in large quantity of CO_2. The world's one of most sensitive are Himalaya is located between these two countries with some part lying in these countries. The increase in CO_2 concentration in the area could also one of the reasons for Himalaya region's rise in temperature being higher than global average temperature rise in past few years.

11.2 Future Emphasis

Climate change is currently occurring at an extraordinary rate and is projected to compound the stress on natural resources and the environment associated with rapid urbanisation, industrialisation, and economic development. Future prediction is important not just because of academic interest but to save nearly more than half of world's population who live in Asia. Any impact to these people will affect global economy.

Climate change will have huge and widespread effects on the availability of, water resources. By the 2050s, access to freshwater in Asia, mainly in large basins, is projected to decrease (Mats et al. 2009).

As per the working group I contribution to fourth assessment report of the Intergovernmental Panel on Climate Change (Christensen et al. 2007) warming is *likely* to be above the global mean in central Asia.

It is predicted that there will be large increase in the Indian summer monsoon season over Arabian Sea, tropical Indian Ocean, south Asia(May 2004; Rupa Kumar et al. 2006). Increase in summer will be observed over south China, Korea and Japan (Gao et al. 2002; Boo et al. 2006; Kimoto 2005; Kitoh et al. 2005; Mizuta et al. 2005). Increase associated with tropical cyclones over Southeast Asia, Japan (Kimoto 2005; Mizuta et al. 2005; Hasegawa and Emori 2005; Kanada et al. 2005)

As per studies conducted by Chen et al., annual precipitation increases by about 20% in coastal areas of northern and Central China whereas increase will be 8% in southern China. The strongest increase in temperature will be up to 4°C occurring in northern China (Chen et al. 2004).

Considering the importance of climate change, developing countries including China and India have undertaken efforts for to reduce their emissions growth. There is international pro-activity to develop and implement new technologies to reduce GHG emissions. Some of them are: the Carbon Sequestration Leadership; the Hydrogen partnership; the Methane to Markets Partnership, and the Asia-Pacific Partnership for Clean Development and Climate (2005), which involves Australia, USA, Japan, China, India and South-Korea (Barker et al. 2007).

References

Barker T, Bashmakov I, Bernstein L, Bogner J.E, Bosch P.R, Dave R, Davidson O.R, Fisher B.S, Gupta S, Halsnæs K, Heij G.J, Kahn Ribeiro S, Kobayashi S, Levine M.D, Martino D.L, Masera O, Metz B, Meyer L.A, Nabuurs G.J, Najam A, Nakicenovic N, Rogner H.H, Roy J, Sathaye J, Schock R, Shukla P, Sims R.E.H, Smith P, Tirpak D.A, Urge-Vorsatz D, Zhou D, 2007: Technical Summary. In: Climate Change 2007: Mitigation. Contribution of Working Group III to the Fourth Assessment Report of the Intergovernmental Panel on Climate Change [Metz B, Davidson O.R, Bosch P.R, Dave R, Meyer L.A (eds)], Cambridge University Press, Cambridge, United Kingdom and New York, NY, USA.

Boo K.O, Kwon W.T, and Baek H.J, 2006: Change of extreme events of temperature and precipitation over Korea using regional projection of future climate change. Geophys. Res. Lett., 33(1), L01701, doi:10.1029/2005GL023378.

Chen, Ming, David Pollard, Eric J. Barron, 2004: Regional Climate Change in East Asia Simulated by an Interactive Atmosphere-Soil-Vegetation Model. J. Climate, 17, 557–572

Christensen J.H, Hewitson B, Busuioc A, Chen A, Gao X, Held I, Jones R, Kolli R.K, Kwon W.T, Laprise R, Magaña Rueda V, Mearns L, Menéndez C.G, Räisänen J, Rinke A, Sarr A and Whetton P, 2007: Regional Climate Projections. In: Climate Change 2007: The Physical

Science Basis. Contribution of Working Group I to the Fourth Assessment Report of the Intergovernmental Panel on Climate Change [Solomon S, Qin D, Manning M, Chen Z, Marquis M, Averyt K.B, Tignor M and Miller H.L(eds.)]. Cambridge University Press, Cambridge, United Kingdom and New York, NY, USA.

Dyurgerov M, and Meier M.F, 2005: Glaciers and the Changing Earth System: A 2004 Snapshot. Occasional Paper 58, Institute of Arctic and Alpine Research, University of Colorado, Boulder, CO, 118 pp.

Gao X.J, Zhao Z.C, and Giorgi F, 2002: Changes of extreme events in regional climate simulations over East Asia. Adv Atmos Sci., 19, 927–942.

Hasegawa A, and Emori S, 2005: Tropical cyclones and associated precipitation over the Western North Pacific: T106 atmospheric GCM simulation for present-day and doubled CO2 climates. Scientific Online Letters on the Atmosphere, 1, 145–148.

Hinzman L.D, Kane D.L, Yoshikawa K, Carr A, Bolton W.R and Fraver M, 2003: Hydrological variations among watersheds with varying degrees of permafrost. Proceedings of the 8th International Conference on Permafrost, 21–25 July 2003, Zurich, Switzerland [Phillips, M., S.M. Springman, and L.U. Arenson (eds.)]. A.A. Balkema, Lisse, the Netherlands, pp. 407–411.

Hewitt K, 2005: The Karakoram anomaly? Glacier expansion and the "elevation effect", Karakoram Himalaya. Mountain Research and Development, 25(4), 332–340.

Kimoto M, 2005: Simulated change of the east Asian circulation under global warming scenario. Geophys. Res. Lett., 32, L16701, doi:10.1029/2005GRL023383.

Kanada S, Muroi C, Wakazuki Y, Yasunaga K, Hashimoto A, Kato T, Kurihara K, Yoshizaki M and Noda A, 2005: Structure of mesoscale convective systems during the late Baiu season in the global warming climate simulated by a non-hydrostatic regional model. Scientific Online Letters on the Atmosphere, 1, 117–120.

Kitoh A.M, Hosaka Y, Adachi and Kamiguchi K, 2005: Future projections of precipitation characteristics in East Asia simulated by the MRI CGCM2. Adv Atmos Sci., 22(4), 467–478

Lemke P, Ren J, Alley R.B, Allison I, Carrasco J, Flato G, Fujii Y, Kaser G, Mote P, Thomas R.H and Zhang T, 2007: Observations: Changes in Snow, Ice and Frozen Ground. In: Climate Change 2007: The Physical Science Basis. Contribution of Working Group I to the Fourth Assessment Report of the Intergovernmental Panel on Climate Change [Solomon S, Qin D, Manning M, Chen Z, Marquis M, Averyt K.B, Tignor M and Miller H.L(eds eds.)]. Cambridge University Press, Cambridge, United Kingdom and New York, NY, USA.

Mats Eriksson, Xu Jianchu, Arun Bhakta Shrestha, Ramesh Ananda Vaidya, Santosh Nepal, Klas Sandström, 2009: The Changing Himalayas, Impact of climate change on water resources and livelihoods in the greater Himalayas, International Centre for Integrated Mountain Development (ICIMOD).

Mizuta R, Uchiyama T, Kamiguchi K, Kitoh A and Noda A, 2005: Changes in extremes indices over Japan due to global warming projected by a global 20-km-mesh atmospheric model. Scientific Online Letters on the Atmosphere, 1 153–156.

May W, 2004: Simulation of the variability and extremes of daily rainfall during the Indian summer monsoon for present and future times in a global time-slice experiment. Clim. Dyn., 22, 183–204.

Ren J, Qin D, Kang S, Hou S, Pu J, Jing Z, 2004: Glacier variations and climate warming and drying in the central Himalayas. Chin. Sci. Bull., 49(1), 65–69.

Rupa Kumar K, Sahai A. K, Krishna Kumar K, Patwardhan S. K, Mishra P.K, Revadekar J.V, Kamala K and Pant G.B, 2006: High-resolution climate change scenarios for India for the 21st century. Curr. Sci. India., 90, 334–345.

Su Z, and Shi Y, 2002: Response of monsoonal temperate glaciers to global warming since the Littel Ice Age. Quat. Int., 97–98, 123–131.

Solomina O, Barry R, and Bodnya M, 2004: The retreat of Tien Shan glaciers (Kyrgyzstan) since the Little Ice Age estimated from aerial photographs, lichonometric and historical data. Geografiska Annaler, 86A(2), 205–215.

Chapter 12
Capacity Assessment: Is Asia Ready to Face Climate Change

The Asia, home to about four billion people, has come a long way in its fight against poverty. Substantial proportion of people living in poverty was cut down in last two decades through remarkable growth. This progress has come at a high environmental cost. Tackling climate change is a priority in Asia due to numerous acute climate change factors and vulnerabilities. The demand for energy will continue to raise so as economic development. A large number of people live along the coast and on low-lying islands. Rising sea levels threaten coastal areas in the continent. Extreme events are expected to rise, and natural habitats are at risk.

The achievement of the any goals set by country depends on capacities of country. While financial resources are important, they are not enough to achieve goals. Without skills, integrity, information, well functioning institutes, countries fail to achieve their goals. This aspect is applicable to coping with climate change as well.

Due to mounting pressures from a changing climate, and increasing population, decision makers, scientists, and resource managers have to understand, obtain, and integrate climate forecasts as well as observational data. Ability to adapt and respond to climate change depends on our understanding of the climate and how to incorporate this into day to day decisions (CCSP 2008).

Capacity assessment helps country by providing knowledge of capacity gap; to prioritize the action; to analyze vulnerability to climate change extremes, and adaptive capacity; to analyze constraints and opportunities for adaptation; to assess greater understanding about local impacts of climate change. But many time assessments itself will be misleading due to hiding the actual facts during capacity assessment process. In such case capacity development will end up in just in a series of workshops, publications and visit to foreign countries by a group of elite officers.

Major latest disasters such as the December 2004 tsunami in the Indian Ocean; October 2005 earthquake of the Pakistani Kashmir; and May 2008 cyclone Nargis that hit Myanmar (Ilan and Tam 2010) have devastated Asian region.

An estimated 62,000 ground water wells were contaminated due to tsunami 2004 by seawater, wastewater and sewage (Hari and Yuko 2008). Indonesia's BAPPENA's (State Ministry of National Development Planning) estimated that 20% of sea grass beds, 30% of coral reefs, 25–35% of wetlands, and 50% of sandy beaches of the west coast was damaged due to Indian ocean tsunami 2004

R. Chandrappa et al., *Coping with Climate Change*,
DOI 10.1007/978-3-642-19674-4_12, © Springer-Verlag Berlin Heidelberg 2011

(BAPPENAS 2005). In Thailand about 15–20% of the coral reefs were affected due to the same tsunami by siltation and sand infiltration (Hari and Yuko 2008). These incidents shows that secondary effect which take long time to subsidize cannot be over ruled during capacity assessment.

The human tragedy can not be trapped in statistics. The statement holds good for other species as well. Many of the current impacts will have implication in future. There is always uncertainty about the location, timings and magnitude of impacts.

12.1 Climate Forecast and Data Products That Support Decision-Making

Forecasting climate has many rules. As shown in Table 12.1 climate forecast would benefit farmers to engineers. Earth's climate is varying and also varying in response to human activity (Ingram et al 2008). Weather forecasts are of three types:

1. Forecasting the next few days – Weather forecasts
2. Forecasting the next few months – Seasonal climate forecasts
3. Projecting the climate in the years to come – Long-range climate forecasts

IPCC's Fourth Assessment Report (IPCC 2007a, b) indicates that there is a more than 90% probability that the observed warming since the 1950s is due to the emission of GHGs. Temperature projections for the twenty-first century indicates a major speeding up of warming over that observed in the twentieth century (Ruosteenoja et al. 2003). In Asia, it is very likely that all areas will warm in this century. Warming is least rapid in Southeast Asia; stronger over south Asia and eastern Asia; and greatest in the Central, Western, and Northern Asia. Warming will be significant in arid regions of Asia and the Himalayan highlands, as well as the Tibetan Plateau (Gao et al. 2003; Yao et al. 2006).

Table 12.1 Stakeholders and forecast application useful to them

Stakeholders	Forecast application
Farmers	Planting and harvesting of crops, application of agrochemicals, planning irrigation
Fishermen	Planning and preparing fish net incase of sea fishing. Protecting fish in the pond by netting the out flow
Agriculture laborers	Searching agricultural employment market price,
Traders	transportation of goods
Manufacturer of warm weather products	To manufacture and store warm weather products like ice cream, cool drinks, etc.
Cold weather products	To manufacture and store warm weather products like sweater, coats, etc.
Rain weather products	To manufacture and store warm weather products like umbrella, raincoat, etc.
Construction engineers/workers	To plan construction activities
Forest managers	To plan planting saplings

12.2 Making Decision-Support Information Responsive to Decision-Maker

Climate impacts are not uniform all over the world. Hence it will not be possible to copy strategy adopted in one place in other places. The decisions related to climate adaptation needs in depth knowledge of the region, understanding of climate science, forecasting capabilities, enlisting probable alternation solution, and implementation capabilities. Thus these chain of activities needs information and shall be made available to decision maker as well as policy formulating authorities in time to enable them to include in national policy so that the information can be converted to action plan with allocation of resources.

Policy makers in Asia shall plan keeping changing climate in the mind instead of planning for past climate. Policy makers must pay attention to a variety of climate futures and promote innovation. Agricultural productivity and water management need to improve to feed future population without further affecting already stressed ecosystems. The decision for future should be long-term based on the information about changing climate and species demography.

Although there is uncertainty about the specific risks associated with climate change in the future, planning now to adapt to a range of different scenarios will help minimize the cost of catastrophes and create communities that can well adapt without loss of life and economy.

12.3 Looking Toward the Future

Twenty-first century is referred as Asian Century as the twentieth century is often called the American Century, and the nineteenth century as the British Century. There is hurry within the region to compete and make as much wealth as possible. The wealthy investors from other continents are also eager to make profit from the situation. As a result there is increased industrial activity resulting in rising consumption. Roads which ware constructed by projecting traffic based on past decades have been choked with vehicles. Drainage system provided considering past population and predicting similar trends in future can't take the load due to population explosion and increased sky scrappers in some of the fast growing cities of Asia. Salaries have been doubled, tripled and multiplied in past decade and so as purchasing power. Everybody is in hurry not waste the time as 'time is money' and every one want to profit 'as much as possible'.

The century also woke-up to new issue – changing climate. In spite of Asia being pioneer in climate related disasters like drought, flood, and cyclone, the continent became alert after 2004 tsunami swept whatever little wealth the countries had created in the shore of Atlantic Ocean.

Until a decade back countries left citizens to cope with the climate. The only help that the affected family received is some meager compensation in case of death of family member, some subsidized food for drought affected area and packets thrown from aero planes on flood hit areas. The citizens had to bribe officials even to get petty compensation. The coping strategy of citizens adopted included walking miles to fetch water; bathing once in a week or month; wetting cloths to keep themselves cool; washing face with a mug of water; and waiting for rain for sowing seeds.

Today countries have action plans for coping with climate change and a good amount of wealth created and barrowed is earmarked for mitigation of climate change. Countries are communicating each other and eager to help each other during disasters. Early warnings systems have either have taken the positions or are in the process of being established. Funds are flowing for the climate related research and hundreds of satellites are watching the planet round the clock. Thousands of scientists all over the world are using past and present knowledge to predict possible climate change in future and recommending suitable action to policymakers.

Future policies and research should not only address current problem but also impact due to change in policy and new policies and technologies. Food security can't be achieved by just growing enough food but also depends on storage and distribution network capability of the nation. The policy should also address fare pricing to farmers. The shifting to new technology should also address long term implications of such technology.

Appendix C gives the country wise scenario of climate vulnerability with respect to individual country. Looking in to past climate related disasters and initial communication of each country under UNFCCC, the countries with snowfall and mountains are highly vulnerable to due to melting of ice and subsequent floods, and mass movements. On the other hand mountains become advantageous to those countries which are adjacent to sea shore as sea level rise will not submerge major portion of the country. The countries which highly depend on agriculture and fishing will face threat due to fall in agriculture output and fish catch. The forest of different countries will face different impacts with respect to phenology, forest fires and loss of species. The countries with high technological capability can withstand impact more compared to those countries with lower technological capability. But still the continent is not fully adopting itself. The conventional agriculture is still practiced in developing Asia instead of adopting high intensity agriculture practiced in Singapore. People still prefer fishing in seas and come back will lesser fish catch instead of growing fish in cages as practiced in Singapore.

Governments have already started budgetary provisions for climate change. Government expenditure in India on adaptation to climate variability surpassed 2.6% of the GDP in the year 2007, with agriculture, water resources, health and sanitation, forests, coastal-zone infrastructure and extreme weather events, being specific areas of concern (GOI 2008). China plans to protect 90% of typical forest ecosystems and national key wildlife effectively by 2010 (NDRCPRC 2007). Armenia Plans to gradual increase of forest cover from present 11.2% of territory up to 20.1% (MNPRA 1998). Forest conservation continues have top priority in Bhutan (RGoB 2000). As per first communication under UNFCCC, Indonesia plans to phase out fuel subsidies and promote use of renewable energy. Sri Lanka has launched programmes to educate, train and promote awareness on climate change (GoS 2000). Similarly most of the countries have plans in place to conserve energy, increase energy efficiency, conserve water resources, and protect flora and fauna.

Countries throughout the world have worked hard to safeguard the people from climate change. Figure 12.1 shows the schematic diagram where in disaster risk can neither be totally eliminated nor adapted to. At the most it can be reduced to certain extent. Like an industrial accident can be minimised and is minimised in many top industrial houses, the people in developed countries are also been safeguarded by their governments. An action in this regards is yet to taken in developed countries that have an action plan prepared but struggling to take action. Figure 12.2 shows poor people warming themselves during winter. Rich people can afford room heaters and warm water but poor people have to adapt for changing climate with available resources. While the rich and influential people have been at the peak of

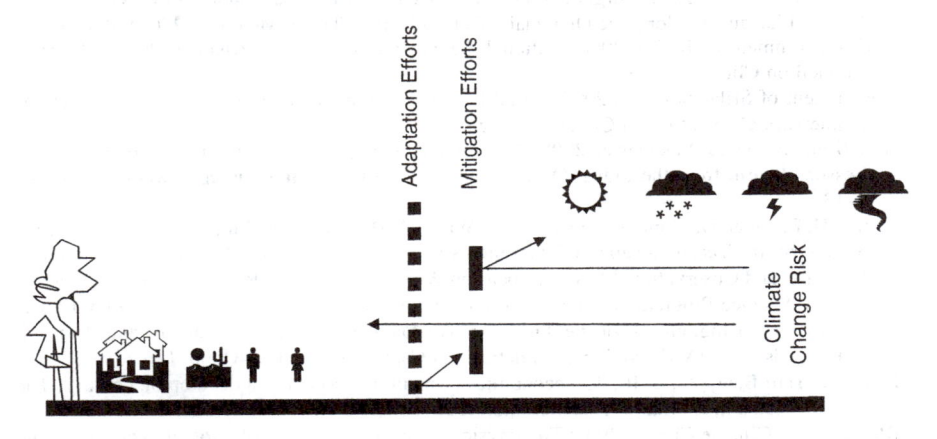

Fig. 12.1 Protecting people by adaptation and mitigation efforts

Fig. 12.2 Adapting to cold wave in Delhi, India

the safety, poor people are at the bottom the safety and are vulnerable to risk as usual. The numbers with respect to risk and vulnerability to disasters has been discussed in **Chaps. 3, 5, 7 and 9** as well as **Appendix C.**

References

BAPPENAS, 2005: Indonesia: preliminary damage and loss assessment. State Ministry of National Development Planning (BAPPENAS), Government of Indonesia.

CCSP, 2008: Decision-Support Experiments and Evaluations using Seasonal-to-Interannual Forecasts and Observational Data: A Focus on Water Resources. A Report by the U.S. Climate Change Science Program and the Subcommittee on Global Change Research [Nancy Beller-Simms, Helen Ingram, David Feldman, Nathan Mantua, Katharine L. Jacobs, and Anne M. Waple (eds.)]. NOAA's National Climatic Data Center, Asheville, NC, 192 pp.

Gao X.J, Li D.L, Zhao Z.C, Giorgi F, 2003: 'Climate change due to greenhouse effects in Qinghai-Xizang Plateau and along the Qianghai-Tibet Railway'. Plateau Meteorol. 22(5): 458–463

GOI (Government of India), 2008: National Action Plan on Climate Change, Prime Ministers Council on Climate Change.

Government of Srilanka (GoS) 2000: Initial National Communication under the United Nations Framework Convention on Climate Change

Hari Srinivas, Yuko Nakagawa, 2008: Environmental implications for disaster preparedness: Lessons Learnt from the Indian Ocean Tsunami, Journal of Environmental Management 89, 4–13

Ingram H, Feldman D, Mantua N, Jacobs K.L, Waple A.M, and Beller-Simms N, 2008: Executive Summary. In: Decision-Support Experiments and Evaluations using Seasonal-to-Interannual Forecasts and Observational Data: A Focus on Water Resources. A Report by the U.S. Climate Change Science Program and the Subcommittee on Global Change Research [Nancy Beller-Simms, Helen Ingram, David Feldman, Nathan Mantua, Katharine L. Jacobs, and Anne M. Waple (eds.)]. NOAA's National Climatic Data Center, Asheville, NC, pp. 1–6.

Ilan Noy, Tam Bang Vu, 2010: The economics of natural disasters in a developing country: The case of Vietnam, Journal of Asian Economics 21, 345–354

IPCC, 2007a: Climate Change 2007: The Physical Science Basis. Contribution of Working Group I to the Fourth Assessment Report of the Intergovernmental Panel on Climate Change [Solomon, S; Qin, D; Manning, M; Chen, Z; Marquis, M; Averyt, KB; Tignor, M; Miller, HL (eds)]. Cambridge and New York: Cambridge University Press

IPCC, 2007b: Climate Change 2007: Impacts, Adaptation and Vulnerability. Contribution of Working Group II to the Fourth Assessment Report of the Intergovernmental Panel on Climate Change [Parry, ML; Canziani, OF; Palutikof, JP; van der Linden, PJ; Hanson, CE (eds)]. Cambridge: Cambridge University Press

Ministry of Nature Protection of the Republic of Armenia (MNPRA), 1998: First National Communication of the Republic of Armenia Under The United Nations Framework Convention on Climate Change

National Development and Reform Commission, People's Republic of China (NDRCPRC) 2007: China's National Climate Change Programme

Royal Government of Bhutan (RGoB), 2000: First Green House Gas Inventory

Ruosteenoja K, Carter T.R, Jylhä K, Tuomenvirta H, 2003: 'Future climate in world regions: an intercomparison of model-based projections for the new IPCC emissions scenarios'. The Finnish Environment 644: 83. Helsinki: Finnish Environment Institute

Yao T.D, Guo X.J, Lonnie T, Duan K.Q, Wang N.L, Pu J.C, Xu B.Q, Yang X.X, Sun W.Z, 2006: 'δ18O Record and Temperature Change over the Past 100 years in Ice Cores on the Tibetan Plateau'. Science in China: Series D Earth Science 49(1): 1–9

Chapter 13
Hazard and Risk Assessment

Development planning usually does not consider climate change. As a consequence, disasters cause economic and ecologic disruption during disasters. Right from the early stages, planners should include probable hazards due to changing climate. Proper planning can reduce damage from the extreme events due to climate change. This chapter discusses some fundamentals which could help planners to save citizens, economy and ecology by consequences of changing climate.

Natural hazards can be defined as "*those elements of the physical environment, harmful to man and caused by forces extraneous to him*" (Burton et al. 1978). But with changing time natural hazards are now not strictly "*Act of God*". The GHG emission and subsequent change in climate has been proved to be case of many extreme natural events.

Even though humans can not change frequency or magnitude of natural hazards, they can bring down the severity of hazard. Human can save forest and mangroves which act as barriers against many disasters. Destruction of endangered species can be avoided by safeguarding them by restricting tourism and development in the locality.

The direct and secondary impacts of disasters increase vulnerability to further disasters. The reports after natural disaster focus only on number of dead and wounded and on the value of property lost and damaged. Natural disasters can pose a significant threat to ecosystems and natural habitats.

Direct effects of disaster are:

- Loss of life
- Loss of property and
- Physical and psychological damage to people

Secondary effects of disaster are:

- Environmental degradation
- Spread of diseases
- Impact on growth of economy and
- Drop in agricultural production

R. Chandrappa et al., *Coping with Climate Change*,
DOI 10.1007/978-3-642-19674-4_13, © Springer-Verlag Berlin Heidelberg 2011

Risk assessment helps planners to:

- Recognize and understand natural hazards in the area
- Plan mitigation
- Identify alternate solution
- Earmark funds needed
- Identify knowledge gaps
- Identify risks to new projects from natural hazards and
- Make decisions about how to deal with those risks

Key information required in Hazard and Risk Assessment is:

- Location and extent of area of interest
- Frequency and probability of hazard occurrence and
- Intensity/severity of Hazards

Project planners should also be aware of:

- Secondary hazards resulting from a hazard event like landslides due to earthquake/flood
- Problems that could occur outside the area of hazards like power disruption due to failure of power station in the hazard area and
- How hazard events occur, including natural and anthropogenic that creates or exacerbate hazards like deforestation and mining which cause slope instability and landslides

Hazard mapping is a tool in hazard identification and assessment as maps can accurately record the location and information clearly and conveniently at appropriate scale or level of detail. Mapping may be based on a range of data sources like existing topo-sheets, remote sensing images, surveying. The subsequent sections deal with mapping hazard analysis.

13.1   Aerial Mapping and Surveying

Aerial photographs are probably the most familiar remote sensing data source used by planners (OAS 1969). Arial mapping was one of the quick methods which were adopted earlier days prior to rapid growth of remote sensing technology. Of all the sensors, aerial photography gives the closest depiction of what the human eye can see. The interpreter who is familiar with photographs can easily interpret the pictures.

In aerial photography a flight was made to run in parallel paths and photographs were taken at redesigned interval with sufficient overlap. The photographs are then arranged properly to obtain clear cut picture of magnitude of the damage.

The most used scales in aerial photography vary between 1:5,000 and 1:120,000 (DRDEESESAOAS 1991). Arial photography can be obtained using either colour photographs or black and white photographs.

Radar can also be used in aerial mapping. But, an interpreter with the knowledge of technology will be required to interpret the correctly. Thermal infrared part of the spectrum can also be used to produce thermal pattern of the terrain. In spite of numerous advantages, photography and radar technology has a disadvantage that it cannot be used at all times and weather.

Currently aerial surveying is still most preferred and used technology for reconnaissance to provide immediate relief to victims of disasters. Remote sensing technology now provides an economically feasible alternative means in place of traditional methods wherein sensor data are digitally transmitted to ground stations and pre-processed to improve quality of data captured.

Landslides or mass movements are important disaster associated with climate change. Analysis of landslide can be easily achieved using Aerial mapping and surveying. Systematic large-scale aerial photography provides a detail for analysis of desertification, volcanic eruption, earthquake and many other disasters which fall in to manmade as well as natural disaster category.

13.2 Riverine Flooding Analyses

Riverine flooding is another way to tell river is flooding. Normal floods are generally welcome in many parts of the world as they supply rich soil. Flooding at unpredicted scale and extreme frequency causes damage to life, livelihoods and the environment. Over the past decades, the pattern of floods across all countries is changing. The Fourth Assessment Report (2007) of IPCC predicts – *'heavy precipitation events, which are very likely to increase in frequency, will augment flood risk'*.

Floodplains are neither static nor stable as river may change its course from time to time. Due to their continuously changing nature, floodplains need to be investigated thoroughly. When a river reaches its flood stage, water rises and spills over the banks of the river. The amount of flooding depends on the amount of precipitation in the river catchment area, terrain, soil characteristics, and land around the river system. The dynamics of riverine flooding vary with topography of the area. In relatively flat areas flooding will be shallow and slow moving, whereas in hilly and mountainous areas, floods will occur and subside at faster rate. Floods with the short notice, large depths, and high velocities are dangerous compared to slow shallow floods.

Common types of riverine flooding are:

Overbank flooding: Increase in volume of water within a river and the overflow of water from the river onto the adjacent floodplain is termed as over bank flooding.

Flash floods: A rapid and extreme flow of high water into a normally dry area is termed as flash floods.

Dam and levee failure: Some of the most significant losses due to the failure of levee or dam can be attributed to poor design and engineering. But lapses in construction standards cannot be over ruled especially in the countries with high corruption.

Alluvial fans: These are deposits of rock and soil that have eroded from mountainsides and accumulated on valley floors in a fan-shaped pattern.

Ice jam flooding: This type of flooding is caused by jamming of ice. Formation of a jam results in a rapid rise of water both at the point of the jam and upstream.

The vulnerability of a community to floods, is determined by a range of factors such as location; physiography and hydrogeomorphology; structural interventions; climate variability; flood management; economic conditions of exposed communities; social conditions; traditional knowledge; general awareness about problems and solutions; and other survival skills like managing a boat or being able to swim (Das et al. 2009).

Floods, flash floods, river-bank erosion, and sand casting are the most frequent water-induced hazards in many parts of world and all of these hazards affect all aspects of the land, lives, and livelihoods of communities living in the region to a significant degree. Floods leave people homeless and displaced, destroy crops, damage public property, and infrastructure. Victims will suffer from trauma and shock. Repeated cycles of hazards cripple people's resilience and deepen the poverty spiral. Thousands of hectares of fertile land will be lost to the river due to frequent shifting in the river course and erosion of river banks.

Floods are the one of the costliest natural hazard in the world and the importance of floods is increasing due to a number of factors, including sea level rise. Apart from government, floods are a important for insurers and reinsurers. Flood modeling and analysis may be required at a very detailed level at different levels (national, regional, or continent).

Floods affect about nine million hectares per year in India and the 40 million hectares is susceptible to floods. Floods occur in almost all rivers basins in India and riverine flooding is the most significant climate-related hazard in the country. Heavy rainfall, along with inadequate capacity of rivers to carry flood discharge is the main causes of floods (IRADe).

Flooding Analysis is multi disciplinary activity and needs following data (both qualitative and quantitative) for thorough understanding of the situation.

Environmental context

- Land use or land cover (LULC) change
- Biodiversity
- Natural resources
- Degradation of land
- Land restoration and
- Soil productivity

Climate and hazard context

- Trends in rainfall
- Temperature, flood history
- Flooding characteristics
- Sand casting
- Flood damage and
- Droughts

Economic context

- Household income
- Poverty indicators
- Below Poverty Level families
- Price fluctuation
- Self Help Group
- Micro-finance and
- Access to credit

Political context

- Opinion on governance
- Support for political parties
- Benefit from supporters
- Voting in election
- Articulation of grievances or demands and
- Protest campaign or movement

Socio-cultural context

- Indigenous knowledge systems
- Cultural traits
- Religion
- Housing type
- Survival skills
- Water and land related conflicts
- Power relationship and
- Food and seed storage

Demographic context

- Population
- Density
- Sex ratio
- Literacy
- Migration (in and out)
- Rehabilitation and
- Settlement area

Livelihood context

- Existing livelihood types
- Preferred livelihoods
- Diversification of livelihoods
- Economic and cultural constraints
- Impact on natural resources and
- Food security

Health and nutrition context

- Water borne diseases
- Drinking water source and quality
- Sanitation
- Traditional healing system and
- Access to health care

Gender context

- Relative role of women and men in normal and flood time
- Family maintenance
- Income generation
- Water management and
- Sanitation and pregnancy during floods

Agricultural context

- Crop variety
- Cropping method
- Crop calendar
- Winter crops
- Innovative cropping practices
- Water harvesting
- Irrigation and
- Crop marketing

External intervention context

- Government agencies
- Nongovernment agencies
- Private agency and
- International agency

Biodiversity context

- Species diversity and population
- Endemic and endangered species and
- Migratory species and migration trends

Analysis of these data shall be made to draw further conclusion.

13.3   Coastal Flooding Analyses and Mapping

Coastal flooding and erosion are serious problems among coastal countries. Increase in sea level due to rise in temperature shall add to frequency and magnitude of the problem.

The severity of the erosion varies considerably and results from storm surges and wave actions.

Storm surge is the increase in water surface height above normal tide in coastal areas due to wind action over a long stretch of open water. Figure 13.1 shows schematic diagram of formation of storm surge. The low pressure inside a storm or hurricane's eye creates a dome of water near the centre of the storm. But near the land, strong winds in the storm push this dome of water toward the shore.

Wave action includes two components – wave set-up and wave run-up. Wave set-up is the super elevation of the water surface greater than normal surge elevation. It is caused due to onshore movement of water mass by wave action alone. Wave run-up is the action of a wave after water runs up the shoreline or any other obstacle. When waves reach a water depth of about 1.3 times the wave height, the wave breaks and dissipate their energy by generating turbulence in the water. As the

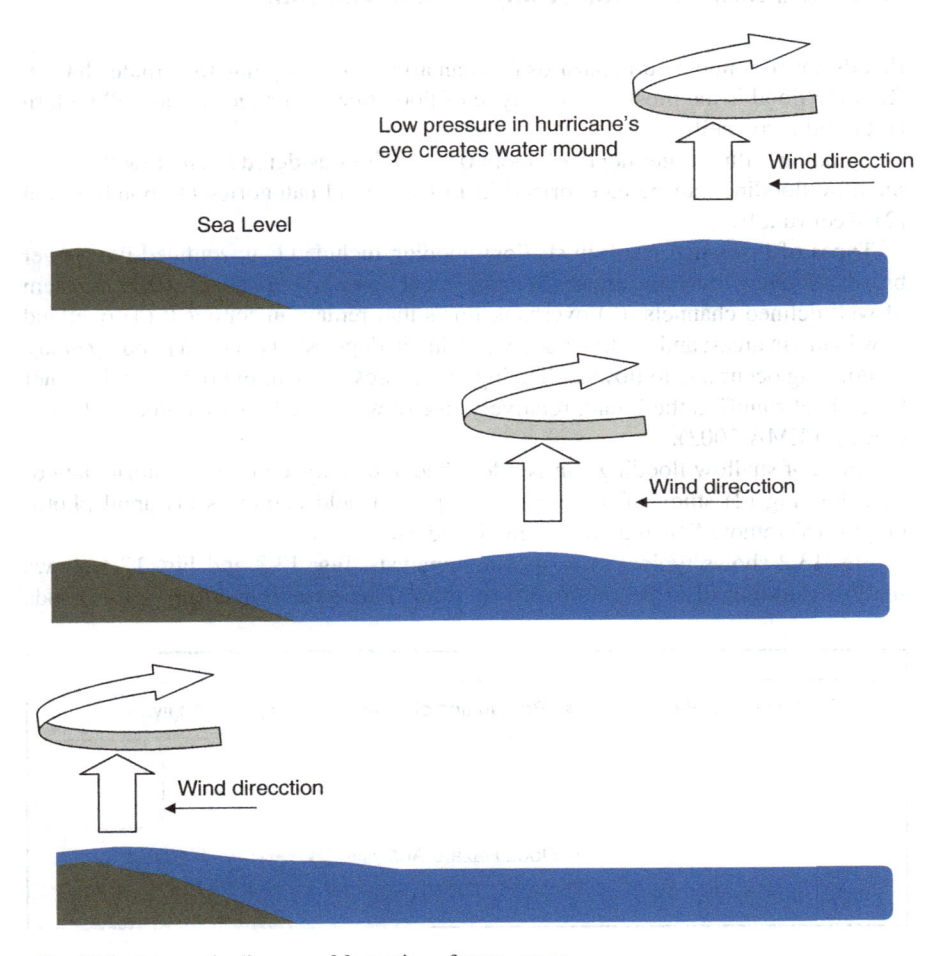

Fig. 13.1 Schematic diagram of formation of storm surge

turbulent water moves further, it spends most of its remaining energy. The beach reacts to changes in wave energy by creating an offshore berm or landward building up of the beach.

Increasing coastal flood risk due to sea-level rise is a major challenge for the twenty-first century.

Coastal flooding analysis and mapping will comprise of following steps:

- Statistical estimation of loads
- Modeling of flood depth and extent and
- Estimation of damage to properties due to flood

13.4 Shallow Flooding Analyses and Mapping

Floods are common natural hazards that can affect a country due to climate change. Riverine flood is the most common type of flood due to changes in rainfall pattern and rapid snowmelt.

Flooding with a water depth of about 0.3–1 m is considered as shallow flooding. Shallow flooding can be categorized in to two broad categories (1) bonding and (2) sheet runoff.

Types of flows that result in shallow flooding include (1) unconfined flows over broad, relatively low relief areas; (2) intermittent flows in arid regions without system of well-defined channels; (3) overbank flows that remain unconfined; (4) overland flow in urban areas; and (4) flows accumulating in depressions to form ponding areas.

Ponding occur due to flows collecting in a depression without outlet on the other hand sheet runoff is the broad, relatively free down slope flow of water in sloping terrain (FEMA 2003).

Areas of shallow flooding can be identified and analyzed by (1) historic data of past flooding (2) study of topographic maps, (3) field enquiries (4) areal photographs, (5) remote Sensing and (6) field studies.

Fig. 13.2 shows typical steps in flood analysis. Fig. 13.3 and Fig. 13.4 shows shallow flooding in Delhi and overflow from Tehri reservoir during 2010 floods

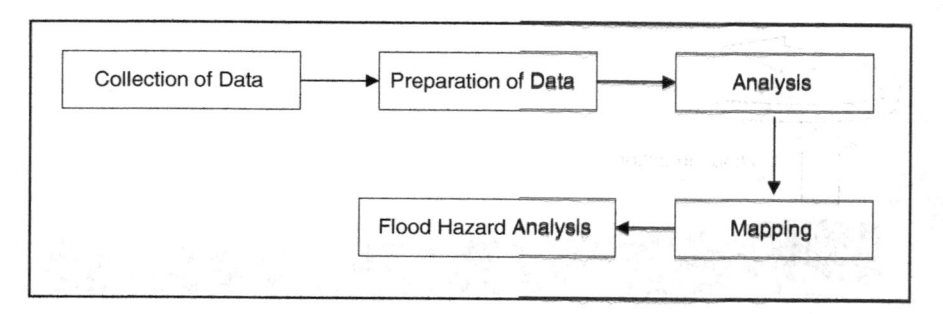

Fig. 13.2 Typical steps in flood analysis

Fig. 13.3 Shallow flooding in Delhi during 2010

Fig. 13.4 Tehri reservoir during 2010 floods

which is typical in many Asian cities. Modern flood analysis involves remote sensing data and conventional Geographic Information System (GIS) data derived from topographic maps. The information is stored in different layers (viz., physiography, land use, watershed, and drainage maps) along with relevant attribution table. The attribution table are updated with flood frequency and inundation depth. The data will be later analysed for different inundation regions using GIS software.

13.5 Ice-Jam Analyses and Mapping

Ice forms in freshwater bodies whenever the surface water cools to 0°C or lower. In cold regions, rivers and lakes are covered with ice for most of the winter season. In late winter and early spring the rise in air and water temperatures result in the weakening of the ice covers leading to ice breakup and the movement. This may lead to ice jams. Constant feeding of ice floes increases the blockage of the river causing flooding of the river banks. Ice jams also cause serious damage to hydropower dams, bridges and other structures across and along rivers where ice jam occurs.

Ice jam analysis and mapping are of use in as glacier-related phenomena and may increase with climate change. The glacial lakes in the Himalayas are growing in both number and size and bigger floods are conceivable (Reynolds 1998; Yamada 1998). Overall, hazards due to flash floods, riverine floods, breached landslide-dam floods, and GLOFs are likely to increase.

There are many types of ice, depending on the mode of formation and evolution (Ashton 1986). They are

Sheet ice: This type of ice is formed in calm water, or in slow-moving River with a velocity of less than 0.5 m/s.

Frazil ice: This type of ice comprises of small particles of ice formed in very turbulent, super cooled water.

Fragmented ice: This type of ice is made up breaking of frazil ice pans or sheet ice growing at the surface of slow-moving water.

Brash ice: This is gathering of ice pieces less than 1.5–2 m dimension formed due to the breakup of an ice cover by escalating water flow or by vessel passage.

There are many types if of ice jams, they are:

Freeze up jams: These jams are made up of mostly frazil ice, with some fragmented ice, and occur during early winter to midwinter.

Breakup jams: These jams occur during periods of thaw, usually in late winter and early spring, and are composed mostly of fragmented ice formed by the breakup of an ice cover or freezeup jam.

Minor breakup jam: These jams occur during midwinter thaw periods due to increase in flow.

Combination jams: Occurrence of both freezeup and breakup jams at a time is refereed as combination jams.

River geometries, weather characteristics, and land-use pattern contribute to the ice jam. The analysis and mapping of the ice-jams is very important to know probable disaster and required mitigation.

As per Integrated Research and Action for Development (IRADe), Ice jams or landslides are among the major causes of flood in addition to typhoons and cyclones.

13.6 Alluvial Fan Flooding Analyses and Mapping

Flooding taking place on the surface of an alluvial fan or similar landform is called *Alluvial fan flooding*. This type of flooding originates at the apex of fan and is distinguished by high-velocity flows. Alluvial fans are gently sloping, fan-shaped landforms. They are created over time by deposition of eroded sediment. Alluvial fan are common at the base of mountain ranges in arid and semiarid regions. The flood is characterized by soil erosion, sediment transport, and deposition;

and, flow paths that can't be predicted easily. The alluvial fan flooding usually begins to occur at the hydrographic apex, and spreads out along paths that are uncertain. During the event it spreads as sheet flood, debris slurries, or in multiple channels characterized by sufficient energy to carry coarse sediment at shallow flow depths. The abrupt deposition may allow flows to initiate new, distinct flow paths. Erosion further influences hydraulic and enlarge the area subject to flooding.

13.7 Mapping of Areas Protected by Levee Systems

A natural or artificial bund or wall to regulate water levels is called as levee. The mapping of levee protected system can be prepared easily by analyzing remote sensing and other conventional data available like topographic maps and survey of the area. The major occurrence of failure of levee and subsequent damage was flood in Asia is that of Pakistan in 2010. About 2.5 billion people were affected and 1,100 people died in the disaster (BBC 2010). The swollen Indus River broke through the levee in southern Sindh province in august 2010. The estimated losses due to flood was $5.1 billion to $7.1 billion due to damages and further $2.12 billion in trade disruption (Michael and Mark 2010).

Floods occurred in this decade in Koshi River in India occurred mainly due to failure of bunds. Floods on the Koshi are unique as river changes its course regularly. These areas suffered direct onslaught of flood, land erosion, sand casting, water logging, scarcity of drinking water, total collapse of health services, disruption of roads and communication systems, engulfing of villages, epidemics, snakebites and large-scale deaths of livestock because of lack of fodder (Singh et al. 2009).

In the year 2010 rainfall was very high in north India. Dehradun got 3,088 mm rain upto September 21, 2010. Tehri reservoir (Fig. 13.3) crossed all time high of 831.05 m which brought rain in plains of Uttar Pradesh.

Bihar state in India has for a long time been a victim of floods and shifts in the courses of the rivers. Civilisation flourished around these rivers due to fertile silt brought by the river was very useful for agriculture. Colonial rulers, tried to check the flooding and meandering of the rivers by embanking them, but accumulated rain resulted in flood. The situation was aggravated by seepage from the embankments water logging after the rains. Failure of sluice gate coupled with seepage, water logging and flood lead to embankment of the tributaries as well. By 2003 the state had has 3,430 km of embankments, 2,952 km in the north and 478 km in the south. But despite these measures about 16.5% of the designated flood-prone area of the country is located in this state which is home to 22.1% of the flood-affected population of the country (GoI 1980).

Considering the above two scenarios it is evident that construction of levees may not solve the problem but it could aggravate it as well. Hence it is necessary to make study prior to constructing structures. The mapping and analysis of the areas planned for levees would give clear picture of the disaster that could be mitigated before investing on structures.

References

Ashton G, ed., 1986: River and lake ice engineering, Water Resources Publications, Littleton, Co

British Broadcasting Corporation (BBC), 2010: '2.5m people affected' by Pakistan floods officials say <http://www.bbc.co.uk/news/world-south-asia-10834414> retrieved on 22. Sept 2010

Burton I, Kates R.W and White G.F., 1978: The Environment as Hazard (New York: Oxford University Press, 1978

Das P; Chutiya D; Hazarika N, 2009: Adjusting to floods on the Brahmaputra plains, Assam, India. Kathmandu: International Centre for Integrated Mountain Development

Department of Regional Development and Environment Executive Secretariat for Economic and Social Affairs Organization of American States, With support from the Office of Foreign Disaster Assistance United States Agency for International Development (DRDEESE-SAOAS), 1991: Primer on Natural Hazard Management in Integrated Regional Development Planning.

FEMA (Federal Emergency Management Agency), 2003: Guidelines and specifications for flood hazard mapping partners, Annexure E: Guidance for Shallow Flooding Analyses and Mapping.

GoI (Government of India), 1980: Report on national commission on floods. New Delhi

IRADe (Integrated Research and Action for Development), National Circumstances of India Geographic, Demographic, Socio-Economic And Environmental Governance Background, downloaded from www.irade.org/natcom.pdf on 22.Sept.2010

IPCC, 2007: Fourth Assessment Report. Geneva: Intergovernmental Panel on Climate Change.

Michael J. Hicks and Mark L. Burton, 2010: Preliminary Damage Estimates for Pakistani Flood, Center for Business And eConoMiC reseArCH, BALL stAte university

Organization of American States (OAS), 1969: Physical Resource Investigations for Economic Development

Reynolds J.M, 1998: 'High-altitude glacial lake hazard assessment and mitigation: A Himalayan perspective.' In Maund, JG; Eddleston, M (eds) Geohazards in Engineering Geology, Engineering Geology special publications No 15, pp 25–34. London: Geological Society

Singh P, Ghose N, Chaudhary N, Hansda R, 2009: Life in the shadow of embankments – Turning lost lands into assets in the Koshi Basin of Bihar, India. Kathmandu: ICIMOD

Yamada, T, 1998: Glacier Lake and its outburst flood in the Nepal Himalaya, Monograph No 1. Tokyo: Japanese Society of Snow and Ice, Data Center for Glacier Research

Chapter 14
Preventing and Mitigation of Disasters

Preventing and mitigation of disasters need basic understanding about the phenomenon/event; knowledge of what to do before the event; knowledge of what to do during the event; and knowledge of what to do after the event. Policy level, approaches to identify and plan for expected changes need to be linked with approaches that allow local populations to evolve their own reaction strategies in response to the opportunities and restrictions they face within their location-specific contexts (Marcus 2010).

All the disasters can not be mitigated. People have to adapt to earthquake and volcanic eruption as science has not evolved to mitigate such disasters. But climate related disasters can be mitigated as the recent trend in climate change is influenced by anthropogenic activities. Mitigation in the context of climate change means the implementation of policies to reduce GHG emissions and enhance processes, activities or mechanisms to GHG's from the atmosphere. Knowledge gaps in the context of mitigation of climate change exist in: (1) lack of accurate data, modelling, and analysis; and (2) lack of knowledge on mitigation technologies.

Within in these constrains the possible mitigation measure that could mitigate climate change are:

Transportation

- Use of electrically driven vehicles
- Use of fuel cells and battery driven electric vehicles
- Use of eco friendly fuel like natural gas and hydrogen
- Increasing conversion efficiency of fuel to energy
- Changing to less carbon-intensive fuel and
- Reducing emissions of non-CO_2 GHGs from vehicle exhausts

Road traffic

- Reducing vehicle load by change in material used for manufacturing vehicles
- Reducing dimensions of vehicles so that total weight of the vehicle will be reduced and material required for making vehicles will be reduced
- Intelligent traffic management to avoid unnecessary traffic jams and to many signals

R. Chandrappa et al., *Coping with Climate Change*,
DOI 10.1007/978-3-642-19674-4_14, © Springer-Verlag Berlin Heidelberg 2011

Fig. 14.1 Mass
transportation could reduce
GHGs provided it covers
entire city. Otherwise citizens
have to depend on private
transport

- Widening of road and increasing underpass and over bridges to avoid traffic
  congestion
- Increasing mass transportation (Fig. 14.1) efficiency, reliability, service hours,
  passenger safety and security and
- Reducing mass transportation costs

  Air traffic

- Management of air traffic
- Flying at optimum cruise altitude
- Minimising taxiing time
- Flying minimum-distance great circle routes and
- Holding and stacking fuel around air ports

  Rail transport

- Improving aerodynamics
- Increasing energy efficiency
- Electrification of railways and
- Reduction in train weight by using low weight material

  Building

- Better thermal envelops
- Improved building methods and building materials
- Reducing emission during manufacturing of building blocks
- Designing building to use efficient use of sun light
- Adopt district heating instead of providing heating system for individual building
- Use of solar water heaters and
- Energy efficient appliances and cooking stoves

Agriculture

- Restoration of cultivated organic soils
- Improved cropland management
- Increase agro-forestry
- Improve grazing land management
- Restoration of degraded land
- Diversify crop in areas of high rice cultivation
- Adopt drip irrigation to avoid stagnant water in sugarcane fields
- Stop cultivating tobacco (Box 14.1) and
- Management of rice field to control CH_4

Box 14.1 Tobacco and climate change

Tobacco is an agricultural product processed from the leaves of plants belonging to genus Nicotiana. It can be used as an organic pesticide and in some medicines. But it is most commonly used as a recreational drug in the forms of smoking, chewing, snuffing, etc. The usage of tobacco is practiced by about 1.1 billion people, and up to 1/3 of the adult population. It currently causes 5.4 million deaths per year. 4.1 million hectares of land were under tobacco cultivation in 1985 (Simon and Wong 1990). Wood for tobacco curing is one of the causes of deforestation in China, Brazil, India, the Philippines and most of Africa. A wood shortage is alarming in Malawi and Tanzania due to deforestation in tobacco growing region (Simon and Wong 1990). Money used for tobacco purchase causes malnutrition in poor families. Cultivation and processing of Tobacco and is lively hood for many. But it does not mean that they have right to destroy others lives. Even though it can be used for growing food items many people grow tobacco due to price it fetches. Even though licensing is required in some countries, corruption over rules the law. Countries whose constitution promise safeguarding of citizens often fails to stop the tobacco farming due to reason best known to people in power. The associated burden due to sick people and loss of work time is often forgotten. The economy of tobacco is attractive to farmers because it will fetch returns higher than conventional food crops like wheat and maize. It is attractive to manufacturers of tobacco products because it brings in profit. Countries will get tax due to sale transport and export of tobacco products as well as cancer curing drugs.

But with changing global climate and consequent food crisis, it is time to say goodbye to tobacco and say yes to food crops. But both farmers and countries are reluctant to stop cultivation. Stopping tobacco cultivation can act both as mitigation and adaptation measure as it saves land for food crops and it mitigates emission due to combustion of wood as well as saves forest which is a carbon sink.

Live stock

- Increase livestock food quality
- Slaughter animals when they attain optimum weight
- Increase research with respect to additives and vaccination to avoid methane generation in digestive track

Industry

- Decrease energy use
- Increase energy efficiency
- Reduce energy loss/theft
- Reduce use of carbonaceous fuel like coal, oil
- Switch over to non carbonaceous fuel like radio active material
- Switch over to non fuel energy sources like solar and wind and
- Use bio-fuel without emitting soot

Behavioural change

- Switchover to vegetarianism as major non-vegetarian ruminants are cause for one third of global anthropogenic emission
- Avoid smoking
- Avoid unnecessary travelling
- Avoid bursting crackers
- Increase use of Information and Communication Technology (ICT) by increasing video conferencing, teleconferencing, electronic communication, etc., to avoid travelling and
- Avoid physical conferences and meetings

Forestry

- Increase forest area and
- Increase site level carbon density

Waste management

- Reduce, recycle and reuse waste
- Avoid combustion for waste management and
- Trap and use methane from waste processing and disposal activities

New polices and legislation

- Incorporation of carbon or energy taxes
- Trade coordination to increase GHG reduction
- Liberal funding towards climate research
- Policies that modify foreign direct investment towards GHG emitting activities
- Stipulating Ambient CO_2 standards
- Voluntary corporate agreement to reduce carbon emission
- Penalties to GHG emitters
- Subsidies and incentives for lowering GHG emission

- Dissipation of climate related information
- Banning crackers and
- Restriction of private vehicle ownership

All the solutions are not acceptable to all the people. The rich people who travel for holidays will recommend fuel efficiency rather than stopping travelling. The people who are fond of meat will advocate butchering animal earliest to avoid methane emission. People who do not use public transports promote fuel efficient private vehicles and biofuel. Public transport depends on local conditions, occupancy rates, and primary energy sources, and built environment. But public will not avail public transport just because it is available. Many public transport services in developed countries are overcrowded resulting in suffocation, infection spreading and pickpockets. The public transportation in some countries is also characterised by sexual harassment, inefficient/infrequent service, impolite staff, lack of adherence to schedule, non availability of service during night and during riots. Unless these issues are addressed the demand for private transport will increase in coming years with zooming number of cars and escalating number of other private vehicles.

The society which depends on energy does not cut down luxury for sake of climate. But ultimately the result of GHG emission will depend on anthropogenic activities and variations of these activities. It has been argued that international market-based approaches applied at a larger scale covering many countries and sectors, can offer a cost-effective means to combat climate change. Policies tailored to fit specific national circumstances may be necessary for reducing emissions across all sectors and gases based on cost-effectiveness; environmental effectiveness; acceptability to citizens; distributional effects; and institutional feasibility. The instruments used fro mitigation can be both mandatory and voluntary in nature. Voluntary instruments help in raising awareness among stakeholders to accelerate the application of best available technologies. But it is not always true that dissipation of information will always result in desired results. The examples of past failure in information dissipation include failure to reduce tobacco consumption in spite of publishing effect on tobacco on cigarette packets. Dissipation of information has also failed in reducing pollution and alcohol consumption. In such scenario one can argue legal instrument will be more effective. But impact of corruption on legal system can not be overruled.

In spite of huge effort after Kyoto protocol, down trend in temperature rise is not visible and policy interventions to curb GHG have not shown reduction in Carbon emission in the world. Possibly legal intervention at national level could result in GHG emission. The possible legal interventions could be stipulating national ambient air quality standard for carbon dioxide; banning energy inefficient technologies like cupola; specifying minimum mileage for each category of vehicle; and fixing maximum weight for vehicles.

In spite of its international importance ambient CO_2 is not measured extensively though out the world and even though the countries through out the world have standards for other pollutants no efforts have been made to stipulate ambient air quality standard with respect to CO_2. The pollutant concentration in a given region is sum of pollutant generated near the point and pollutant travelled to the point from different place. By stipulating CO_2 one can or track CO_2 by running trajectory models reach possible source to take remedial action.

Climate change mitigation and adaptation have several common elements which may be complementary, substitutable, independent or competitive. Mitigation and adaptation also have different features and time scales. Different societies have different capacities for mitigation and adaptation, depending on economic development, access to resources, markets, finance, information, and other governance issues. Mitigation and adaptation policy has a global and national/regional dimension. Adaptation can be both reactive and proactive. Technology plays important role in mitigation as well as adaptation to disasters. The technology change has two phases. In the first phase technology visualized, created, and developed and in the next it is commercialised, disseminated and used.

14.1 Coping with Natural Hazards

Natural hazards have always been a concern to the human population around the world. In spite of technical achievements, people still continue to suffer all over the world due to disasters. Disaster endangers human life, and damage infrastructure. Earth climate system functions in a way that high latitudes and tropical regions experience higher precipitation, whereas other area would get lower precipitation (IPCC 2001). Option for dealing with natural hazards depends on the physical characteristics of the land, land-use patterns, susceptibility to particular hazards, income level, and cultural characteristics. Hence climate change has the potential increase climate induced disasters. But at the same time it cannot be overruled that non climatic disasters continue to occur and some time co-occur with climate induced disasters (like simultaneous occurrence of flood and earthquake) increasing more damage to life and property.

Possible causes of variability in disasters are: (1) density of population; (2) magnitude of the disaster; (3) poor building technology; (4) lack of planning; (5) insufficient rescue and emergency capability; (6) inadequate communication infrastructure; and (7) lower disaster prediction capability.

Overcoming the above problems need following strategies and efforts by: (1) discouraging population accumulation by proper urban planning; (2) improve building technology and build the capacity of civil engineers; (3) improve rescue and emergency capabilities; (4) improve communication infrastructure; and (5) enhance disaster prediction capability.

The effects of a disaster will last a long time and affected country will need more funds and international aid. Hence it is prudent not to exhaust funds and use all international aid immediately after disaster.

Social factors account for vulnerability to climate impacts. Major factors that account for vulnerability are poverty; gender and age. Poor people have greater exposure to disasters as most of them live in makeshift houses or unsafe houses. Developing countries often lack roads allowing emergency vehicle access. It can be observed in many cities even a moderate rain can disrupt traffic in peak hour and cause traffic congestion. Further the poor people are unable to move to less dangerous sites and will not have insurance for the houses. Many times houses are built in sites not owned by them and hence will not have legal ownership to make an insurance policy. People living in illegal property will not get legal and financial protection from state.

14.2 Planning and Preparation for Different Time Scales

Duration of disaster varies from disaster to disaster. While floods in river occur at a frequency of ever few years, events like tsunamis can occur once in a few decades or century. Stochastic information and prediction technology helps in being prepared to such disaster to mitigate impact. In addition the new challenge of changing climate also needs to be addressed in different time scales.

Planning and preparation could be in terms of structural, social, legal and economic measures. Structural measures include flood protection by construction of dams, water storage reservoirs, assembly point during emergency; improved water storage and supply scheme. Social measures include advanced alarming and warning systems; publicity and education; and capacity building through formation of self help groups. Legal measures include protection of groundwater aquifers through restriction in ground water extraction, and surface water use. Economic measures can come by adopting insurance for public property, and health care insurance for economically weaker section. A cost-benefit analysis helps to quantify the damage due to climate change but it is difficult to estimate the social cost of carbon (SCC) due to the large uncertainties involved.

Kyoto Protocol legally binds commitments for Annex I Countries to reduce GHG emissions globally by 5.2% below the 1990 levels at the end of 2012. Climate change will adversely affect the MDGs (Millennium Development Goals) agreed to achieve by the year 2015 by 192 United Nations member states and at least 23 international organizations. The eight MGDs are: (1) eradicate extreme hunger and poverty; (2) achieve universal primary education; (3) promote gender equality and empower women; (4) reduce child mortality; (5) improve maternal health; (6) combat HIV/AIDs, malaria and other diseases; (7) ensure environmental sustainability; and (8) develop a global partnership for development. The impacts of climate change will affect the poorest, which have the least resources and the least capacity to adapt.

Stringent stabilization scenarios would be required for minimizing the probability of dangerous anthropogenic interference' with the earth's climate systems to start declining before 2015 and reach less than 50% of today's emission level by 2050 (TERI 2009).

14.3 Planning for Environmental Emergencies and Climate Change

Emergency can occur at different level so as climate impact. The preparedness should be done according to anticipated scenarios at international, regional, national, and local level. The institutions involved in international emergency are different from that of local emergency and so as at different level. Emergency management must be comprehensive, progressive, risk-driven, integrated, collaborative, coordinated, flexible, and professional (IAEM 2007). These qualities form the principles of environmental management. Table 14.1 gives actors at different level during emergency and these agencies shall meet and communicate each other and prepare emergency preparedness plan at different level.

Emergency preparedness should be aimed at minimizing the damage. Preparedness includes measures taken in expectation of the disaster. Disaster preparedness include: (1) public safety information; and (2) disaster awareness planning. Disaster awareness planning addresses improving the ability of a particular area to natural disasters. Disaster preparedness supports the development of a system which deals with activities that should take place in case of disaster.

The international agencies anticipate emergency and natural calamities that are likely to affect different countries at a time and be prepared accordingly. It is not uncommon to extend/receive help from other countries and international institutions even if the disaster occurs within the jurisdiction of a country.

The national emergency plans should take into account the needs of the emergency services of defense, paramilitary force, different department, ministries and neighboring countries.

The regional emergency plans involves jurisdiction of more than one local bodies or district administration.

The local emergency plans should take into account the needs of the emergency services (the police, health and fire services), people and property affected, utility companies (electricity, gas, water and telephone) and neighboring local authorities.

Table 14.1 Actors at different level during emergency

Level	Actors
International	International governments
National	Different agencies within the government such as defence, police, medical department, agency responsible for environmental protection; NGOs; and expert's in different fields. Actors involved depend on individual c country
Region within a country	Regional institutions
Local	Actors involved in local emergency are differ from country to country and institutional setup and involve different agencies within the government such as defence, police, medical department, agency responsible for environmental protection; NGOs; and experts in individual fields
Industry/institution	Pre-identified personnel from the organisations and government agencies responsible for service during emergency

The Industry/institution should plan in a way that it identifies teams and fix up responsibilities. It should take into account the external assistance required (the police, health and fire services), people and property affected, utility companies (electricity, gas, water and telephone) during the planning.

All agencies that deal with emergency shall aim to save lives; prevent the disaster escalating; relieve suffering; safeguard the environment; protect property; assist with criminal, judicial, public, technical or other inquiries; restore a normal situation as soon as possible.

Asia has witnessed numerous disasters in recent years which has lead to cooperation at the regional levels, such as ASEAN, SAARC (South Asian Association for Regional Cooperation); Central Asia. Cooperation efforts have been promoted also between Japan, China and the Republic of Korea to combat disaster (ACDR 2010). In 2005, ASEAN members signed the first treaty in the world for interstate disaster management and in 2010 the Asia-Pacific Economic Cooperation forum decided to form permanent working group by elevating the existing task force on emergency preparedness (CSIS 2010).

On December 1997, a Regional Haze Action Plan (RHAP) was completed with objectives to prevent forest fires by better management policies and enforcement; operational mechanisms for monitoring land and forest fires; Improve regional land and forest fire-fighting capability and other alleviation measures (ADB 2001).

In order to plan environmental emergencies and climate change *South Asian Regional Policy Dialogue on Disaster Risk Reduction and Management* was held in New Delhi, India on 21–22 August 2006. The SAARC initiated a 'Regional Study on the Causes and Consequences of Natural Disasters and the Protection and Preservation of Environment', and was completed in 1991. A study on 'Greenhouse Effect and its Impact on the Region' finalized in 1992, SAARC Plan of Action on Environment was accepted in 1997. The SAARC Meteorology Research Centre (SMRC) was established in Dhaka, Bangladesh in 1995 and the SAARC Coastal Zone Management Centre (SCZMC) was established in Male, Maldives in 2004.

Taking into consideration the *Disaster Management in South Asia a Comprehensive Regional Framework for Action 2006–2015* was adopted at the expert group meeting held at Dhaka, Bangladesh in February 2006.

SAARC Disaster Management Centre Management (SDMC) was set up in October 2006 for regional cooperation for preparedness and mitigation of national disasters.

The Asia-Pacific Economic Cooperation (APEC)'s Energy Security Initiative (ESI), is the principal mechanism through which the member countries addresses long- and short-term energy security.

14.4 Being Prepared Saves Lives

The prevention and reduction of risks needs application of engineering solutions, public policy decisions, insurance policies, and regulatory actions. The risk tolerance for disasters depends on numerous factors. Losses due to disasters are

measured in dollars but the environmental damages can't be completely estimated. Reducing and being prepared for disasters comes at increased cost.
But such expenditure of tax payer's money is justified as it saves their lives and
property.

Beyond the efforts of regional institutions to combat climate impact, a few
significant multilateral and bilateral activities are occurring such as Mekong
River Commission, the Coral Triangle Initiative (CTI). Such positive indicators
of are quite helpful within the region and out side the region to fight against climate
change.

Since the 2004 Indian Ocean tsunami, regional initiatives to combat natural
disasters have proliferated. Since the 2004 tsunami, there has been extraordinary
global momentum for reducing exposure to disaster risk. A landmark moment in
this regard is signing of the Hyogo Framework for Action by 168 nations in January
2005. Despite the many common disaster risk factors, all Asian countries are not
homogenously vulnerable. Hence there are now more than ten important regional
arrangements engaged in disaster management (CSIS 2010).

Apart from the efforts visible, there are many issues which are not addressed
completely due to apathy from governments and public. Many cities are not prepared for disasters and there is indiscriminate growth of urban settlements by
compulsory acquisition of rural property adjacent to mega cities. Sky scrapers
with least safety measures pose risk to life and increase human density in an area.
The increase in private vehicle will increase traffic jams. As could be seen in
Fig. 14.2, the crowd near the door in public transport service implies shortage of
public transport facility in many cities encouraging sale of private vehicles. On the
other hand eco-friendly people who avail the service of public transportation has to
hang in the doors in peak hours subjecting themselves for risk.

The high density of population, formation of slums (Fig.14.3) and occupying
footpaths by street vendor (Fig. 14.4) makes cities more vulnerable to the impacts
of natural disasters. Poor governance will only increase haphazard settlement and

Fig. 14.2 Poor public
transportation system will
encourage sale of private
vehicle. Those who opt for
public transport will be
subject to risk in overcrowded
vehicles

Fig. 14.3 Formation of slums

Fig. 14.4 Street vendor occupying footpaths

business establishments making it difficult to clear them afterwards. In many cities government build new houses to slum dwellers in taxpayer's money. There are numerous slums in developing world wherein occupants pay rent to people who have grabbed the precious space in the city. On the other hand middle class people find difficulty in owning or renting houses in urban locality legally due to non-availability of space. In such situation people do not cooperate and live discretely without forming a community. Hence there will be lot of pressure on government to solve every problem due to natural disasters.

Sparsely populated areas rely on the government to only a limited extent for disaster warning and aftermath assistance in dealing with it. Hence building the capacity of local community in sparsely populated area will improve hazard management in such area effectively.

References

ADB (Asian Development Bank), 2001: Fire, Smoke, and Haze The ASEAN Response Strategy As viewed in <www.aseansec.org> on 14th Sept. 2010

ACDR (Asian Conference on Disaster Reduction), 2010: Hyogo, Japan, Conference Summary

CSIS (Centre for strategic and International Studies), 2010: Asia's Response to climate Change and Natural Disasters, Implications for an Evolving Regional Architecture

IAEM (International Associations of Environmental Managers), 2007: Principles of Emergency *Management Supplement*, <http://www.iaem.com/publications/documents/PrinciplesofEmer gencyManagement.pdf> downloaded on 4th September 2010

IPCC, 2001: Third Assessment Report on Climate Change, Cambridge Univ. Press.

Marcus Moench, 2010: Responding to climate and other change processes in complex contexts: Challenges facing development of adaptive policy frameworks in the Ganga Basin, Technological Forecasting & Social Change 77, 975–986

Simon Chapman and Wong Wai Leng, 1990: Tobacco in the 3rd World, Clarion <http://www.nsma.org.au/world3.html> retrieved on 15 November 2010

TERI (The Energy and Resources Institute) 2009: Simplifying Climate Change, based on the findings of the IPCC Fourth Assessment report, pp140

Chapter 15
Sea Level Change and Asia

One of the most marked effects of climate change is melting of masses of ice around the world. The melting of the glaciers and ice sheets has two main impacts: (1) variation in runoff from the melting of mountain glaciers; and (2) sea level rise.

There are three major processes by which climate change affects sea level: (1) thermal expansion; (2) melting of glaciers and ice caps; (3) loss of ice mass from Greenland and Antarctica.

The impacts of sea level rise are many folds: (1) thermohaline (or deep ocean overturning) circulation will lead to change in global flow of ocean currents; (2) destruction of property built on coastal area; (3) damage to flora and fauna in ocean; and (4) sea water intrusion to groundwater as well as surface water near coasts

River deltas vulnerable to rise in sea level are compounded by land subsidence and human intervention such as sediment trapping by dams (Church et al. 2008).

Global warming is a complex natural disaster with a long slow beginning that is likely to cause noteworthy impacts on large portion of this planet's population (Joshua 1999). Current sea level rise has occurred at a mean rate of 1.8 mm per year for the past century(Bruce 1997; Church et al. 2006), and more recently rates estimated near 2.8 ± 0.4 mm (Chambers et al. 2003) to 3.1 ± 0.7 mm (Bindoff et al. 2007) 2 mm per year (measured during the period 1993–2003). Current sea level rise is due significantly to global warming (Bindoff et al. 2007), which will increase sea level over the coming century and longer periods (Meehl et al. 2007). Studies by Unnikrishnan and Shankar (2007) revealed the sea level rise was between 1.06 and 1.75 mm per year. At the end of the twentieth century, thermal expansion and melting of land ice contributed roughly equally to sea level rise, while thermal expansion is expected to contribute more than half of the rise in the upcoming century (Bindoff et al. 2007). Values for predicted sea level rise over the course of this century typically range from 90 to 880 mm, with a central value of 480 mm. Models of glacier mass balance (the difference between melting and accumulation of snow and ice on a glacier) give a theoretical maximum value for sea level rise in the current century of 2 m (and a "more plausible" one of 0.8 m), based on limitations on how quickly glaciers can melt (Pfeffer et al. 2008).

R. Chandrappa et al., *Coping with Climate Change*,
DOI 10.1007/978-3-642-19674-4_15, © Springer-Verlag Berlin Heidelberg 2011

15.1   History and Reality of Sea level Change

Sea level has changed drastically over time. Some of the changes in sea level over time are discussed in the following paragraphs.

550 Million Years Ago

The Cambrian lasted from 542 ± 0.3 million years ago to 488.3 ± 1.7 million years ago (Gradstein et al. 2004). During this period sea level raised gradually (Haq and Schutter 2008) with maximum sea level around 150 m higher than the present day sea level.

500 Million Years Ago

Ordovician covers the time between 488.3 ± 1.7 and 443.7 ± 1.5 million years ago. The Katian Age of the Late Ordovician period varied between 456 and 446 million years ago. This era was characterised by relatively stable sea level with a large drop associated with the end-Ordovician glaciations. Maximum sea level in this period was about 225 m higher than the present day (Haq and Schutter 2008) sea level.

450 Million Years Ago

The Silurian period extended from the end of the Ordovician Period to about 443.7 ± 1.5 million years ago. Relative stability of sea level occurred at the lower level during the Silurian with maximum sea level around 200 m higher than present day (Haq and Schutter 2008).

400 Million Years Ago

The Devonian is a geologic period spanned from 416 to 359.2 million years ago. Sea level saw a gradual fall through the Devonian, continuing through the Mississippian to long-term low at the Mississippian/Pennsylvanian boundary with sea level was about 100–200 m higher than present day (Haq and Schutter 2008).

350 Million Years Ago

The Carboniferous spanned from about 359.2 ± 2.5 million years ago. The sea level at this period varied between 0 and 100 m higher than present day (Haq and Schutter 2008).

300 Million Years Ago

The Permian Period spanned from 299.0 ± 0.8 to 251.0 ± 0.4 million years ago. A gradual rise in sea level occurred until the beginning of the Permian followed by a gentle decrease lasting until the Mesozoic with maximum sea level of around 80 m higher than present day (Haq and Schutter 2008).

250 Million Years Ago

The Mesozoic Era spanned from about 250 to 67 million years ago. Sea level today is very near the lowest level ever attained about 250 million years ago (Haq and Schutter 2008).

20,000 Years Ago

The world's sea level was about 130 m lower than today, due evaporation of large amount of sea water that had been deposited as snow and ice.

10,000 Years Ago

The majority of ice that had deposited 20,000 years ago melted about 10,000 years ago.

3,000 Years Ago

Global sea level rose by about 120 m until ice age (about 21,000 years ago), and became stable between 3,000 and 2,000 years ago. Global sea level did not change considerably from then until the late nineteenth century. Global average sea level rose at a rate of about 1.7 mm per year during twentieth century (Bindoff et al. 2007).

100 Years Ago

The pace of sea level rise was estimated as 1.8 mm per year for the past 70 years (Douglas 2001; Peltier 2001). Miller and Douglas (2004) find a range of 1.5–2.0 mm per year for the twentieth century. Holgate and Woodworth (2004) estimated sea level change at pace of 1.7 ± 0.4 mm per year averaged along the global coastline during the period 1948–2002.

50 Years Ago

Church et al. (2004) observed a global rise of 1.8 ± 0.3 mm yr per year during 1950–2000, and Church and White (2006) observed a change of 1.7 ± 0.3 mm per year for the twentieth century.

Sea ice in the Sea of Okhtosk is most vulnerable to climate, and it influences climate, ecosystems, fisheries, and transportation around the area (Tokioka et al. 1995). Since this area is the southernmost ocean in the Northern Hemisphere, early disappearance of sea ice result in a large sea-surface temperature (SST) rise over the Sea of Okhotsk. Sea level rise along China's coasts in the past 50 years was 2.5 mm per year which was slightly higher than the global average (NDRCPRC 2007).

15.2 Impact on Flora and Fauna

Changes in climate change threaten biological species thereby adding stress to the terrestrial and marine ecosystems that are already strained by land use change, exploitation of resources, oil spillage, pollution, over-harvesting, and introduction of alien species.

The link between biodiversity and climate change is two way. Climate change alters the state of biodiversity, changes in biodiversity alters the climate change. The coral reefs of Southeast Asia are the most gravely vulnerable, with 40% of reefs effectively lost, 45% under threat, and 15% at low threat (Hoegh-Guldberg et al. 2009).

The stress on coral reefs, mangroves, and sea grasses is of two broad categories. The first one is originate directly from activities within the region like declining water quality, overexploitation of resources, sewage discharge, deforestation, change in land use, and destructive fishing. The second threat is global phenomena like rising carbon dioxide in the earth's atmosphere, acidification of the world's oceans.

The Southeast Asian region is a major hub for shipping traffic and several mega-ports. A number of threats arise from this intense shipping activity including oil spills, pollution from ports, ballast/bilge discharge, garbage disposal, groundings and anchor-damage, leading to the direct destruction of coral reefs formations.

Expansion of anthropogenic activities has affected coastal ecosystems. Land use to build human dwellings, airports and tourist venues has removed coral reefs and mangroves from coastal areas. Corals have been mined to make cement and building material. Mangroves, sea grass beds, salt marsh areas and coral reefs were affected by dredging and port development. About 21% of the coral reefs of Southeast Asia are threatened by run-off. Thirty-five percent of coral reefs are endangered by nutrients and sediments (Burke et al. 2002).

Coral reefs and sea grasses that flourish in clear and low nutrient waters disappear once coastal waters become clouded with sediment and laden with nutrients. Increase in turbidity in sea water restricts light falling on coral and other photosynthetic organisms. While increase in nutrients favors seaweeds, sediment will smother coral reefs.

The increase in population in has altered natural forests and landscapes in many Asian countries. Large river systems in the Philippines have added huge quantities of sediments oppressing coral reefs.

Increase in extreme weather events in the most severe categories, such as hurricanes or typhoons, occur with the global climate (Goldenberg et al. 2001; Webster et al. 2005), may cause significant loss/erosion of or damage to shorelines. Successful reproduction of marine turtles depends primarily on available terrestrial habitat. Female turtles emerge onto beaches to lay eggs in nesting season (Miller 1997). Studies reveal that loss of nesting beach due to hurricanes, cyclones and storms, decreased hatching success and hatchling emergence success (Martin 1996; Ross 2005; Pike and Stiner 2007; Prusty et al. 2007; Van Houton and Bass 2007).

However, susceptibility to storm-related threats may vary by species (Pike and Stiner 2007).

Sea turtle nests at several sites in the western Indian Ocean of Southeast Asia will be threatened due to climate change and sea level rise. The impacts of tidal flooding include physical damage to infrastructure, lost working time and reduced productivity at work place. The tides also cause traffic jams, illness, and disrupt activities of schools.

Ecologically, the Asian mega deltas (Ayeyarwady delta of Myanmar; Chao Phraya delta of Thailand; Mekong Delta of Vietnam; Song Hong delta of Vietnam; Zhujiang delta of China; Changjiang delta of China; Huanghe delta of China; Lena Delta of Russia; Ganges-Brahmapura Delta of India and Bangladesh; Indus Delta of Pakistan; Shatt-el-Arab Delta of Iraq and Iran) are important diverse ecosystems of unique plants and animals located in different climatic regions (IUCN 2003; ACIA 2005; Macintosh 2005; Sanlaville and Prieur 2005). These mega deltas of Asia are exposed to climate change and sea level rise resulting in higher frequency and level of inundation due to storm surges and floods (Nicholls 2004; Woodroffe et al. 2006)

Mangrove forests at present occupy 14,650,000 ha of coastline worldwide (Wilkie and Fortuna 2003). Mangrove ecosystems are important, especially in developing countries, as they play an important role in human sustainability and livelihoods (Alongi 2002). These forests are used traditionally for food, timber, fuel, and medicine (Saenger 2002). Mangroves are breeding sites for birds, mammals, fish, crustaceans, shellfish, and reptiles; and sites for accretion of sediment, nutrients, and contaminants (Twilley 1995; Kathiresan and Bingham 2001; Manson et al. 2005). Mangroves offer protection from waves, tidal bores, and tsunamis and can dampen shoreline erosion (Mazda et al. 1997). Some mangroves will survive and even thrive with the forecasted changes in climate whereas others won't survive (Daniel 2008).

15.3 Impact on Asian Society

Change in climate is expected to aggravate current stresses on water resources due to population growth and change in economy as well as land-use change. Melting of glaciers and reductions in snow cover are likely to accelerate water availability, hydropower potential, and changing seasonality of flows. Changes in precipitation and temperature will change runoff and water availability.

Runoff is anticipated with high confidence to rise by 10–40% by mid-century at higher latitudes and in some wet tropical areas, and decline by 10–30% over some dry regions at mid-latitudes and dry tropics (IPCC 2007) resulting in subsequent impact to economy and ecosystem of the region.

Productive ecosystems yielding sea food and coastal recreation together support the economies of many coastal communities. In Asia fishing is gender specific and each gender has special roles which vary from area to area, culture to culture, season to season and country to country. Complementing the fishing women and children engage themselves in the collection of shellfish. The collection of shellfish contributes to some of diet needs of the community, especially in times when fishing activity cannot be carried out due to extreme weather conditions. The abundance of reef fish is depending on the health of the local coral reefs and the extent of fishing in the area. Fishing is usually seasonal activity to and sometimes part time activity to supplement other activities. Fishing is often a social activity, strengthening bonds between people. Some people hesitate to join conservation programs despite the concerns of over fishing, loss of marine diversity, coastal zone degradation and coral reef destruction (Minura 2008). On the other hand some of the wealthy importing nations continue to import sea products in spite of adequate knowledge about marine diversity and threat due to unscientific and uncontrolled marine activities.

Critical issues of coastal community are population explosion, the lack of suitable agricultural land, and the decline in fish and shellfish landings. The lack of arable land is mostly due to the narrow coastal flat areas along the coast. Poor agricultural practices in some area have increased the amount of soil erosion from the hill slopes.

Increased sea-level rise will reduce the habitat suitable for shellfish which plays a very important part in food security for some families. Climate change will flood narrow rocky line and the cliffs will experience more erosion, increasing the sediments into the coastal waters making some of the species vulnerable. The level of impact of sea-level rise on coastal communities will depend on the access to other sources of diet from agriculture and animal husbandry.

Several locations within South-East Asia have experienced sea water rise. Tidal floods are an everyday event in many parts of Asia. Worse tidal flooding occurs during rainy season and west monsoon between October and January. The flood not only affects coastal villages but also affects cities (Hoegh-Guldberg et al. 2009).

Increase in drought-affected areas will affect agriculture, water supply, energy production and health. The increase in flood will damage physical infrastructure and water quality.

Table 15.1 gives possible impact to urban settlement in sea shore, adaptation principle and adaptation strategy.

Asia is the most densely inhabited and hence the most vulnerable continent in the world. It faces high risk of climate impacts and low adaptive capacity. In addition to rising sea levels, much of Asia experience sinking of the land due to heavy extraction of groundwater, tectonic activity, and natural ground movement. Some of these climate impacts directly interact with the urban area, whilst other impacts like heat waves and precipitation pattern act indirectly.

Climate change impacts on Asian mega cities in high-risk zones, such as main river delta regions, are at the top of the international research agenda (Cruz et al. 2007). Vietnam ranks high amongst the affected countries (DasGupta et al. 2007).

Many of the largest cities in Asia are situated on the coast and more susceptible to sea water rise. Cities are hotspots of innovation and provide solution to many of world's problems due to presence of highly educated people. Dhaka in Bangladesh located just meters above current sea levels, is frequently impacted by tropical

Table 15.1 Possible impact to urban settlement in sea shore, adaptation principle and adaptation strategy

Possible impact to urban settlement in seashore	Adaptation principle	Adaptation strategy
Sea level rise, tidal flooding, fluvial flooding, urban drainage flooding, heat, human comfort and health	Precautionary principle	Spatial planning of different land use areas, transport, flood hazard areas
	Preventive principle	Maintenance of natural protective like mangrove forests, hilly areas. Providing adequate infrastructure, Planned evacuation by moving government and private institutions to less vulnerable area. Education and awareness to local community. Removing illegal buildings
	International and international cooperation	Fast and efficient information among institution of other countries and within the country
	Transparency and public consultation	Frequent consultation with public of affected area, publication of policies, reduction of corruption
	Disaster management	Prepare disaster preparedness plan and conduct mock drill

cyclones and flooding, and has very limited adaptive capability. Most of the vulnerable cities are highly vulnerable due to frequent flooding and have relatively low adaptive ability.

Calcutta in India, Manila in Philippines, Phnom Penh in Cambodia, Jakarta in Indonesia, Ho Chi Minh City in Vietnam and Shanghai in China are some of important cities which are very susceptible to sea-level rise. Historically these cities have densely built-up area in a low lying region and are sensitive to climatic impacts. Position of Ho chi Minh City in an estuarine region of the Dong Nai River system and northeast of the Mekong River Delta contribute to this sensitivity (Nguyen 2006).

Coastal oil production areas are likely to be affected by storm surges (IPCC 1996). But not are oil producing countries of Asia will be affected due to rise in sea level as seven of the 21 countries in arid and semi arid region of Asia are landlocked and coastal zones in this region are not significantly affected by sea-level changes.

To conclude international community are already working on the challenges like sea water intrusion, coastal submergence, loss of biodiversity and loss of economy. These concerns should also become priority to all the countries located in coastal area to protect its citizens.

References

ACIA (Arctic Climate Impact Assessment), 2005: Impacts of a Warming Arctic: Arctic Climate Impact Assessment. Cambridge University Press, Cambridge, 140 pp.
Alongi D.M, 2002: Present state and future of the world's mangrove forests. Environmental Conservation 29, 331–349.

Bindoff N.L, Willebrand J, Artale V, Cazenave A, Gregory J, Gulev S, Hanawa K, Le Quéré C, Levitus S, Nojiri Y, Shum C.K, Talley L.D and Unnikrishnan A, 2007: Observations: Oceanic Climate Change and Sea Level. In: Climate Change 2007: The Physical Science Basis. Contribution of Working Group I to the Fourth Assessment Report of the Intergovernmental Panel on Climate Change [Solomon, S., D. Qin, M. Manning, Z. Chen, M. Marquis, K.B. Averyt, M. Tignor and H.L. Miller (eds.)]. Cambridge University Press, Cambridge, United Kingdom and New York, NY, USA.

Burke L, Selig L, Spalding M, 2002: Reefs at Risk in Southeast Asia. World Resources Institute, Washington DC

Bruce C. Douglas, 1997: "Global Sea Rise: A Redetermination". Surveys in Geophysics 18: 279–292. doi:10.1023/A:1006544227856 DOI:dx.doi.org.

Chambers D.P, Ries J.C, Urban T.J, 2003: "Calibration and Verification of Jason-1 Using Global Along-Track Residuals with TOPEX". Marine Geodesy 26: 305. doi:10.1080/714044523 DOI: dx.doi.org .

Church J.A, White N.J, Aarup T, Wilson W.S, Woodworth P.L, Domingues C.M, Hunter J.R, and Lambeck K, 2008: 'Understanding global sea levels: past, present and future'. Sustainability Science, 3 (1): 1–167.

Church John A, White Neil J, Coleman Richard, Lambeck Kurt, Mitrovica, Jerry X, 2004: Estimates of the regional distribution of sea level rise over the 1950 to 2000 period. J. Clim., 17(13), 2609–2625.

Church John, White, Neil, 2006: "A 20th century acceleration in global sea-level rise". Geophysical 33: L01602. January 6, 2006. doi:10.1029/2005GL024826 DOI:dx.doi.org . L01602.

Church J.A, White N.J, and Hunter J.R, 2006: Sea-level rise at tropical Pacific and Indian Ocean islands. Global Planet. Change, 53, 155–168.

Cruz R.V, Harasawa H, Lal M, Wu S, Anokhin Y, Punsalmaa B, 2007: "Asia". In: Climate Change 2007: Impacts, Adaptation and Vulnerability. Contribution of Working Group II to the Fourth Assessment Report of the Intergovernmental Panel on Climate Change (IPCC).Cambridge, UK: Cambridge University Press, pp. 469–506.

Daniel M Alongi, 2008: Mangrove forests: Resilience, protection from tsunamis, and responses to global climate change, Estuarine, Coastal and Shelf Science 76, 1–13

DasGupta S, Laplante B, Meisner C, Wheeler D, & Yan J, 2007: "The Impact of Sea Level Rise on Developing Countries: A Comparative Analysis". World Bank Policy Research Working Paper 4136. Washington, DC, USA: World Bank.

Douglas B.C, 2001: Sea level change in the era of the recording tide gauges. In: Sea Level Rise: History and Consequences [Douglas, B.C., Kearney, M.S., and S.P. Leatherman (eds.)]. Academic Press, New York, pp. 37–64.

Goldenberg S.B, Landsea C.W, Mestas-Nunez A.M, Gray W.M, 2001: The recent increase in Atlantic hurricane activity: causes and implications. Science 293:474–479

Gradstein Felix M, Ogg J.G, Smith A.G, 2004: A Geologic Time Scale 2004. Cambridge: Cambridge University Press. ISBN 0521786738.

Haq B.U, Schutter S.R 2008: "A Chronology of Paleozoic". Science 322 (5898): 64. doi:10.1126/ science.1161648 DOI:dx.doi.org . PMID 18832639. http://www.sciencemag.org/cgi/content/ full/322/5898/64.

Hoegh-Guldberg O, Hoegh-Guldberg H, Veron J.E.N, Green A, Gomez E.D, Lough J, King M, Ambariyanto, Hansen L, Cinner J, Dews G, Russ G, Schuttenberg H. Z, Peñafl or E.L, Eakin C.M, Christensen T.R.L, Abbey M, Areki F, Kosaka R.A, Tewfik A, Oliver J, 2009:. The Coral Triangle and Climate Change: Ecosystems, People and Societies at Risk. WWF Australia, Brisbane, 276 pp.

Holgate S.J and Woodworth P.L, 2004: Evidence for enhanced coastal sea level rise during the 1990s. Geophys. Res. Lett., 31, L07305, doi:10.1029/2004GL019626.

IPCC, 1996: Climate change 1995: the IPCC second assessment report, Vol 2: scientific-technical analyses of impacts. Adaptations and mitigation of climate change. Watson RT, Zinyowera MC, Moss RH (eds). Cambridge University Press, Cambridge

IPCC, 2007: Climate Change 2007, Synthesis report, pp49

IUCN (The World Conservation Union), 2003: The lower Indus river: balancing development and maintenance of wetland ecosystems and dependent livelihoods. Water and Nature Initiative, 5 pp. Accessed 24.01.07: www.iucn.org/themes/wani/flow/cases/Indus.pdf.

Joshua D. Lichterman, 1999: Disasters to come, Futures 31 (1999) 593–607

Kathiresan K, Bingham B.L, 2001: Biology of mangroves and mangrove ecosystems. Advances in Marine Biology 40, 81e251.Leatherman (eds.). Academic Press, San Diego, pp. 65–95.

Martin R.E, 1996: Storm impacts on loggerhead turtle reproductive success. Mar Turtle Newsl 73:10–12

Manson R.A, Loneragan N.R, Skilleter G.A, Phinn S.R, 2005: An evaluation of the evidence for linkages between mangroves and fisheries: a synthesis of the literature and identification of research directions. Oceanography and Marine Biology: An Annual Review 43, 483e513.

Mazda Y, Wolanski E, King B, Sase A, Ohtsuka D, Magi M, 1997: Drag force due to vegetation in mangrove swamps. Mangroves and Salt Marshes 1, 193–199.

Meehl G.A, Stocker T.F, Collins W.D, Friedlingstein P, Gaye A.T, Gregory J.M, Kitoh A, Knutti R, Murphy J.M, Noda A, Raper S.C.B, Watterson I.G, Weaver A.J and Zhao Z.C, 2007: Global Climate Projections. In: Climate, Change 2007: The Physical Science Basis. Contribution of Working Group I to the Fourth Assessment Report of the Intergovernmental Panel on Climate Change [Solomon, S.,D. Qin, M. Manning, Z. Chen, M. Marquis, K.B. Averyt, M. Tignor and H.L. Miller (eds.)]. Cambridge University Press, Cambridge, United Kingdom and New York, NY, USA. Projections of Global Average Sea Level Change for the 21st Century Chapter 10, p 820

Miller J.D, 1997: Reproduction in sea turtles. In: Lutz PL, Musick JA (eds) The biology of sea turtles, Vol 1. CRC Press, Boca Raton, FL, p 51–81

Miller L, and Douglas B.C, 2004: Mass and volume contributions to 20th century global sea level rise. Nature, **428**, 406–409.

Minura N, Ed., 2008: Asia -Pacific coasts and their management. Dordrecht, Springer.

National Development and Reform Commission, People's Republic of China (NDRCPRC), 2007: China's National Climate Change Programme

Nicholls, R.J., 2004: Coastal flooding and wetland loss in the 21st century: changes under the SRES climate and socio-economic scenarios. Global Environ. Chang., **14**, 69–86.

Nguyen Huu Nhan, 2006: "The Environment in Ho Chi Minh City Harbours". In: Wolanski, Eric: The Environment in Asia Pacific Harbours, Amsterdam: Springer Netherlands, pp. 261-291.

Peltier, W.R., 2001: Global glacial isostatic adjustment and modern instrumental records of relative sea level history. In: Sea Level Rise: History and Consequences [Douglas, B.C., M.S. Kearney, and S.P. Leatherman (eds.)]. Academic Press, San Diego, pp. 65–95.

Pfeffer Wt, Harper Jt; O'Neel S, 2008: "Kinematic constraints on glacier contributions to 21st-century sea-level rise". Science (New York, N.Y.) **321** (5894): 1340–3. doi:10.1126/science.1159099 DOI:dx.doi.org . ISSN 0036-8075. PMID 18772435.

Pike D.A, Stiner J.C, 2007: Sea turtle species vary in their susceptibility to tropical cyclones. Oecologia 153:471–478

Pratiwo 2004: The City Planning of Semarang 1900-1970. The 1st International Urban Conference, Surabaya, 23rd-25th 2004

Prusty G, Dash S, Singh M.P, 2007: Spatio-temporal analysis of multi-date IRS imageries for turtle habitat dynamics characterisation at Gahirmatha coast, India. Int J Remote Sens 28:871–883

Ross J.P, 2005: Hurricane effects on nesting Caretta caretta. Mar Turtle Newsl 108:13–14

Saenger P, 2002: Mangrove Ecology, Silviculture and Conservation. Kluwer, Dordrecht.

Sanlaville P. and A. Prieur, 2005: Asia, Middle East, coastal ecology and geomorphology. Encyclopedia of Coastal Science, M.L. Schwartz, Ed., Springer, Dordrecht, 71–83.

Twilley R.R, 1995: Properties of mangrove ecosystems related to the energy signature of coastal environments. In: Hall C.A.S. (Ed.), Maximum Power: The Ideas and Applications of H.T. Odum. University of Colorado Press, Boulder, pp. 43–62.

Tokioka T, Noda A, Kitoh A, Nikaidou Y, Nakagawa S, Motoi T, Yukimoto S, and Takata K, 1995: A transient CO_2 experiment with the MRI CGCM - quick report. J. Meteor. Soc. Japan, 73, 817–826.

UNDP. 2004: Human Development Report: 2004. United Nations Development Program (UNDP), Oxford University Press, New York.

Unnikrishnan A.S and Shankar D, 2007: Area Sea Levels trends along the North Indian Ocean Coasts consistent with global estimates? Global and Planetary Change

Van Houton K.S, Bass O.L, 2007: Stormy oceans are associated with declines in sea turtle hatching. Curr Biol 17:R590

Webster P.J, Holland G.J, Curry J.A, Chang H.R, 2005: Changes in tropical cyclone number, duration, and intensity in a warming environment. Science 309:1844–1846

Wilkie M.L, Fortuna S, 2003: Status and trends in mangrove area extent worldwide. Forest Resources Assessment Working Paper 63. Forest Resources Division, FAO, Rome. http://www.fao.org/docrep/007/j1533e/J1533E00.htm.

Woodroffe C.D, Nicholls R.J, Saito Y, Chen Z and Goodbred S.L, 2006: Landscape variability and the response of Asian megadeltas to environmental change. Global Change and Integrated Coastal Management: The Asia-Pacific Region, Harvey N, Ed., Springer, 277–314.

Chapter 16
Impact on Biodiversity: Asian Scenario

Biodiversity is the degree of variation of life forms in an ecosystem, biome, or entire planet. With rapid civilisation the diverse species was hunted and over exploited for food as well as recreation. Civilisation is also cause for clearing diverse fauna in many countries. Rapid environmental degradation has been the cause for extinction of many species. Natural barriers such as large rivers, seas, mountains and deserts have contributed to diversity by enabling independent evolution on both side of the barrier. On the other hand artificial barriers like international borders, roads, settlement and farm land, military activities has contributed to fragmentation of ecosystem and destruction of many species.

The conservation of natural ecosystems and biodiversity are critical to fulfilling poverty alleviation and sustainable development. Biodiversity is the foundation of agriculture, health, food, industry, forests, and fisheries, soil conservation and water quality. Biological resources provide the raw materials for livelihoods, industries, medicines, trade, tourism and food. Genetic diversity is the basis for improved crops, improved agricultural production, and food security.

The effects of climate change on forests comprise both positive (e.g. increases in forest vitality and growth from CO_2 fertilization, and enhanced growing seasons) and negative effects (e.g. increase in insects and pathogens) (Ayres and Lombardero 2000; Bachelet et al. 2003; Lucht et al. 2006; Scholze et al. 2006; Lloyd and Bunn 2007). Climate change will also alter many soil processes, affecting the entire ecosystem. Increase in temperature trigger stomata opening enhancing the sensitivity of plants to Air Pollutants (Andrzej et al. 2007). The Millennium Ecosystem Assessment (MEA) identified climate change as one of the major reasons for having adverse affects on biodiversity (MEA 2005).

Based on climatic features Asia can be divided into four sub-regions: boreal; arid and semi arid; temperate; and tropical. Boreal forest also known as Taiga is a biome characterised by coniferous forests. Except boreal forest other forest has been degraded by anthropogenic activities.

Temperate forests in Asia are important because of their high degree of endemism and biological diversity. Tropical moist forests are important source of wood in many regions.

Biodiversity all over the world depends on soil and water. Soil organic carbon is an critical component of the agro ecosystem due to its role in the dynamics of GHGs

R. Chandrappa et al., *Coping with Climate Change*,
DOI 10.1007/978-3-642-19674-4_16, © Springer-Verlag Berlin Heidelberg 2011

(Kirschbaum 2000). Soils comprise of the largest proportion of the terrestrial reserve of carbon (Batjes and Sombroek 1997). Soil up to one meter depth under tropical forests, temperate forests and croplands hold, 216, 100 and 128 Gt of global carbon Stocks respectively (George et al. 1985). Soil Organic Carbon (SOC) in the Indian Himalayan mountain region has depleted due to changes in climate and land use in past few decades leading to a declining trend in productivity (Martin et al. 2010).

Freshwater aquatic ecosystem in Asia has high flora and fauna diversity. With projected change in temperature and precipitation, water quality will get affected leading to eutrophication.

Twenty-five 'biodiversity hotspots' in the world were identified as areas of high concentrations of endemic species and experiencing immense habitat loss (Myers et al. 2000). Southeast Asia has with four of these hotspots with rich and unique biota (Mittermeier et al. 1999).

16.1 Changing Species Demography

Climate change is forecasted to drive species ranges toward the poles. Species-specific responses vary. Climate change may strongly influence distribution and abundance of fishes (Wood and McDonald 1997), through changes in growth, survival, reproduction, or responses to changes at other trophic levels (Beaugrand et al. 2002, 2003).

Rise in the frequency, extent, and/or severity of drought and heat stress connected with climate change could fundamentally change the composition, structure, and biogeography of forests in many areas (Craig et al. 2010). Migratory species act as indicators of the interdependence between ecosystems and ecological health. Shifts in timing of important life cycle events, and range, are ways that birds and their ecological communities are displaying response to climate change. Such shifts enhance threats and risks to the community (Root and Hughes 2005).

Several studies have reveled that demographic parameters of seabirds were associated with either El Nino-Southern Oscillation (Barbraud and Weimerskirch 2003) or the North Atlantic Oscillation (Sandvik et al. 2005), or more local variance in sea ice extent/concentration (Jenouvrier et al. 2005) or sea surface temperature (Frederiksen et al. 2004; Nevoux et al. 2007). Climate change and air pollution affect forests by changes in soil processes, tree growth, species composition and distribution, increased plant susceptibility to stressors, increased fuel built-up and fire danger, water resources, etc. (Andrzej et al. 2007). Climate change will have significant impact on migratory species. These species are more susceptible to climate impact as they depend on multiple habitats for breeding and feeding. Further they are exposed to climatic variations during their migration. Changing water temperature also affects breeding time of some of the species.

Temperature change also has an effect on the gender of some species. Warmer water temperatures lead to greater female populations in certain turtles and fish.

Temperature change also lead to diseases such as increase in tumors in Green Sea Turtles. Invasive species are moving into areas where they have never been before, disturbing the balance of local ecosystems (UNEP/CMS 2006).

The shifts in life cycle events fall out of step with other living organisms they interact with. Long-distance migratory birds face greater risks due to climate change than resident birds. In several cases, shifts in range and habitats are not possible due to fragmented landscapes.

Some birds feed primarily on flying arthropods whose abundance is directly related to air temperature (Roeder 1953). As a result, the timing of lying by these birds is closely correlated with the abundance of arthropods (Winkler and Allen 1996). Altered temperature, precipitation and moisture, directly affect birds (WWF 2006). Shifts in timing of important life cycle events, and shifts in ranges, are strong response to climate change displayed by birds to climate change (Root and Hughes 2005).

Considerable evidence now shows that bird species are shifting pole-ward, or to higher elevation in tropical mountains (Parmesan and Yohe 2003)

Rising sea temperatures caused by climate change may also contribute to jellyfish blooms, as jellyfish. Climate influences has been cause for increase in jellyfish populations in the North Sea (Lynam et al. 2004). Vast numbers of jellyfish were harvested in Southeast Asia for the global market (Omori and Nakano 2001).

Uye and Kasuya (1999) suggest that numbers of indigenous ctenophores may be rising in some Japanese coastal waters.

Studies on plankton have been systematically recorded as they are indicators of climate change for following reasons: (1) few species of plankton are commercially exploited; hence, any long-term changes can be attributed to climate change; (2) most species are short lived and hence population size is less influenced by the existence of individuals from previous years; (3) plankton can show changes in distribution quickly as they are free floating and can respond easily to climate change; and (4) nonlinear responses of plankton can amplify slight environmental disturbances (Taylor et al. 2002; Graeme et al. 2005).

Antarctic krill, which are diet of whales and other marine mammals, have declined in the past 25 years (Atkinson et al. 2004). This decline is due to reduced plankton which is food for krill in the form of phytoplankton in summer and ice algae in winter (Graeme et al. 2005).

The change in species demography is also influenced by migration of human population due to climate change. The migration and increased activity in new location by immigrated humans will either extinguish the local habitat or force to migrate to new habitat. Managing existing protected areas is highlighted as Adaptive capacity of migratory species and the ecosystems they depend on may be limited (MEA 2005).

Removal of vegetation cover increases the rate of permafrost melt in cold regions. Thus ecosystems in cold area, which are under stress and constant degradation, will expose ground contributing to increase in climate and in turn degradation of ecosystem.

16.2   Vanishing Species

Climate change is one of the most documented causes of reported population declines in marine predators (Hughes 2000; Croxall et al. 2002; Weimerskirch et al. 2003). Even moderate climate change will exceed the ability of several plants and animals to migrate or adapt (Both et al. 2006). Studies form Thomas et al. (2004) indicates that Climate change will put large numbers bird species at risk of extinction with extinction rates varying from 2 to 72% depending on the region, climate circumstances and prospective for birds to shift to new habitat. Long-distance migratory birds face elevated climate change risk (WWF 2006).

Biodiversity in mountain areas where migration of species is physically restricted will be threatened due to rapid changes in climate resulting in rapid losses of habitat and genetic diversity from the mountain ecosystem. Extinction is the most severe of all climate impact for biodiversity. Climate change has caused the extinction of 70 harlequin frog species (Pounds et al. 2006). Study made by Allison et al. (2005) demonstrate that climate change is having noticeable impacts on marine fish distributions, and noticed rates of boundary movement with rise in temperature indicate that in future distribution shifts could be obvious.

Species of fresh-water ecosystems, with a low resistance to high temperature which include fish in the salmon family, Daphnia and other large Cladocera, Mysidacea, and Gammaridae may suffer heat damage (Government of Japan 1997)

Carbon dioxide emitted by anthropogenic activities will be absorbed by rain precipitated over oceans and by ocean water making it acidic. This directly impacts marine animals like corals and molluscs which have calcareous shells or skeleton. Ocean acidification weakens coral skeletons, reduce growth rates of corals.

Staghorn corals are hard or 'stony' corals belonging to the genus *Acropora* generally located between 25°N and 25°S. In Asia pacific region they are located abundantly in 'Coral Triangle' region of the Solomon Islands, Papua New Guinea, Indonesia, East Timor, Philippines and Malaysia. Staghorn corals have a symbiotic relationship with photosynthetic algae called zooxanthellae. Climate change could result in disruption of such symbiotic relations threatening both the species.

Corals and zooxanthellae usually live only 1–2°C below their upper temperature tolerance. Seawater temperature rise leads to faster photosynthesis by zooxanthellae and increases the amount of oxygen which can increase to toxic levels within the corals' tissues. To stay alive, corals expel most, of the algae from their tissues, thus losing their source of energy. Corals appear white or 'bleached' due to loss of pigmented algae making white calcareous skeleton visible through the transparent coral tissues.

Several factors are influential in the choice of an optimal nesting site for turtles. These include low salinity, high humidity and well ventilated area with near shore oceanography conducive to dispersal of hatchlings into oceanic currents (Miller 1997; Foley et al. 2006). Turtles require sufficient space above the high tide line for nesting. Some species require adequate beach vegetation for clutch shading (Naro-Maciel et al. 1999; van de Merwe et al. 2005; Kamel and Mrosovsky 2006).

The seven existing species of marine turtle have survived paleo-climatic regimes (Hamann et al. 2007). It is likely that many current nesting beaches, and migratory routes are completely different 10,000 yr ago (FitzSimmons et al. 1999; Hamann et al. 2007).

Green turtles are largely herbivorous and are significant regulators of sea grass productivity and biomass in coastal marine habitats (Thayer et al. 1984; Williams 1988; Moran and Bjorndal 2005, 2007; Kuiper-Linley et al. 2007). Changes in sea surface temperature, sediment disturbance, altered penetration of ultra violet light, eutrophication and acidification of coastal waters (Sabine et al. 2004; Hall-Spencer et al. 2008), cause changes in the distribution and types of macro-algal species present in coastal habitats (Lapointe 1999; Bjork et al. 2008), favouring sea grass dominated communities (Harley et al. 2006; Hall-Spencer et al. 2008). Such an effect could be beneficial for green turtles (Chaloupka and Limpus 2001; Balazs and Chaloupka 2004; Broderick et al. 2006; Chaloupka et al. 2008). But sea grasses themselves could ultimately be negatively affected by increased temperatures, and other stress factors altering growth rates, physiology and distribution (Short and Neckles 1999; Bjork et al. 2008; Ehlers et al. 2008).

Nesting sites of The Leatherback Turtle (*Dermochelys coriacea*), largest of all the living turtles are found in many countries, including those in the Americas, Africa, Asia and Australasia. They are found in great numbers where nutrient-laden water moves upwards from lower depths. Sometimes these sites are thousands of kilometres away from the turtles' nesting sites. Gender of this species is determined by embryonic temperature and hence increase in global temperature will result in increase in the number of females relative to males affecting the stability of their populations. An increase in storm frequency is likely to result in beach erosion and degradation, washing away turtle nests and declining nesting habitat. Juveniles use ocean currents to aid dispersal after hatching and adults use them for long-distance migration. Changes to oceanic currents are will affect the quantity and distribution of jellyfish and other prey species.

A study by Baker et al. (2006) predicted that up to 40% of green turtle *Chelonia mydas* nesting beaches could be flooded with 0.9 m of sea level rise, while studies by Fish et al. (2005, 2008) suggested similar losses of hawksbill turtle *Eretmochelys imbricata* nesting habitat (means 50 and 51% decrease, respectively).

Marine species are affected by the growth of coastal populations, unsustainable fishing methods, urbanisation, climate change, offshore oil extraction and mining, industrial pollution, and tourism. Eighty-eight percent of the reefs of South East Asia were at risk from human damage in 2002. About 16% of the world's corals were damaged by temperature-induced coral bleaching and El Niño. Eight fish species are listed in the most threatened category of the Convention on International Trade in Endangered Species of Wild Fauna and Flora (CITES). Among these species are sturgeon (*Acipenser* spp.) and the regionally important Caribbean queen conch (*Strombus gigas*). Overfishing and poor land use practices have led to freshwater fishes being the most highly threatened group of animals, with 20% extinct, threatened, or vulnerable (World Bank 2004). Global warming may result in drop in quantity of cold-water seaweed which may lead to a drop in populations of abalone, turbos, sea urchins, and other sessile creatures that feed on this type of seaweed (Government of Japan 1997).

Some of the important species that are threatened due to climate change are discussed below,

Seals: These are mammals that have evolved from carnivorous land-mammal and adapted to living and feeding in water. They remain more land-bound than whales by coming ashore to produce their pups. Seals are members of the family Phocidae. They live in the oceans of both hemispheres but are mostly confined to polar, sub-polar, and temperate climates. Ringed Seal breeding is depends on availability of sufficient ice, and food nearby. Warmer ocean temperatures will be favorable for Seal parasites and pathogens. Rains or warm temperatures can cause the roofs of lairs to collapse leaving Ringed Seals exposed to predators.

Clown fish: Clownfish has 28 species and are found in tropical and subtropical areas of the Pacific and Indian Oceans. Climate change is affecting clownfish due to climate impact on coral reef habitat and water temperature. All fish are 'cold-blooded' or ectothermic, and life-history is highly influenced by the immediate water temperature. Warmer ocean temperatures will cause eggs to perish.

Arctic fox: The Arctic Fox has a distribution, occurring in Alaska, Canada, Greenland, Iceland, Russia and Scandinavia. Arctic Foxes prey mainly on lemmings and voles. Those at coastal locations feed more on seabirds, Ringed Seal pups, Arctic Hare, fish and carrion. Decline in prey species are likely to have impact on Arctic Fox populations.

Salmon fish: Pacific salmon live in coastal and river waters of Alaska, Russia, Japan and Mexico. Atlantic salmon inhabit in the North Atlantic Ocean and Baltic Sea, including associated rivers in USA, Canada, Iceland, Norway, Finland, Sweden and the United Kingdom. Salmon lays eggs in small pits (called 'redds') that are excavated in gravel-based freshwater streams. These nesting sites are selected based on temperature, currents and oxygen levels. Salmon eggs hatch after about 3 months, and juveniles begin their downstream migration. Young salmon could stay in fresh water up to 4 years, before entering the ocean which coincides with planktonic blooms, upon which the juveniles feed. Rise in water temperatures affects salmons by physiological stress, depletion of energy reserves, increased susceptibility to diseases; and disruptions to breeding efforts.

Beluga whales: Beluga whales live in Arctic and sub-Arctic waters. Extensive ice cover and winter conditions has limited human activities in the Arctic till date safe guarding these animals and invasion of other species. The reduced sea ice play due to global warming will increase human Activities affecting Belugas.

The climate triggered by human activity has been problematic to many species discussed is previous paragraphs. But current knowledge about climate change and its consequences is still incomplete. While countries are already working on humans settlements there still lot of gaps in knowledge with respect to loss of species during climate change and disasters. Since the disaster evaluation prioritizes loss of life followed by people affected and economic damages, it often misses species census in elaborate way after disaster.

References

Allison L. Perry, Paula J. Low, Jim R. Ellis, John D. Reynolds, 2005: Science, Vol 308, PP 1912–1915

Andrzej Bytnerowicz, Kenji Omasa, Elena Paoletti, 2007: Integrated effects of air pollution and climate change on forests: A northern hemisphere perspective, Environmental Pollution 147, 438–445

Atkinson Angus, Volker Siegel, Evgeny Pakhomov and Peter Rothery, 2004: Long-term decline in krill stock and increase in salps within the Southern Ocean. Nature 432, 100–103

Ayres M.P, Lombardero, M.J, 2000: Assessing the consequences of global change for forest disturbances for herbivores and pathogens. The Total Science of the Environment 262, 263–286.

Baker J.D, Littnan C.L, Johnston D.W, 2006: Potential effects of sea level rise on the terrestrial habitats of endangered and endemic megafauna in the Northwestern Hawaiian Islands. Endang Species Res 2:21–30

Bachelet D, Neilson R.P, Hickler T, Drapek R.J, Lenihan J.M., Sykes M.T, Smith, B, Sitch S, Thonicke K, 2003: Simulating past and future dynamics of natural ecosystems in the United States. Global Biogeochemistry Cycles 17, 1045 doi:10.1029/2001GB001508.

Balazs G.H, Chaloupka M, 2004: Thirty-year recovery trend in the once depleted Hawaiian green sea turtle stock. Biol Conserv 117:491–498

Barbraud C, Weimerskirch H, 2003: Climate and density-shape populations dynamics of a marine top predator. Proceedings of the Royal Society B 270, 2111–2116.

Batjes N.H, Sombroek, W.G, 1997: Possibilities for carbon sequestration in tropical and subtropical soils. Global Climate Change Biology 3, 161–173.

Beaugrand G, Brander K.M, Lindley J.A, Souissi S, Reid P.C, 2003: Nature 426, 661.

Beaugrand G, Reid P.C, Ibanez F, Lindley J.A, Edwards M, 2002: Science 296, 1692.

Bjork M, Short F, McLeod E, Beers S, 2008: Managing seagrasses for resilience to climate change. IUCN, Gland

Both C, Bouwhuis S, Lessells C.M. & Visser M.W, 2006: Climate change and population declines in a long-distance migratory bird. Nature 441: 81

Broderick A.C, Frauenstein R, Glen F, Hays G.C and others, 2006: Are green turtles globally endangered? Glob Ecol Biogeogr 15:21–26

Chaloupka M.Y, Kamezaki N, Limpus C.J, 2008: Is climate change affecting the population dynamics of the endangered Pacific loggerhead sea turtle? J Exp Mar Biol Ecol 356:136–143

Chaloupka M.Y, Limpus C.J, 2001: Trends in the abundance of sea turtles resident in Southern Great Barrier Reef waters. Biol Conserv 102:235–249

Christopher P Lynam, Stephen J Hay, Andrew S Brierley, 2004: Interannual variability in abundance of North Sea jellyfish and links to the North Atlantic Oscillation. *Limnology and Oceanography* 49(3), 637–643

Craig D. Allen, Alison K. Macalady, Haroun Chenchouni, Dominique Bachelet, Nate McDowell, Michel Vennetier, Thomas Kitzberger, Andreas Rigling, David D. Breshears, E.H. (Ted) Hogg, Patrick Gonzalez, Rod Fensham, Zhen Zhang, Jorge Castro, Natalia Demidova, Jong-Hwan Lim, Gillian Allard, Steven W. Running, Akkin Semerci, Neil Cobb, 2010: A global overview of drought and heat-induced tree mortality reveals emerging climate change risks for forests, Forest Ecology and Management 259, 660–684

Croxall J.P, Trathan P.N, Murphy E.J, 2002: Environmental change and Antarctic seabird populations. Science 297, 1510–1514.

Ehlers A, Worm B, Reusch T.B.H, 2008: Importance of genetic diversity in eelgrass *Zostera marina* for its resilience to global warming. Mar Ecol Prog Ser 355:1–7

Fish M.R, Cote I.M, Gill J.A, Jones A.P, Renshoff S, Watkinson A.R, 2005: Predicting the impact of sea-level rise on Caribbean sea turtle nesting habitat. Conserv Biol 19: 482–491

Fish M.R, Cote I.M, Horrocks J.A, Mulligan B, Watkinson A.R, Jones A.P, 2008: Construction setback regulations and sea level rise: mitigating sea turtle nesting beach loss. Ocean Coast Manag 51:330–341

FitzSimmons N, Moritz C, Bowen B.W, 1999: Population identification. In: Eckert K.L, Bjorndal K.A, Abreu-Grobois F.A, Donnelly M (eds) Research and management techniques for the conservation of sea turtles. IUCN/SSC Marine Turtle Specialist Group Publication No. 4, IUCN, Gland, p 72–79

Foley A.M, Peck S.A, Harman G.R, 2006: Effects of sand characteristics and inundation on the hatching success of loggerhead sea turtle (*Caretta caretta*) clutches on low-relief mangrove islands in southwest Florida. Chelonian Conserv Biol 5:32–41

Frederiksen M, Wanless S, Harris M.P, Rothery P, Wilson L.J, 2004: The role of industrial fisheries and oceanographic change in the decline of North Sea blacklegged kittiwakes. Journal of Applied Ecology 41, 1129–1139.

George R, Francis A.P, Lino Grima, Henry A.R, Thomas H.W, 1985: A prospectus for the management of the Long Point ecosystem. Great Lakes Fishery Commission, Ann Arbor, Technical Report No. 43 (WGBR).

Government of Japan, 1997: Japan's Second National Communication under the United Nations Framework Convention on Climate Change

Graeme C. Hays, Anthony J. Richardson and Carol Robinson, 2005: Climate change and marine plankton. Trends in Ecology and Evolution Vol.20 No.6, 337–344

Hall-Spencer J.M, Rodolfo-Metalpa R, Martin S, Ransome E and others, 2008: Volcanic carbon dioxide vents show ecosystem effects of ocean acidification. Nature 454:96–99

Hamann M, Limpus C.J, Read M.A, 2007: Chapter 15 Vulnerability of marine reptiles in the Great Barrier Reef to climate change. In: Johnson JE, Marshall PA (eds) Climate change and the Great Barrier Reef: a vulnerability assessment, Great Barrier Reef Marine Park Authority and Australia Greenhouse Office, Hobart, p 465–496

Harley C.D.G, Hughes A.R, Hultgren K.M, Miner B.G, Sorte C.J, Thornber C.S, Rodriguez L.F, Tomanek L, Williams S.L, 2006: The impacts of climate change in coastal marine systems. Ecol Lett 9:228–241

Hughes L, 2000: Biological consequences of global warming: is the signal already. Trends in Ecology and Evolution 15, 56–61.

Jenouvrier S, Barbraud C, Cazelles B, Weimerskirch H, 2005: Modelling population dynamics of seabirds: importance of the effects of climate fluctuations on breeding proportions. OIKOS 108, 511–522.

Kamel S.J, Mrosovsky N, 2006: Deforestation: Risk of sex ratio distortion in hawksbill sea turtles. Ecol Appl 16:923–931

Kuiper-Linley M, Johnson C.R, Lanyon J.M, 2007: Effects of stimulated green turtle regrazing on seagrass abundance, growth and nutritional status in Moreton bay, south-east Queensland, Australia. Mar Freshw Res 58:492–503

Kirschbaum M.U.F., 2000: Will changes in soil organic carbon act as a positive or negative feedback on global warming? Biogeochemistry 48, 21–51.

Lapointe B.E, 1999: Simultaneous top-down and bottom-up forces control macroalgal blooms on coral reefs. Limnol Oceanogr 44:1586–1592

Lloyd A.H, Bunn A.G, 2007: Responses of the circumpolar boreal forest to 20th century climate variability. Environmental Research Letters 2, 045013, doi:10;1088/1748-9326/2/4/045013.

Lucht W, Schaphoff S, Erbrecht T, Heyder U, Cramer W, 2006: Terrestrial vegetation redistribution and carbon balance under climate change. Carbon Balance and Management 1, 6 doi:10.1186/1750-0680-1-6.

Myers Norman, Russell A. Mittermeier, Cristina G. Mittermeier, Gustavo A. B. da Fonseca & Jennifer Kent, 2000: Biodiversity hotspots for conservation priorities. Nature 403, 853–858

Martin D, Tarsem Lal, Sachdev C.B, Sharma J.P, 2010: Soil organic carbon storage changes with climate change, landform and land use conditions in Garhwal hills of the Indian Himalayan mountains, Agriculture, Ecosystems and Environment 138, 64–73

MEA (Millennium Ecosystem Assessment), 2005: Ecosystems and human wellbeing: Synthesis. Washington DC: Island Press

Miller J.D, 1997: Reproduction in sea turtles. In: Lutz PL, Musick JA (eds) The biology of sea turtles, Vol 1. CRC Press, Boca Raton, FL, p 51–81

Mittermeier A Russell, Norman Myers, and Cristina Goettsch Mittermeier, 1999: Hotspots: Earth's Biologically Richest and Most Endangered Terrestrial Ecoregions, Cemex, Conservation International

Moran K.L, Bjorndal K.A, 2005: Simulated green turtle grazing affects structure and productivity of seagrass pastures. Mar Ecol Prog Ser 305:235–247

Moran K.L, Bjorndal K.A, 2007: Simulated green turtle grazing affects nutrient composition of the seagrass *Thalassia testudinum*. Mar Biol 150:1083–1092

Naro-Maciel E, Mrosovsky N, Marcovaldi MA, 1999: Thermal profiles of sea turtle hatcheries and nesting areas at Praia do Forte, Brazil. Chelonian Conserv Biol 3:407–413

Nevoux M, Weimerskirch H, Barbraud C, 2007: Environmental variation and experience-related differences in the demography of the long-lived blackbrowed albatross. Journal of Animal Ecology 76, 159–167.

Omori M and Nakano E, 2001: Jellyfish fishery in southeast Asia Hydrobiologia 451 (Dev. Hydrobiol. 155): 19–26.

Parmesan C & Yohe G, 2003: A globally coherent fingerprint of climate change impacts across natural systems. Nature 421: 37.

Pounds A.J, Bustamante M.R, Coloma L.A, Consuegra J.A, Fogden M.P.L, Foster P.N, La Marca E, Masters K.L, Merino-Viteri A, Puschendorf R, Ron S.R, Sánchez-Azofeifa G.A, Still C.J & Young B.E, 2006: Widespread amphibian extinctions from epidemic disease driven by global warming. Nature 439: 161.

Roeder K.D. 1953: Insect physiology. New York: Wiley.

Root T & Hughes L, 2005: Present and future phenological changes in wild plants and animals. In: Lovejoy T.E. and Hannah. L. (Eds.) Climate Change and Biodiversity, Yale University Press, New Haven & London. pp. 61.

Sabine L Christopher, Richard A. Feely, Nicolas Gruber, Robert M. Key, Kitack Lee, John L. Bullister, Rik Wanninkhof, C. S. Wong, Douglas W. R. Wallace, Bronte Tilbrook, Frank J. Millero, Tsung-Hung Peng, Alexander Kozyr, Tsueno Ono and Aida F. Rios, 2004: The oceanic sink for anthropogenic CO_2. Science 305:367–371

Sandvik H, Erikstad K.E., Barbett R.T, Yoccoz N.G, 2005: The effect of climate on adult survival in five species of North Atlantic seabirds. Journal of Animal Ecology 74, 817–831.

Scholze M, Knorr W, Arnell N.W, Prentice I, 2006: A climate-change risk analysis for world ecosystems. Proceedings of the National Academy of Sciences of the United States of America 103, 13116–13120.

Short F.T, Neckles H.A, 1999: The effects of global climate change on seagrasses. Aquat Bot 63:169–196

Taylor H Arnold, Icarus Allen J & Paul A Clark 2002: Extraction of a weak climatic signal by an ecosystem. Nature 416, 629–632

Thayer G.W, Bjorndal K.A, Ogden J.C, Williams S.L, Zieman J.C, 1984: Role of larger herbivores in sea grass communities. Estuaries 7:351–376

Thomas C.D, Cameron A, Green R.E, Bakkenes M, Beaumont L.J, Collingham Y.C, Erasmus B.F.N, De Siquiera M.F, Grainger A, Hannah L, Hughes L, Huntley B, Van Jaarsveld A.S, Midgley G.F, Miles L, Ortega-Huerta M.A, Peterson A.T, Phillips O. & Williams S.E, 2004: Extinction risk from climate change. Nature 427: 145.

UNEP/CMS, 2006: Migratory Species and Climate Change: Impacts of a Changing Environment on Wild Animals Secretariat, Bonn, Germany. 68 pp

Uye S & Kasuya T, 1999: Functional roles of ctenophores in the marine coastal ecosystem. In Okutani, T., S. Ohta & R. Ueshima (eds) Update Progress in Aquatic Invertebrate Zoology. Tokai University Press, Tokyo: 57–76 (in Japanese with English Abstract).

van de Merwe J, Ibrahim K, Whittier J, 2005: Effects of hatchery shading and nest depth on the development and quality of *Chelonia mydas* hatchlings: implications for hatchery management in Peninsular, Malaysia. Aust J Zool 53: 205–211

Williams SL, 1988: *Thalassia testudinum* productivity and grazing by green turtles in a highly disturbed seagrass bed. Mar Biol 98:447–455

Weimerskirch H, Inchausti P, Guinet C, Barbraud C, 2003: Trends in bird and seal populations as indicators of a system shift in the Southern Ocean. Antarctic Science 15, 249–256.

Winkler D.W and Allen P.E, 1996: The seasonal decline in tree swallow clutch size: physiological constraint or strategic adjustment? Ecology 77, 922–932.

Wood C. M, McDonald D.G, Eds., 1997: Global Warming: Implications for Freshwater and Marine Fish, Cambridge Univ. Press, Cambridge.

World Bank, 2004: Saving Fish and Fishers toward Sustainable and Equitable Governance of the Global Fishing Sector

WWF (World Wild Fund), 2006: Bird Species and Climate Change. PP 12

Chapter 17
Changing Livelihood Due to Climate Change: Asian Perspective

Livelihoods vary in different social, ecological, geographical and institutional settings. The definition of 'livelihood' is widely discussed.

The most widely accepted definition of livelihood was defined by Chambers and Conway (1992):

> *a livelihood comprises the capabilities, assets (including both material and social resources) and activities required for a means of living.*

According to Ellis (2000) livelihood is

> *the activities, the assets, and the access that jointly determine the living gained by an individual or household.*

Livelihood deals with people, their resources and social relationships. While gaining a livelihood, people may have to cope with risks and uncertainties, such as climate change, resource depletion, and pressure on the land, changing life style, epidemics, market risk, increasing food prices, inflation, and competition.

Changing climate will affect livelihood positively as well as negatively. The climate impact will be experienced intensely by poor and vary from region to region and house to house within region. The extent to which a society is vulnerable to climate change depends on: (1) the ways and speed of climate changes; (2) the degree to which the wellbeing of people depends on climate sensitive systems; (3) and the capacity of the society to adapt to climate-induced changes (Jon et al. 2003).

It would be very difficult to assess vulnerability to climate change of community based on per capita income as some people are paid by food/shelter/clothing by employer. Hence the income level is segregated based on nutritional/clothing/shelter affordability and summarised in Table 17.1.

As per Holling (2001), adaptive cycle, links different time and spatial frameworks. He identifies three core characteristics that shape the cycle: (1) inherent potential; (2) internal controllability; and (3) adaptive capacity.

The inherent potential of a system defines the range of possible options for the future. As could be seen in Fig. 17.1 the internal controllability of the system reflects the degree of connectedness among internal controlling variables and processes, along with the degree of rigidity or flexibility of these controls. Box 17.1 gives principles governing livelihood change due to climate change.

R. Chandrappa et al., *Coping with Climate Change*,
DOI 10.1007/978-3-642-19674-4_17, © Springer-Verlag Berlin Heidelberg 2011

Table 17.1 Income level and climate impact

Income level	Example	Climate impact vulnerability	Carbon footprint T of CO_2 equivalent per year
Ultra low: Income less than cost of nutritional requirement of family	Rag pickers, street vendors, beggars, agricultural labours	Very high: Loss of income likely to push them towards disease and death	Less than 0.1
Very low: Income equal to cost of nutritional requirement family	Agricultural labours in high/medium yielding area, micro entrepreneurs	Very high: Loss of income likely to push them towards disease and death	0.1–0.2
Low: Income exceeds nutritional requirement and can afford minimum clothing requirement	Agricultural labours in high yielding area, house hold servants, labours in unorganised sectors, micro entrepreneurs	Very high: Loss of income likely to push them towards disease and death. Likely to exhaust micro savings and approach for loan	0.2–1
Lower medium: Income exceeds nutritional requirement and can afford minimum clothing requirement and medicine	Agricultural labours in high yielding area, house hold servants, servants in unorganised sectors, micro entrepreneurs	Very high: Loss of income likely to push them towards disease and death. Likely to exhaust micro savings and approach for loan	1–2
Upper medium: Income exceeds nutritional requirement and afford clothing, medicine, afford low cost housing and public transportation. This group will have low savings and disposable income	Agricultural labours, farmer will very low land ownership, fishing community, servants in unorganised sectors, micro entrepreneurs, government servants in lower hierarchy	High: Likely to exhaust micro savings and approach for loan. Rural community may migrate to urban area in search of better opportunity	2–5
High: Can afford housing which can withstand minor natural calamities, can afford private transportation. This group will have sufficient savings and disposable income	People in organised sector	Medium: May adopt climate change up to some extent based on severity of impact exposure	5–25
Very high: Housing which can withstand medium natural calamities, afford private transportation. This group will have high savings and disposable income	People in organised sector, medium income entrepreneurs, medium income farmers	Medium: May adopt climate change up to some period based on severity of impact exposure	25–50

(continued)

Table 17.1 (continued)

Income level	Example	Climate impact vulnerability	Carbon footprint T of CO_2 equivalent per year
Ultra high: Housing which can withstand high natural calamities, Owns house in more than one location/country, private transportation. This group will have high savings disposable income	High income entrepreneurs; high income politicians; high income sports people; high income entertainers; high income movie stars	Low: Can modify living condition to changing climate, and withstand natural calamities. Will get affected only if exposed to major calamities/accidents. They can relocate themselves to other location/country	>100

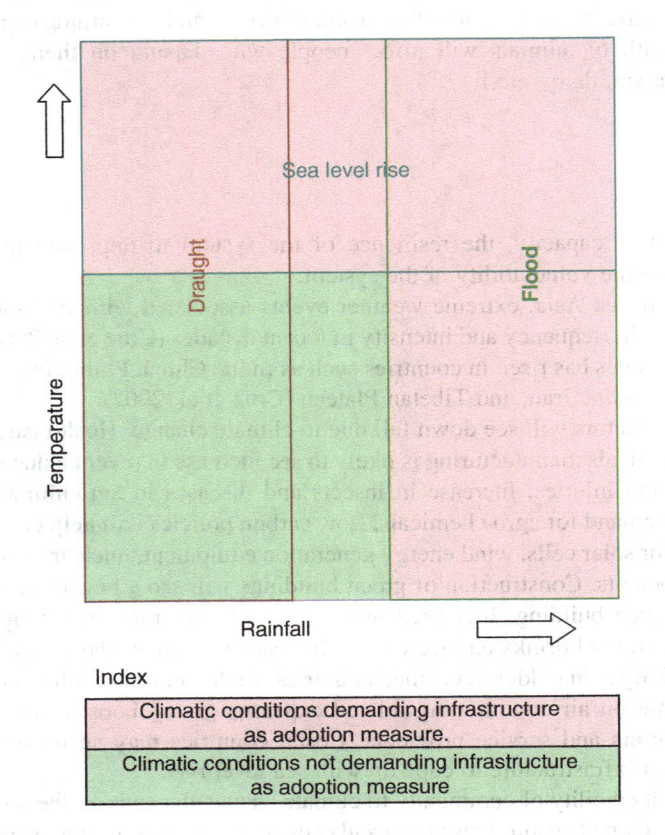

Fig. 17.1 Climate change and infrastructure requirement

> **Box 17.1 Principles governing livelihood change due to climate change**
> Lively hood is being affected by climate change and will affect in feature.
> While there are many reasons for it and some of them are still being invented
> some governing reasons are listed below:
>
> 1. Changing temperature will alter the crop yield affecting people who
>   depend on agriculture
> 2. The soil will get affected due to climate change that will affect crops
> 3. The population of insects will increase and affect on crops
> 4. Rise in sea level will destroy the habitation, crops and habitats of species
>   which thrive on sea shores
> 5. Increase in temperature which leads to lower precipitation and draught
>   will lead to drying of water bodies affecting agriculture and inland
> 6. Climate change and diminishing forest will affect lively hood of people
>   who depend on the forest
> 7. Increase in flood storms and cyclone will affect trade and commerce
> 8. Increase in disease will affect health of people and the earning capacity and
> 9. Health of animals will affect people who depend on them (wild life
>   tourism, dairy, etc.)

The adaptive capacity, the resilience of the system to unpredictable shocks is opposite of the vulnerability of the system.

In Southeast Asia, extreme weather events associated with El Niño have also augmented in frequency and intensity in recent decades (Cruz et al. 2007). Damage due to cyclones has risen in countries such as India, China, Philippines, Japan, Viet Nam, Cambodia, Iran, and Tibetan Plateau (Cruz et al. 2007).

Not all sectors will see down fall due to climate change. Health care sector and pharmaceuticals manufacturing is likely to see increase in revenue due to increased diseases and injuries. Increase in insects and diseases in agricultural corps will increase demand for agro chemicals. Low carbon policies will help creation of new markets for solar cells, wind energy generation equipment, nuclear energy generating equipments. Construction of green buildings will see a boost in raw materials used for such building. Increased warm days will see more consumption of sun cream, beer, cool drinks and ice cream. Increase in warmer days may bring down heating charges in colder developed countries. Rich countries with hot climate will spend more on air conditioning related expenses giving boost to air conditioner manufacturers and service providers. Costal countries may spend more on construction of infrastructure to cope up with sea level rise.

The vulnerability of community to climate change depends on the level to which they depend on (1) natural resources and ecosystem services; (2) the degree to which the resources and services they rely on are sensitive to climate impact; and (3) their capacity to adapt to alteration in these resources and services (Jon and Neil 2007).

Climate change is not just an environmental issue; it is question of life and death for many. Impact on lively hood is different to different people. Affected people, who have suffered losses of their livelihood through natural disaster or conflict, have the right to protect, recover, improve and develop their livelihoods. In the case of Nepal, measures to overcome climate impact include resource conservation in the agricultural sector; efficient water supply management; establishment of a hydrological forecasting system; and extensive plantations (UNFCCC 2005).

Figure 17.2 shows the occupations that are likely to see boom and down fall due to climate change. Table 17.2 shows impact on trade and commerce due to climate change. Not all occupations will get affected. Doctors may get more patients and there could be increase in sale of green building material. Agriculture and fishing will definitely get affected. But adaptation to new technology can sustain agriculture where as fishing will see a downfall if people do not stop/reduce eating fish in near future.

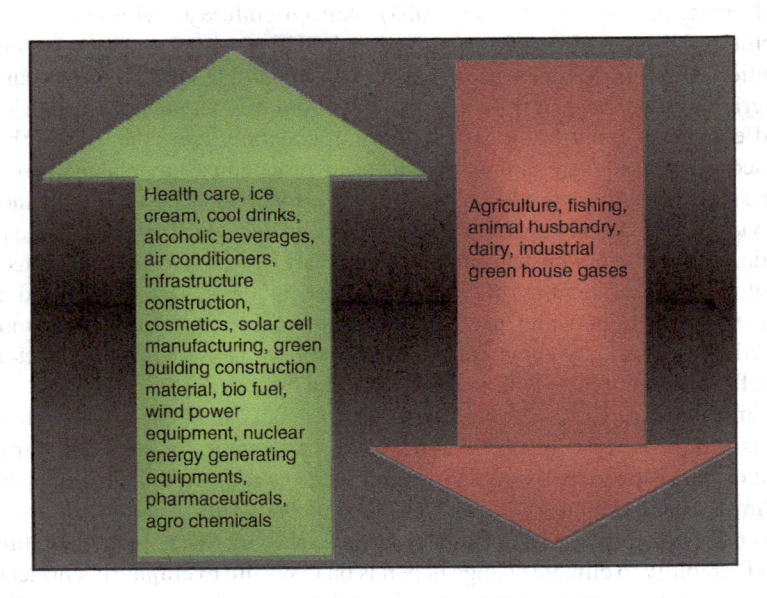

Fig. 17.2 Occupations that are likely see boom and downfall due to climate change

Table 17.2 Impact on trade and commerce due to climate change

Sector	Impact	Adaptation
Trade and commerce	Destruction of infrastructure and industry due to climate induced disasters; loss of raw material due to drop in agricultural production and fresh water availability; drop in economy due to increased because burden on working population	Climate-proof infrastructure, improved early warning systems, improved erosion control through public works, diversify education to impart new skill to citizens

17.1 Difficulties in Agriculture

Asia's fast-growing economies remain home to more than 600 million rural populance living in severe poverty. In spite of enormous rural-urban migration, rural poverty will stay dominant for several more decades (World Bank 2008). Modern agriculture has sharply increased yields of crops, but has caused ecological damage as well. Global agriculture is under significant pressure to meet the demands of rising populations using finite land resource. Intensive animal husbandry has raised meat, paltry and dairy products but has also resulted in cruelty towards animals. Developing countries in Asia are likely to get affected by agricultural failure due to climate change. Table 17.3 shows impact on agriculture due to climate change. Slow agricultural productivity and food security pose challenges to many countries in the Asia.

Agriculture is a form of natural resource management and agricultural products can be classified into foods, fibres, fuels, and raw materials. Ninety-eight percent of the rural inhabitants in South Asia, 96% in East Asia and the Pacific are in transforming nations (World Bank 2008). Non agriculture activities have overtaken agriculture in terms of people employed and GDP in countries with economies in transition. But this does not mean that world can survive without agriculture.

Agriculture operates in three divergent economies – nations having agriculture-based economy, countries with transforming economy and countries with industrialised economy. Agricultural practice and procedures has changed over time increasing agricultural productivity. Use of agrochemicals has greatly increased crop yields. Increase in supply of grains has led to cheaper livestock. In the present decade several factors pushed up the price of grains consumed by humans due to drought in several regions of world; increase in demand for grain-fed animal products; retail boom; and use of food grain to bio-fuel production and trade restrictions imposed by some countries. Agriculture along with other sectors produces faster economic growth and help in reducing poverty.

Climate change is threatening agricultural systems and hence the livelihoods as well as food security of people who depend on agriculture in the Asia. Agriculture is one of the sectors most vulnerable to climate change due to its high dependence on climate. Unlike industrial operation most of the farming is done in open fields and hence both crop as well as animal farming activity depends on the climate.

Vulnerability to climate change depends on exposure to climate events as well as physical, environmental, socioeconomic, and political factors. Low-income populations will feel the effects of climate change more strongly than others. Mountainous areas in Asia experience extreme cold apart from flooding, land degradation,

Table 17.3 Impact on agriculture due to climate change

Sector	Impact	Adaptation
Agriculture	Loss of crop due to floods, droughts, storms, sea water intrusion, rise in temperature; increase in livestock and crop disease; and loss of live stock due to climate induced disasters	Improved early warning systems, construction of dam, better agricultural practice, train agricultural labours and farming community with alternate lively hood skills

soil erosion, and deforestation, which affects crop production. The dams receive increased mud flows in the areas of higher precipitation from erosion resulting in top soil erosion and lowered dam capacity. Extreme cold in some regions is already affecting livestock. South Asia is affected by flood, unreliable rainfall, drought, and rise in sea levels.

Not all crops can grow in all climatic conditions. Dates need hot and dry climate where as apples need cold climate. Rice crops demand ensured water but needs dry weather during harvesting. Wheat will fail if the water stagnates in wheat field. Cotton balls will be spoiled due to rain. Mangos grow well in dry climate and rain will make flowers fall and bring down yield during flowering season. Individual crops demand varying macro and micro nutrients as well as differing caring. Carrot and potato will rot at high water supply. The future of agriculture is tied to better stewardship of natural resources.

Different climatic regions will have different constrains with respect to agriculture.

The major crops in highland terraces are vegetables, millets, tea, medicinal plants, and horticultural crops (like apple, apricot, pineapple, citrus fruits, apple, peer, peach, and banana). Crops like potato, maize, mustard, millets and paddy can be observed in valleys. Mule, sheep, goat and yak major animals reared in the hilly area. The major constraints in hilly area are: low choice of crops; heavy runoff and severe erosion due to steep slope, soil degradation due to shifting agriculture; intense leaching during high runoff, water stagnation in valleys during rainy season. Drying up rivers, melting glaciers, GOLF, landslides are direct effects of climate change which are adding to agricultural constraint in the region.

The major crops in arid region are Groundnut, sunflower, chillies, food grains and cotton. The major constraints in arid region are: high water scarcity, soil salinity, and Nutrient imbalance.

The main crops which are grown in Semi-Arid Region are millets, wheat, pulses, Rice and sugarcane, major constraints are: Over exploitation of groundwater, soil salinity, imperfect drainage, alkalinity, and nutrient imbalance.

The major corps in tropical climate is plantation crops, such as coconut, areca nut, and oil palm. Other crops include rice, pulses, oil seeds, pineapple, tapioca and spices. The major constraints in the region are: water logging, runoff, soil erosion.

Agriculture is a type of natural resource management for the producing food, fuel, and fiber. Crops respond positively to elevated CO_2 in the absence of climate change (Ainsworth and Long 2005; Kimball et al. 2002; Jablonski et al. 2002), the associated effects of high temperatures, changed patterns of precipitation, and varying frequency of drought and floods, will depress yields. These will increase the gap among rich and poor countries (IPCC 2001).

Agricultural labour market and the rural nonfarm economy in Asia and Latin America engage between 45 and 60% of the rural labour force (World Bank 2008). This means Agriculture is major lively hood for enormous population in Asia.

Agriculture is successful in meeting the world's demand for food but still more than 800 million people remain food insecure, and agriculture has left a enormous environmental footprint. Agriculture's environmental footprint can be diminished and farming systems can be made less vulnerable to climate change (World Bank

2008). Developing countries are more vulnerable to climate impact than developed countries, because of the prevalence of agriculture in their economies, the shortage of capital for adaptation measures, their warmer climates, and their increased exposure to extreme events (Parry et al. 2001). Hence climate change is likely to have serious consequences in the developing countries, where about 800 million people are presently undernourished (UN Millennium Project 2005). Agriculture is a source of livelihood for billions of poor people. Importance of agriculture to GDP remains high in South Asia and declined only slightly between 1995 and 2006. Mean temperatures influence crop duration (e.g. Challinor et al. 2005), whereas temperature extremes during flowering can reduce the grain or seed number (Wheeler et al. 2000).

Agriculture contributes to development as an economic activity; as a livelihood; and as a provider of environmental services (World Bank 2008).

The key bio-physical processes may vary with crop species. But both the direct (Long et al. 2006) and the indirect (Mearns et al. 1999) impact of climate change on crops are important in impacts assessments. Melting of glaciers in the Himalayas due to climate change will increase risk of flooding, erosion, mudslides and Glacial Lake Outburst Flood (GLOF) in Nepal, Bangladesh, Pakistan, and north India. Abiotic stress like drought, high temperatures, and flood and wind velocity can severely disturb the crops. Once plants are weakened from abiotic stresses, biotic stresses like pest and diseases can set in (Rosegrant et al. 2007).

Climate change is likely to result in a shift in the boundary of the farming-pastoral transition area to the south in Northeast China, which may augment grassland areas and provide more favourable conditions for livestock (Mats et al. 2009). But, the transition area of the farming-pastoral region is also an area of potential desertification (Li and Zhou 2001; Qiu et al. 2001). Some rangelands may suffer from degradation due to the warmer climate (Dirnbock et al. 2003). Rising temperatures and water stress are expected to result in 30% decrease in crop yields in Central and South Asia by the mid-twenty-first century (UNDP 2006). Augment in agricultural water demand by 6–10% or further is projected for every 1°C rise in temperature (IPCC 2007a). As a result, the net cereal production in South Asia is projected to reduce by at least between 4 and 10% by the end of this century (IPCC 2007a).

More rain is likely to fall in already rain abundant area creating increased risk of flooding while other area will receive less rain fall. Micro level studies shows that Andra Pradesh state in India will be affected by 8–9% decline in yield of water intensive crops due to rise in temperature of 3.5°C by 2050 (World Bank 2006; UNDP 2007).

Falling crop will reduce food for self consumption in farming families and also cut supplies to local market. The competition by wealthy industry like alcoholic beverage manufacturing will further cut down entry of food grain to market for consumption as food. The observed and forecasted increase in alcohol consumption due to increase in temperature will take away food from poor. There is also increase in alcohol consumption within developing countries. Further use of productive agricultural land for growing tobacco would also mean drop in productive land which could otherwise is used for food grains. There is also tremendous pressure on agriculture land around fast growing cities like Bangalore and Hyderabad of India where in entrepreneurs has acquired land for real estate business and construction of industry.

Eastern Terai of Nepal faced rain deficit in the year 2005/06 by early monsoon leading to crop production reduced by 12.5% on national basis. Nearly 10% of agriland were left fallow due to rain discrepancy but mid western Terai witnessed heavy rain with floods, resulting in reduced production by 30% in the year (Regmi 2007).

Cold wave in Nepal in 1997/98 had impacts on agricultural and showed reduction in the production of crops by 27.8, 36.5, 11.2, 30, 37.6 and 38% in potato, toria, sarson, rayo, lentil and chickpea respectively (source: NARC annual reports from 1987/88 to 1997/98).

Top 20 agricultural commodity in Asia are: Rice; pig meat; buffalo milk; cow milk; wheat; vegetables; hen eggs; chicken meat; cattle meat; cotton; potatoes; tomatoes; sugar cane; apples; groundnuts; garlic; oil palm; asparagus; onions; and grapes (FAO 2008).

Following paragraphs summarize impacts of climate change on major crops and livestock:

17.1.1 Rice

Long periods of sunshine are favourable for high rice yields and growth is optimal when the daily air temperature is between 24 and 36°C. The variation between day and night temperatures must be minimal during flowering and grain production. An irrigation water temperature of not less than 18°C is favoured. In south Asia, rice production has to be doubled by the year 2020 (IRRI 2000). Increase in the CO_2 level will help increase the production. But impact due to droughts and storms and floods can not be over seen. Past studies revealed that panicle initiation, flowering, heading, milking stage and crop maturity period has decreased due to the increase in temperature (Malla 2008). Grain yields may decline by 9–10% for each 1°C rise in temperature. Flooded rice ecosystem cannot be sustained if the drought conditions are prolonged and it may be necessary to develop non flooded and dry land rice cultivation. Increase in temperature will increase yield due to increase in photosynthesis. But, the flavour of the rice might be affected as the quantity of magnesium in the rice grains may decline.

17.1.2 Potato

Potato needs relatively mild temperature during early growth and cool weather conditions during tuber development. Humidity and rains lead to insect-pest, disease, viruses and epidemics. Poor drainage and lack of aeration restricts the tuber development and leads to rotting of tubers. Potato needs about 25°C at the time of germination, about 20°C for vegetative growth and between 17 and 20°C for tuberization and tuber development. Higher temperature has an adverse effect on the tuber growth, and temperature above 30°C stops tuber formation.

17.1.3 Wheat

Climatic parameters like rain and temperature strongly affect the crop. Physiological growth stages like panicle initiation, heading, flowering, milking and physiological maturity decreases due to increase in temperature. Increase in the CO_2 level will help increase the production. The study conducted in India showed that, there will decrease in potential yield by 1.5–5.8% in subtropical region and 17–18% in tropical zone (Agrawal and Kalra 1994). But impact due to droughts, storms and floods can not be over seen.

17.1.4 Maize

Being a C4 photosynthetic pathway plant, grain productivity is less responsive to increase in atmospheric CO_2 level. But impact due to winds, droughts, storms and floods can not be over seen.

17.1.5 Tomato

The tomato is sensitive to frost and does not thrive at low temperatures. Both high and low temperatures interfere with the setting of fruit. Fruits will crack if moisture supply follows drought. Rainfall in flowering stage will lead to loss of crop. Shortage of tomato supply created crisis in India during December, 2010.

17.1.6 Cassava

Cassava is a shrubby, tropical, perennial plant. It requires a minimum temperature of 25°C to grow and many varieties are drought resistant. But storms and floods will affect the crop.

17.1.7 Rubber

Rubber is best suited in a tropical climate. Even distribution of rainfall with no dry seasons exceeding 1 month along with high mean daily air temperature of between 25 and 28°C, and high rainfall exceeding 2,000 mm/year are ideal conditions for growing rubber. Increase in rainfall results in loss of tapping days and crop washout. A crop decrease of 3–15% will occur due to drought conditions (Government of Malaysia 2000).

17.1.8 Sugar Cane

Sugar cane needs warm growing season along with proper irrigation, long hours of bright sun shine and high relative humidity to yield more tonnage. The crop needs a ripening season of around 2–3 months with warm days, clear skies, cool nights and a dry weather without rainfall for build up of sugar. A higher temperature leads to reversion of sucrose into fructose and glucose resulting in less accumulation of sucrose. Rain in ripening period will result in poor juice quality, higher vegetation growth, formation of water shoots and rise in tissue moisture. Severe cold weather slow down bud sprouting and stop cane growth. Cane leaves and meristem (tissue with undifferentiated cells) tissues are killed at temperature −1 to 2°C. High wind velocity leads to cane breakage and leave leaf damage. Transpiration rate will decrease on cloudy days resulting in reduction in uptake of nutrients (Ikisan 2010).

17.1.9 Groundnut

Groundnut is grown in tropical region and demands a good irrigation as well as warm temperature. The groundnut crop can not stand frost for long and water stagnation. Cool and wet climate results in slow germination, plantlet surfacing, seed rot and diseases.

17.1.10 Garlic

Garlic a major tropical crop that will survive well in varying climate. But it will fail during extreme climatic events like drought and floods.

17.1.11 Onion

The crop grows well in warm temperature with soils having good drainage. High soil moisture, humidity during rains leads to diseases and rotting of crop. Unseasonal rains in India, the second largest onion producer resulted in shooting of onion process three to five times during December 2010. The issue became national crisis and authorities have to suspend export of the commodity to fulfil the local demand. The situation was not new but is repetition of similar shortage in 1980 and 1998.

17.1.12　Oil Palm

Oil palm flourishes in humid tropical climate in which rain occurs mostly at night and days are bright and sunny. Minimum monthly rainfall required is around 1,500 mm with absence of dry seasons, and an evenly distributed sunshine exceeding 2,000 h per year will result in optimum yield. A mean maximum temperature of about 29–33°C and a mean minimum temperature of 22–24°C results in highest bunch production. Increase in drought and flooding will affect the crop (Government of Malaysia 2000).

17.1.13　Asparagus

During periods of high temperatures, the tips tend to open and hence, the spears must be harvested more frequently in hot weather. The crop requires a vegetative growth period of adequate length to renew the carbohydrate reserves in the crown, which is source of new spears. Rust and other diseases will be less severe during dry season. Sprouting of the buds in the crown begins at temperatures of 10°C or above. Studies by Bill (1999) revealed that the optimal temperature for asparagus spear emergence was from 24.5 to 33°C and no spears emerged above 35°C.

17.1.14　Cocoa

Cocoa flourishes in areas where annual rainfall is 1,500–2,000 mm with three or less number of dry months. It should not be planted in areas with annual rainfall less than 1,250 mm unless irrigation is provided. Areas with annual rainfall more than 2,500 mm are also not preferred as it reduces yield by 10–20% due to water logging. The excessive rainfall causes high disease incidence, particularly *Phytophthora* and pink diseases (Government of Malaysia 2000).

17.1.15　Horticultural Crops

Climate induced impacts on horticultural crops are gaining importance. Past studies revealed that Oleic acid concentration increased and linolenic acid decreased in soybean seed with rise in temperature (Thomas et al. 2003). Seventy-five percent increase in air CO_2 content increased sourness in orange and Vitamin C (antioxidant) by about 5% (Malla 2008). Apple trees develop their vegetative and fruiting buds in the summer and buds go dormant as winter approaches. These buds remain dormant until they have accumulated sufficient chilling units. Insufficient receipt of chilling temperatures by buds results in delayed foliation; reduced fruit set; and reduced fruit quality (Ranbir et al. 2009).

17.1.16 Livestock

Livestock includes poultry, dairy production and rearing animals such as cattle, buffaloes, sheep, goats and pigs. Live stock sector plays the major role in developing counties. Meat and milk products are perishable goods, demanding energy to conserve the products. Livestock sector generates about 1.4% of the world's GDP and 40% of agricultural GDP. Live stocks are considered one of the leading stressors of ecosystem. But they stress ecosystem only if they are reared in fragile ecosystem by people who own them. Livestock which are cause of climate change are also victim of climate change. As shown in Fig. 17.3 not all live stocks are reared in pastures and animal houses. It is some time part of small house hold providing only income. They can be seen on streets of roads in many Indian cities. Rise in temperature by 2°C would decrease the meat and milk quality; hatchability of poultry; and enhances the possibility of disease in the livestock (Malla 2008). Livestock is usually raised under shade to avoid direct heat load from solar radiation. Air temperature, relative humidity and airflow affect production. Since livestock maintain a constant deep body temperature, heat generated must be lost to the atmosphere. Animals experience heat stress due to rise in temperature (Government of Malaysia 2000). Large animals die due to fall of trees and smaller animals are vulnerable to the strong wind.

Fig. 17.3 Livestock which are cause of climate change are also victim of climate change

17.1.17   Pests and Diseases

Increase in temperature and CO_2 will lead to an increase in population of pests due to increases the rate of reproductive cycle of insect and pest. The rise in temperature increases severity of diseases in presence of host plant. Pest and disease of plain ecosystem may gradually shift to hills and mountains affecting the ecosystem over there.

Other events which crisscross climate related events that affect fall in food crops are:

- Air Pollution
- Water Pollution
- Drop in surface water and ground water resources
- Human-Animal conflict near forest area where in wild animals like elephants often consume and destroy crops due to shortage of food in forest and
- Short-circuit and sparks in live wires running across agricultural lands during heavy rainfall

The climate impact of climate varies between the C3 and C4 plants. C3 plants convert carbon dioxide and ribulose bisphosphate (a 5-carbon sugar) into 3-phosphoglecerene. Example of C3 plants includes rice and barley. These plants thrive in area with moderate sunlight intensity and temperature.

C4 plants possess mechanisms to raise intercellular CO_2 concentration at the site of fixation. Examples of C4 plants include corn and sugarcane. C4 plants are more efficient in hot climate. Examples of C4 plant include sugarcane, maize etc.

Photosynthetic pathways in C3 and C4 plants lead to different physiological behaviour and ecological performance. C4 grasses tend to have higher water use efficiency and higher photosynthetic capacity than C3 grasses. C3 grasses are more efficient at lower temperatures (10–20°C) than C4 grasses. However C3 grasses suffer yield losses at higher temperatures, due to photorespiration losses as previously fixed CO_2 is lost.

The contribution of snow and glacial melt from Himalayas to the major rivers in the region ranges from 2 to 50% of the average flow. Snow and ice melt contribute about 70% of the flow in the summer monsoon to the Ganges, Indus, Tarim, and Kabul rivers (Kattelmann 1987; Singh and Bengtsson 2004; Barnett et al. 2005). The rivers of Nepal contribute about 40% of the average annual flow in the Ganges Basin, where 500 million people live. Himalayas also contribute about 70% of the flow in the dry season (Alford 1992). In western China, glacial melt contribute to 25% of the population in dry season (Xu 2008). The Indus Irrigation Scheme in Pakistan depends on runoff originating from Himalayas (Winiger et al. 2005). The agriculture which depends on water from above source may face discrepancies due to climate change.

During mid-1980s, cereal yields were comparably low. But 15 years later yields increased by more than 50% and poverty had declined by 30% in South Asia (World Bank 2008). Rise in population, Air Pollution, declining farm size and water scarcity is major challenges in Asia with respect to Agriculture. Overcoming widespread poverty in Asia requires tackling widening of rural-urban income disparities and this task needs adaptation with changing climate.

Key adaptation measures in agricultural practice to climate change include:

(i) Changes in practices to adopt to climate impact – change in sowing dates, change in crops, use of drought resistant variety, etc.

(ii) Changes in water management for efficient water use – rainwater harvesting, drip irrigation, mulching

(iii) Increase awareness in rural area with respect to new technology

(iv) Agricultural diversification – multiple cropping like planting horticultural/ commercial trees with cereal and vegetable crops, crop rotation

(v) Decreasing the bureaucratic hurdles for farming community

(vi) Increase in the production and distribution of improved seeds and other agricultural inputs like agro chemicals and agro machineries

(vii) Agricultural information systems – free and ready information on internet, toll free telephonic advice, and SMS advise

(viii) Adaptation of new technologies

(ix) Crop insurance and

(x) Communities training for adaptation of natural calamities through self help groups

Thirty-nine percent of area in production in South Asia and 29% of area in production in East Asia is under irrigation (World Bank 2008). Due high level of uncertainties in rain fed agriculture and reduced glacial runoff, investment in water is essential. With rising costs of large-scale irrigation schemes and environmental damage associated with it is prudent to implement small-scale schemes and rain water harvesting.

Box 17.2 Intensive agriculture and climate change

Agriculture is of very limited in Singapore economy. Agricultural products in Singapore include vegetables, eggs, fish, milk, ornamental plants and ornamental fish. With limited land and sea Singapore's agricultural developments take place in Agro technology Parks on land and Marine Parks at sea. Food fish production happens in floating cages at sea.

Hydroponic cultivation adapted by Singapore may become widely accepted in developing countries if the conventional agriculture becomes too difficult. This is a method of growing plants using mineral nutrient solutions, in the mineral nutrient solution only or in an inert medium, such as gravel, mineral wool, or coconut husk.

Productivity in hydroponics is high due control of pests and optimum supply of nutrients.

17.2 Dropping Fish Reserves

Man is depending on fishing even before dawn of civilization and millions still rely on fishing for their income and nutrition. But fish resource is under threat of collapse endangering the livelihoods of these millions and degrading the health of the oceans. Fisheries are over fished and exhausted by some people of few generations at the cost of future generations and natural ecosystems. The fisheries sector, environmental and agricultural sectors has many features in common. Current fishing practice has overstressed resource base.

Fishing was first industrialized in the early nineteenth century with the invension of steam trawlers. The sector quickly expanded in 1950s and 1960s. Invention and use of radar, and acoustic fish finders added to mechanised fishing.

According to estimates by FAO that 25% of the world's major fisheries are overfished and 40% are fully fished (World Bank 2004). The result is a diminishing fish catch and net income in the sector. Studies show reductions in the size and value of fish caught (Pauly et al. 1998, 2002; Myers and Worm 2003). Since the 1950s, the total number of people fishing and fish farming worldwide has at least quadrupled resulting in 270% average fishing power increase between 1965 and 1995 (World Bank 2004).

It is now common and vital for those fishing communities already vulnerable to the impacts of present day climate changes that successful adaptation must be accomplished through actions that reduce the vulnerabilities. While big fisher men travel further looking for new location to fish, poor become more prevalent as the climate changes.

Climate change is an additional problem to already existing fishing pressure, loss of habitat, pollution, disturbance, and introduced species. The reasons which govern the biological response are: (1) Changing temperature; (2) light supply (determined by ice cover, cloudiness and surface mixed-layer thickness); (3) alteration in nutrient supply in marine ecosystem due to change in vertical stability; (4) Increase in population of rivals in food chain like sardines.

By mid of this century, yearly average river runoff are projected to diminish by 10–30% over some area at mid-latitudes and in the dry tropics (IPCC 2007b). This means some part of Asia will face drop in fresh water resources by directly affecting aquatic ecosystem in this region. The drop in freshwater resource and added pollution due to increased population and industrialisation will further deteriorate aquatic ecosystem.

Trophic-level marine species are highly vulnerable to anthropogenic pressure (Dulvy et al. 2003). Exploitation of many marine fishes and mammals has reduced their abundance (Baker and Clapham 2004; Baum et al. 2003; Christensen et al. 2003; Myers and Worm 2003). These species maintain ecosystem structure and function (Estes et al. 1998; Jeremy et al. 2001), and are dependent on lower profile forage species (Pauly et al. 1998; Skewgar et al. 2007). In addition to providing humans with food, they provide prospect for wildlife viewing (Loomis and Larson 1994; Rudd and Tupper 2002). Further many people derive satisfaction knowing that they exist and will continue to do so (Loomis and White 1996; Pearce and Moran 1994; Van Kooten and Bulte 2000).

Figure 17.4 shows factors affecting fish catch and population. Table 17.4 shows impact on fisheries due to climate change. In coastal areas, volume of phytoplankton will increase with rise of water temperatures. This increases food supply, but the production efficiency may decrease due to motion and other factors. The quantity of cold-water seaweed may decline, resulting in populations creatures that feed on this type of seaweed.

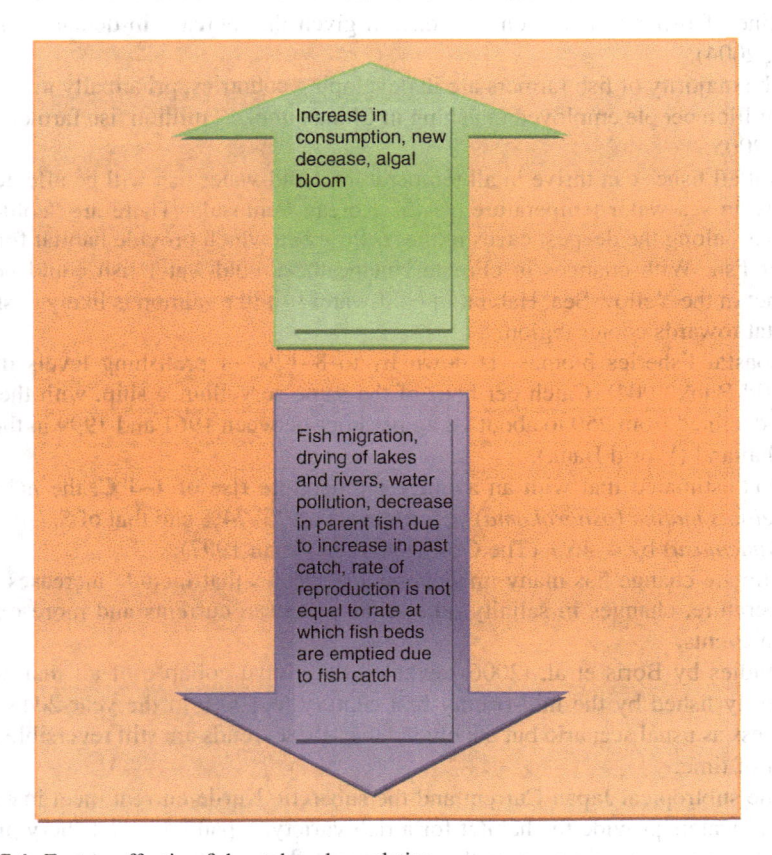

Fig. 17.4 Factors affecting fish catch and population

Table 17.4 Impact on fisheries due to climate change

Sector	Impact	Adaptation
Fishing	Sea level rise, Increased salinity in estuaries, reduced fluvial fisheries. It is not possible upto 4–7 days after the cyclone as trees and other objects float around the water, which gets caught on the nets	Control of Water Pollution, training fishermen with alternate lively hood skills, encourage vegetarian food

Marine biodiversity has changed due to exploitation, pollution, habitat destruction, and climate change (Dulvy et al. 2003; Heike et al. 2006; John et al. 2003; Jeremy et al. 2001; Boris et al. 2005). Marine extinctions are slowly revealed at the global scale (Dulvy et al. 2003). Ecosystems in estuaries (Heike et al. 2006), coral reefs (John et al. 2003), and coastal (Jeremy et al. 2001) are rapidly losing populations, species, along with oceanic fish communities (Boris et al. 2005).

Since the 1980s, average catches of north-east Atlantic cod have remained at 200,000 tonnes per year (O'Brien et al. 2000) despite variations in recruitment, and there has been a subsequent decline in abundance (Yoneda and Wright 2004). Decline of fisheries is reason for concern given the increase in demand (Briones et al. 2004).

The majority of fish farmers are in developing countries, principally in Asia and 4.5 million people employed in fishing in China out of 12 million fish farmers in the year 2005.

Not all fishes can thrive in all temperature. Cold-water fish will be affected due to rise in sea water temperature off the Korean Peninsula. There are "cold-water masses" along the deepest caves in the Yellow Sea which provide habitat for cold-water fish. With changes in climate change these cold-water fish could become extinct in the Yellow Sea. Habitat of cold-water fish like salmon is likely to shift its habitat towards cooler region.

Coastal fisheries biomass is down by to 8–12% of prefishing levels in Asia (World Bank 2004). Catch per hour of the same surveillance ship, with the same gear declined from 250 to about 18 kg per hour between 1961 and 1999 in the Gulf of Thailand (World Bank).

It is estimated that with an average temperature rise of 1–4°C, the habitat of *Salvelinus malma* (*oshorokoma*) will decrease by 25–74% and that of *S. leucomaenis* (*amemasu*) by 4–46% (The Government of Japan 1997).

Climate change has many impacts on sea species that include increases in sea temperature, changes in salinity, alterations in ocean currents and more extreme storm events.

Studies by Boris et al. (2006) revealed the global collapse of all marine taxa currently fished by the mid-twenty-first century to 100% in the year 2048 in the business as usual scenario but concluded that, these trends are still reversible at this point of time.

The subtropical Japan Current and the subarctic Kurile current meet in Japan's coastal waters provide the habitat for a rich variety. Japan's future fishery production depends upon changes in the course and flow of the Japan Current due to climate change (The Government of Japan 1997).

In summary the Asia and Pacific region which is the world's largest producer of fish, will observe decline in production of fish due to changes in climate and anthropogenic stress. With future changes in sea surface temperature; ocean currents; sea levels; salinity; wind speed and direction; and other associated changes, world's fish population will be under great stress.

17.3 Diminishing Forest

We are now positioned in a situation inevitable to changes to climate patterns. Adaptation to climate change is therefore not a '*spare part in the tool kit*' to be used '*if some thing happens*' but a tool that need to be used every day.

The two important aspects to consider in assessing the impact of climate change on forests are: (1) They form an essential component of the global carbon cycle, acting as a sink and reservoir of carbon; (2) The forest ecosystem is a fundamental part of the global biological system, continuously reacting to deviation in climate.

The forest's response to climate change is biologically complex and the effects are not quickly observed. Over the past several years, researchers have explored a wide range of issues that influence how forests are affected and may adapt to climate change. First investigations on climate impacts were on vegetation composition (Emanuel et al. 1985; Smith et al. 1992; Woodward 1992; Leemans and Vandenborn 1994; Cramer 1996) and forest succession (Solomon 1986; Pastor and Post 1988; Kienast 1991; Prentice et al. 1991; Bugmann 1997; Price et al. 1999).

Climate change has shown to influence tree phenology (Kramer et al. 1996; Menzel and Fabian 1999; Linkosalo et al. 2000), forest productivity (McGuire et al. 1993; Joyce 1995; Aber et al. 2001; Coops and Waring 2001), and growth (Woodbury et al. 1998; Hasenauer et al. 1999). Tree responses may vary among regions (Billington and Pelham 1991; Beuker et al. 1998; Persson 1998; Bigras 2000). Possible consequences due to warmer and more variable climate include frequent insect and disease outbreaks (Fleming and Volney 1995; Volney 1996; Ayres and Lombardero 2000; Simberloff 2000) and/or wild fires (Kasischke et al. 1995; Wein and Groot 1996; Flannigan et al. 1998).

Varying growth patterns and disturbance is likely to have implications for forest management and timber markets (Mills and Haynes 1995; Perez-Garcia et al. 1997; Sohngen and Mendelsohn 1998). Secondary effects of climate change on the forest industry (see Nabuurs and Moiseyev 1999; Mills et al. 2000), and necessity to adjust forest policies and resource planning (Binkley and Van Kooten 1994) need further research. Effects of climate changes are not limited to the industry but also on heating and cooking (Solberg 1996). Poor people who depend on free fuel wood, vegetable and fruits may not get sufficient raw material for cooling.

Non-timber forest products such as fuel, honey, flowers, forest foods or medicinal plants, are important for the livelihood of the rural communities. Many products and services depend on forest resources but there are no reasonable estimates of the future demand for these products and services (Easterling et al. 2007). Although climate change will affect the availability of forest resources, land-use change and deforestation in tropical zones, is also extremely important (Zhao et al. 2005). Droughts combined with deforestation increase fire danger (Laurance and Williamson 2001).

Numerous studies consider mainly the impact of climate impact on forest resources, industry and economy. Some experts have analyzed impact on ecological system (e.g., Sohngen and Sedjo 2005). Communities in developing countries are

Table 17.5 Impact on forestry due to climate change and adaptation measures

Sector	Impact	Adaptation
Agro forestry	Loss of vegetation, ecosystem change, drought, increased physical vulnerability and species change, reduction in trees due to climate change and anthropogenic demand	Reforestation, discourage use of wood, create market for substitutes for wood, research for substitutes of wood, produce paper by agro resides and used paper. Management of forest fires

especially vulnerable because of the limited ability of resource dependent communities to react to risk in a proactive manner (Davidson et al. 2003; Lawrence 2003).

Table 17.5 shows impact on forestry due to climate change and adaptation measures. The possible adaptation would be policy and practice which ensures following:

- Reforestation
- Discourage use of wood
- Create market for substitutes for wood
- Research for substitutes of wood
- Protect wild life
- Production of paper by agro resides and used paper and
- Management of forest fires

References

Aber J.A, Neilson R.P, McNulty S, Lenihan J, Bachelet D, Drapek R, 2001: Forest processes and global environmental change: Predicting the effects of individual and multiple stresses. Bioscience 51, 735–751.

Ainsworth E.A, Long S.P, 2005: What have we learned from 15 years of free-air CO_2 enrichment (FACE)? A meta-analysis of the responses of photosynthesis, canopy properties and plant production to rising CO2, New Phytol. 165, 351–372.

Agrawal P.K and Kalra N, 1994: Analysing the limitation set by climate factors, genotype, and water and nitrogen availability on productivity of wheat II. Climatically potential yield and optimal management strategies. Field crop res. 38:93–103

Alford D, 1992: Hydrological Aspects of the Himalayan Region. Kathmandu: ICIMOD

Ayres M.P, Lombardero M.J, 2000: Assessing the consequences of global change for forest disturbance from herbivores and pathogens. Science Total Environ. 262, 263–286.

Baker C.S, Clapham P.J, 2004: Trends in Ecology and Evolution 19, 365.

Barnett T.P, Adam J.C, Lettenmaier D.P, 2005: 'Potential impacts of a warming climate on water availability in a snow-dominated region'. Nature 438(17): 303–309

Baum K Julia, Ransom A Myers, Daniel G Kehler, Boris Worm, Shelton J Harley and Penny A. Doherty, 2003: Collapse and Conservation of Shark Populations in the Northwest Atlantic, 299, 389–392

Beuker E, Valtonen E, Repo T, 1998: Seasonal variation in the frost hardiness of Scots pine and Norway spruce in old provenance experiments in Finland. For. Ecol. Manage. 107, 87–98.

Bigras F.J, 2000: Selection of white spruce families in the context of climate change: heat tolerance. Tree Phys. 20, 1227–1234.

Billington H.L, Pelham J, 1991: Genetic variation in the date of budburst in Scottish birch populations: implications for climate change. Funct. Ecol. 5, 403–409.

Bill B. Dean, 1999: The Effect of Temperature On Asparagus Spear Growth And Correlation of Heat Units Accumulated In The Field With Spear Yield, ISHS Acta Horticulturae 479: IX International Asparagus Symposium

Binkley C.S, Van Kooten G.C, 1994: Integrating climate change and forests: economic and ecological assessments. Climate Change 28, 91–110.

Briones M., Dey M.M and Ahmed M, 2004: The future for fish in the food and livelihoods of the poor in Asia. NAGA, 50 World Fish Center Quarterly, 27, 48.

Boris Worm, Marcel Sandow, Andreas Oschlies, Heike K. Lotze, Ransom A. Myers, 2005, Global Patterns of Predator Diversity in the Open Oceans, Science 309, 1365; DOI: 10.1126/science.1113399

Boris Worm, Edward B. Barbier, Nicola Beaumont, Emmett Duffy J, Carl Folke, Benjamin S. Halpern, Jeremy B.C. Jackson, Heike K. Lotze, Fiorenza Micheli, Stephen R. Palumbi, Enric Sala, Kimberley A. Selkoe, John J. Stachowicz, Reg Watson, 2006: Impacts of Biodiversity Loss on Ocean Ecosystem Services, Science 314, 787; DOI: 10.1126/science.1132294, PP 787–790

Bugmann H, 1997: Sensitivity of forests in the European Alps to future climatic change. Climate Res. 8, 35–44.

Challinor A.J, Wheeler T.R, Craufurd P.Q, Slingo J.M, 2005: Simulation of the impact of high temperature stress on annual crop yields. Agric. For. Meteorol. 135 (1–4), 180–189.

Chambers R and Conway G, 1992: Sustainable rural livelihoods: practical concepts for the 21st century, Brighton, Institute of Development Studies

Coops N.C, Waring R.H, 2001: Assessing forest growth across southwestern Oregon under a range of current and future global change scenarios using a process model, 3-PG. Global Change Biol. 7, 15–29.

Cramer W, 1996: Modelling the possible impacts of climate change on broad-scale vegetation structure: examples from Northern Europe. In: Oechel, W., Holten, J.I. (Eds.), Global Change and Arctic Terrestrial Ecosystems. Springer, Berlin, pp. 312–329.

Christensen V, Guenette S, Heymans J.J, Walters C.J, Watson R, Zeller D, Pauly D, 2003: Hundred year decline of North Atlantic predatory fishes, Fish and Fisheries 4, 1–24

Cruz R.V, Harasawa H, Lal M, Wu S, Anokhin Y, Punsalmaa B, Honda Y, Jafari M, Li C and Huu Ninh N, 2007: Asia. Climate Change 2007: Impacts, Adaptation and Vulnerability. Contribution of Working Group II to the Fourth Assessment Report of the Intergovernmental Panel on Climate Change, M.L. Parry, O.F. Canziani, J.P. Palutikof, P.J. van der Linden and C.E. Hanson, Eds., Cambridge University Press, Cambridge, UK, 469–506.

Davidson D.J, Williamson T and Parkins J.R, 2003: Understanding climate change risk and vulnerability in northern forest-based communities. Can. J. Forest Res., 33, 2252–2261.

Dirnbock T, Dullinger S, Grabherr G, 2003: 'A regional impact assessment of climate and land-use change on alpine vegetation'. Journal of Biogeography 30: 401–417

Dulvy N.K, Sadovy Y & Reynolds J.D, 2003: Extinction vulnerability in marine populations. Fish and Fisheries 4: 25–64.

Easterling W.E, Aggarwal P.K, Batima P, Brander K.M, Erda L, Howden S.M, Kirilenko A, Morton J, Soussana J.F, Schmidhuber J and Tubiello F.N, 2007: Food, fibre and forest products. Climate Change 2007: Impacts, Adaptation and Vulnerability. Contribution of Working Group II to the Fourth Assessment Report of the Intergovernmental Panel on Climate Change, Parry M.L, Canziani O.F, Palutikof J.P, van der Linden P.J and Hanson C.E, Eds., Cambridge University Press, Cambridge, UK, 273–313.

Ellis F, 2000: Rural Livelihoods and Diversity in Developing Countries, Oxford, Oxford University Press

Emanuel W.R., Shugart H.H., Stevenson M.P, 1985: Climatic change and the broad-scale distribution of terrestrial ecosystem complexes. Climate Change 7, 29–43.

Estes J.A, Tinker M.T, Williams T M, Doak D.F, 1998: Killer Whale Predation on Sea Otters Linking Oceanic and Nearshore Ecosystems, Science 282, 473–476.

FAO, 2008: <http://faostat.fao.org/site/339/default.aspx> retrieved on 20, September, 2010

Flannigan M.D, Bergeron Y, Engelmark O, Wotton B.M, 1998: Future wildfire in circumboreal forests in relation to global warming. J. Veg. Sci. 9, 469–476.

Fleming R.A, Volney W.J.A, 1995: Effects of climate change on insect defoliator population processes in Canada's boreal forest some plausible scenarios. Water Air Soil Pollut. 82, 445–454.

Government of Malaysia, 2000: Initial National Communication Submitted to The United Nations Framework Convention on Climate Change, Ministry of Science, Technology and The Environment

Hasenauer H, Nemani R.R., Schadauer K, Running S.W, 1999: Forest growth, forest ecology response to changing climate between 1961 and 1990 in Austria. For. Ecol. Manage. 122, 209–219.

Heike K. Lotze, Hunter S. Lenihan, Bruce J. Bourque, Roger H. Bradbury, Richard G. Cooke, Matthew C. Kay, Susan M. Kidwell, Michael X. Kirby, Charles H. Peterson, Jeremy B, Jackson C, 2006: Depletion, Degradation, and Recovery Potential of Estuaries and Coastal Seas, Science, 312. 1806. DOI: 10.1126/science.1128035. pp. 1806–1809

Holling C.S, 2001: "Understanding the complexity of economic, ecological and social systems." Ecosystems. 4 : 390–405.

Hutchings, J.A. and Reynolds, J.D. 2004: Marine fish population collapses: consequences for recovery and extinction risk. BioScience 54: 297–309

Ikisan, 2010 <http://www.ikisan.com/links/ap_sugarcaneSoil%20and%20Climate.shtml> retrieved on 20, September, 2010

IPCC, 2001, Climate change 2001: impacts, adaptation, and vulnerability, Contribution of Working Group II to the Third Assessment Report of the Intergovernmental Panel on Climate Change, Cambridge University Press, Cambridge,

IPCC, 2007a: Climate Change 2007: The Physical Science Basis. Contribution of Working Group I to the Fourth Assessment Report of the Intergovernmental Panel on Climate Change [Solomon S, Qin D, Manning M, Chen Z, Marquis M, Averyt K.B, Tignor M, Miller H.L (eds)]. Cambridge and New York: Cambridge University Press

IPCC, 2007b: Summary for Policymakers. In: Climate Change 2007: Impacts, Adaptation and Vulnerability. Contribution of Working Group II to the Fourth Assessment Report of the Intergovernmental Panel on Climate Change, Parry M.L, Canziani O.F, Palutikof J.P, van der Linden P.J and Hanson C.E, Eds., Cambridge University Press, Cambridge, UK, 7–22.

IRRI, 2000: world rice statistics 1993-1994. International rice research Institute, manila, The Philippines

Jablonski L.M, Wang X, Curtis P.S, 2002: Plant reproduction under elevated CO_2 conditions: a meta-analysis of reports on 79 crop and wild species, New Phytol. 156 (1), 9–26.

Jeremy B.C. Jackson, Michael X. Kirby, Wolfgang H. Berger, Karen A. Bjorndal, Louis W. Botsford, Bruce J. Bourque, Roger H. Bradbury, Richard Cooke, Jon Erlandson, James A. Estes, Terence P. Hughes, Susan Kidwell, Carina B. Lange, Hunter S. Lenihan, John M. Pandolfi, Charles H. Peterson, Robert S. Steneck, Mia J. Tegner, Robert R. Warner, 2001: Historical Overfishing and the Recent Collapse of Coastal Ecosystems, Science: 293, 5530; DOI: 10.1126/science.1059199, pp. 629–637

John M. Pandolfi, Roger H. Bradbury, Enric Sala, Terence P. Hughes, Karen A. Bjorndal, Richard G. Cooke, Deborah McArdle, Loren McClenachan, Marah J. H. Newman, Gustavo Paredes, Robert R. Warner, Jeremy B. C. Jackson, 2003: Global Trajectories of the Long-Term Decline of Coral Reef Ecosystems, Science 301, 955;DOI: 10.1126/science.1085706, PP 955–958

Jon Barnett, Suraje Dessai, and Roger Jones, 2003: Climate Change In Timor Leste: Science, Impacts, Policy and Planning, Briefing to Government, civil society, and donors, República Democrática de Timor-Leste University of Melbourne and CSIRO

Joyce, L.A. (Ed.), 1995: Productivity of America's Forests and Climate Change. US Department of Agriculture, Forest Service, Rocky Mountain Forest and Range Experiment Station, Fort Collins, Gen. Tech. Rep. RM-271, 70 pp.

Jon Barnett, Neil Adger W, 2007: Climate change, human security and violent conflict, Political Geography, 26, 639–655,

Kasischke E.S, Christensen N.L, Stocks B.J, 1995: Fire, global warming and the carbon balance of boreal forests. Ecol. Appl. 5, 437–451.

Kattelmann R. 1987: 'Uncertainty in assessing Himalayan water resources'. Mountain Research and Development 7(3): 279–286

KellomaÈki S, Karjalainen T, Mohren F, LapvetelaÈinen T. (Eds.), 2000: Expert Assessment on the Likely Impacts of Climate Change on Forests and Forestry in Europe. European Forest Institute, Joensuu, Finland, 120 pp.

Kienast F, 1991: Simulated effects of increasing atmospheric CO_2 and changing climate on the successional characteristics of Alpine forest ecosystems. Landscape Ecol. 5, 225–238.

Kimball B.A, Kobayashi K, Bindi M, 2002: Responses of agricultural crops to free-air CO_2 enrichment, Adv. Agron. 77, 293–368.

Kramer K, Friend A, Leinonen I, 1996: Modelling comparison to evaluate the importance of phenology and spring frost damage for the effects of climate change on growth of mixed temperate-zone deciduous forests. Climate Res. 7, 31–41.

Laurance W.F. and Williamson G.B, 2001: Positive feedbacks among forest fragmentation, drought, and climate change in the Amazon. Conserv. Biol., 15, 1529–1535.

Lawrence A., 2003: No forest without timber? Int. For. Rev., 5, 87–96.

Leemans R., Vandenborn G.J, 1994: Determining the potential distribution of vegetation, crops and agricultural productivity. Water Air Soil Pollut. 76, 133–161.

Linkosalo T, Carter T.R, HaÈkkinen R, Hari P, 2000: Predicting spring phenology and frost damage risk of Betula spp. under climatic warming: a comparison of two models. Tree Phys. 20, 1175–1182.

Li B.L, Zhou C.H, 2001: 'Climatic variation and desertification in West Sandy Land of Northeast China Plain'. Journal of Natural Resources, 16:234–239

Long S.P, Ainsworth E.A, Leakey A.D.B, Nösberger J, Ort DR, 2006: Food for thought: lower-than-expected crop yield stimulation with rising CO2 concentrations. Science 312, 1918–1921

Loomis J., Larson D. M., 1994: Marine Resource Economics 9, 275.

Loomis J. B., White D. S., 1996: Ecological Economics 18, 197.

Malla G, 2008: Climate Change And Its Impact On Nepalese Agriculture, The Journal of Agriculture and Environment Vol:.9, pp 62–71

McGuire A.D, Joyce L.A, Kicklighter D.W, Melillo J.M, Esser G, VoÈroÈsmarty C.J, 1993: Productivity response of climax temperate forests to elevated temperature and carbon dioxide: a North American comparison between two global models. Climate Change 24, 287–310.

Mats Eriksson, Xu Jianchu, Arun Bhakta Shrestha, Ramesh Ananda Vaidya, Santosh Nepal, Klas Sandström, 2009: The Changing Himalayas, Impact of climate change on water resources and livelihoods in the greater Himalayas, International Centre for Integrated Mountain Development (ICIMOD).

Mearns L.O, Mavromatis T, Tsvetsinskaya E, 1999: Comparative response of EPIC and CERES crop models to high and low spatial resolution climate change scenarios. J. Geophys. Res. 104(D6), 6623–6646.

Menzel A, Fabian P, 1999: Growing season extended in Europe. Nature 397, 659.

Mills J.R, Alig R, Haynes R.W, Adams D, 2000: Modelling climate change impacts on the forest sector. In: Joyce L.A, Birdsey R. (Eds.), The Impacts of Climate Change on America's Forests: A Technical Document Supporting the 2000 USDA Forest Service RPA Assessment. US Department of Agriculture, Forest Service, Rocky Mountain Research Station, Fort Collins, Gen. Tech. Rep. RMRS-GTR-59, pp. 69–78.

Mills J.R, Haynes R.W, 1995: Influence of climate change on supply and demand for timber. In: Joyce L.A. (Ed.), Productivity of America's Forests and Climate Change. US Department of Agriculture, Forest Service, Rocky Mountain Forest and Range Experiment Station, Fort Collins, Gen. Tech. Rep. RM-271, pp. 46–55.

Myers R. A, and Worm B. 2003: "Rapid Worldwide Depletion of Predatory Fish Communities." Nature, 423 :280–3.

Nabuurs G.J, Moiseyev A, 1999: Consequences of accelerated growth for the forests and forest sector in Germany. In: Karjalainen H, Spiecker, Laroussinie, O. (Eds.), Causes and Consequences of Accelerating Tree Growth in Europe, Nancy, France, 17–19 May 1998. European Forest Institute, Joensuu, Finland, pp. 197–206.

O'Brien C.M, Fox X.J, Planque B & Casey J, 2000: Climate variability and North Sea cod. Nature 404: 142.

Parry M.L, Arnell N.W, McMichael A.J, Nicholls R.J, Martens P, Kovats R.S, Livermore M.T.J, Rosenzweig C, Iglesias A, Fischer G, Millions at risk 2001: defining critical climate change threats and targets, Glob. Environ. Change 11, 181–183.

Pastor J, Post W.M., 1988: Response of northern forests to CO_2 induced climate change. Nature 334, 55–58.

Pauly D, Christensen V, Dalsgaard J, Froese R, and Torres F. 1998: "Fishing Down Marine Food Webs." Science, 279 :860–63;

Pauly D, Christensen V, Guanette S, Pitcher T.L, Sumaila U.R, Walters C.J, Watson R and Zeller D, 2002: "Towards Sustainability in World Fisheries." Nature, 418: 689–95

Pearce D.W, Moran D, 1994: The Economic Value of Biodiversity (James & James/Earthscan).

Persson B, 1998: Will climate change affect the optimal choice of Pinus sylvestris provenances? Silva Fennica 32, 121–128.

Perez-Garcia J, Joyce L.A, Binkley C.S, McGuire A.D, 1997: Economic impacts of climatic change on the global forest sector: an integrated ecological/economic assessment. Crit. Rev. Environ. Sci. and Technol. 27, S123–S138.

Prentice I.C, Sykes M.T, Cramer W, 1991: The possible dynamic response of northern forests to greenhouse warming. Gl. Ecol. Biogeogr. Lett. 1, 129–135.

Price D.T, Halliwell D.H, Apps M.J, Peng C, 1999: Simulating effects of climate change on boreal ecosystem carbon pools in central Canada. J. Biogeogr. 26, 1237–1248.

Qiu G.W, Hao Y.X, Wang S.L, 2001: 'The impacts of climate change on the interlock area of farming-pastoral region and its climatic potential productivity in Northern China'. Arid Zone Research 18: 23–28

Ranbir Singh Ranaa, Bhagata R.M, Vaibhav Kaliaa and Harbans Lalb, 2009: Impact Of Climate Change on Shift of Apple Belt in Himachal Pradesh, ISPRS Archives XXXVIII-8/W3 Workshop Proceedings: Impact of Climate Change on Agriculture

Regmi H. R., 2007: Effect of unusual weather on cereal crops production and household food security. The Journal of Agriculture and Environment, pp 20–29

Rosegrant M, Ringler C, Msangi S, Zhu T, Sulser T, Valmonte-Santos R, Rosegrant M.W, and Wood S, 2007: Agriculture and food security in Asia: The role of agricultural research and knowledge in a changing environment. Journal of Semi-Arid Tropical Agricultural Research. 4(1). pp. 1–35.

Rudd M. A, and Tupper M. H, 2002: The impact of Nassau grouper size and abundance on scuba diver site selection and MPA economics. Coastal Management 30: 133–151.

Simberloff D, 2000: Global climate change and introduced species in United States forests. Sci. Total Environ. 262, 253–261.

Smith T.M, Shugart H.H, Bonan G.B, Smith J.B, 1992: Modelling the potential response of vegetation to global climate change. Adv. Ecol. Res. 22, 93–116.

Solomon A.M, 1986: Transient response to CO_2-induced climate change: simulation modelling experiments in eastern North America. Oecologia 68, 567–579

Singh P and Bengtsson L, 2004: Hydrological sensitivity of a large Himalayan basin to climate change. Hydrological Processes 18: 2363–2385

Skewgar E, Boersma P.D, Harris G, & Caille G, 2007: Anchovy Fishery Threatens Patagonian Ecosystem. Science 315:45.

Sohngen B, Mendelsohn R, 1998: Valuing the market impact of large-scale ecological change: the effect of climate change on US timber. Am. Econ. Rev. 88, 689–710.

Solberg B. (Ed.), 1996: Long-term Trends and Prospects in World Supply and Demand for Wood and Implications for Sustainable Forest Management. European Forest Institute, Joensuu, 150 pp.

Sohngen B and Sedjo R, 2005: Impacts of climate change on forest product markets: implications for North American producers. Forest Chron., **81**, 669–674.

The Government of Japan, 1997: Japan 's Second National Communication Under the United Nations Framework Convention on Climate Change

Thomas J.M.G, Boote K.J, Allen L.H, Gallo-Meagher M and Davis J.M, 2003: Elevated Temperature and Carbon Dioxide Effects on Soybean Seed Composition and Transcript Abundance doi:10.2135/cropsci2003.1548. Crop Science 43:1548–1557

UNDP, 2006: Human Development Report: Beyond Scarcity: Power, Poverty and the Global Water Crisis. New York: United Nations Development Programme

UNDP, 2007: Human Development Report 2007/2008, Fighting Climate Change: Human solidarity in a divided world, pp 93

UNFCCC, 2005: Report of the Asian Regional Workshop on Adaptation. http://maindb.unfccc.int/library/view_pdf.pl?url=http://unfccc.int/resource/docs/2007/sbi/eng/13.pdf retrieved on September 30, 2010.

UN Millennium Project, 2005, Investing in development. A practical plan to achieve the millennium development goals, Report to the UN Secretary General, New York

Van Kooten G.C, Bulte E.H, 2000: The Economics of Nature (Blackwell Scientific,).

Volney W.J.A, 1996: Climate change and management of insect defoliators in boreal forest ecosystems. In: Apps, M., Price, D.T. (Eds.), Forest Ecosystems, Forest Management and the Global Carbon Cycle. Springer, Berlin, pp. 79–87.

Wein R.W, Groot W.J, 1996: Fire-climate change hypotheses for the taiga. In: Goldammer, J.G., Furyaev, V.V. (Eds.), Fire in Ecosystems of Boreal Eurasia. Kluwer Academic Publishers, London, pp. 505–512.

Wheeler T.R., Craufurd P.Q, Ellis R.H, Porter J.R, Vara Prasad, P.V, 2000. Temperature variability and the annual yield of crops. Agric. Ecosyst. Environ. 82, 159–167.

Winiger M, Gumpert M, & Yamout H, 2005: Karakoram-Hindu Kush-Western Himalaya: assessing high-altitude water resources. Hydrological Processes 19(12): 2329–2338

Woodbury P.B, Smith J.E, Weinstein D.A, Laurence J.A, 1998: Assessing potential climate change effects on loblolly pine growth: a probabilistic regional modelling approach. For. Ecol. Manage. 107, 99–116.

Woodward F.I, 1992: A review of the effects of climate on vegetation: ranges, competition and composition. In: Peters, R.L., Lovejoy, T.E. (Eds.), Global Warming and Biological Diversity. Yale University Press, New Haven, pp. 105–123.

World Bank, 2004: Saving Fish and Fishers Toward Sustainable and Equitable Governance of the Global Fishing Sector

World Bank, 2006: Overcoming Drought: Adoptin Strategies for Andra Pradesh

World Bank, 2008: World Development report 2008, Agriculture for Development

Xu J.C, 2008: The highlands: a shared water tower In a changing climate and changing Asia. Working Paper No. 67. Beijing: World Agroforestry Centre, ICRAF-China

Yoneda, M., and Wright, P. J. 2004: Temporal and spatial variation in reproductive investment of Atlantic cod *Gadus morhua* in the northern North Sea and Scottish west coast. Marine Ecology-Progress Series 276: 237–248.

Zhao Y, Wang C, Wang S and Tibig L, 2005: Impacts of present and future climate variability on agriculture and forestry in the humid and sub-humid tropics. Climatic Change, **70**, 73–116.

Chapter 18
Accelerating Innovation and Decelerating Climate Change

Since 1980s "sustainable development" has been the key phrase of environmental protection. 1987 Report of the World Commission on Environment and Development defines sustainable development as:

> *Development that meets the needs of the present without compromising the ability of future generations to meet their own needs.*

The common concern of society is determined within states by the constitution and the legislation within the state. The international community does not have constitution and central authority for enforcement. Common concern of humankind results from treaties, customary rules and non-binding principles. Innovation over the past decades was used to fulfil human demands which have now resulted in stress on climate. Energy security is an important issue facing Asian countries, which currently import more than one-third of all global oil supplies. During the period 1990–2004, the highest growth rate in energy was in Asia (Sims et al. 2007). During the period from 1990 to 2005, China's CO_2 emissions increased from 676 to 1,491 $MtCO_2/yr$ to befall 18.7% of global emissions (IEEJ 2005; BP 2006). The area with the lowest per-capita utilization has changed from Asian developing countries in 1972 to African countries today. Mitigating climate impacts requires significantly stepping up international efforts to dissipate existing technologies and develop new ones. Dissipating climate friendly technology requires capacity building and enhancing the ability of countries.

Strengthening national innovation and technology capacity can become a powerful catalyst for development (Metcalfe and Ramlogan 2008; Metcalfe and Ramlogan 2008). International harmonization of regulatory incentives must be scaled up to provide more financing and to adopt/formulate policy that stimulate demand for climate-smart innovation.

Article 3(1) of the Framework Convention on Climate Change declares:

> *The parties should protect the climate system for the benefit of present and future generations of humankind.*

Major international environmental legislation namely Basel Convention on the Control of Transboundary Movements of Hazardous Wastes and Their Disposal, 1989; Montreal Protocol on the Protection of the Ozone Layer 1987 as amended in

R. Chandrappa et al., *Coping with Climate Change*, DOI 10.1007/978-3-642-19674-4_18, © Springer-Verlag Berlin Heidelberg 2011

1992; Convention on Biological Diversity 1992; and Convention on Climate Change 1992 insist to provide for transfer of technology or for financial assistance. Hence it becomes duty of parties not only to help themselves but should also help others in accelerating the innovation and decelerating climate change.

18.1 The Tools, Technologies, and Institutions

Tools, technology and institutions play great role in encouraging and dissipating climate smart technologies.

Some of the technology areas are: energy efficiency; carbon capture and storage; biomass, wind and solar power; and nuclear power (Edmonds et al. 2007; Stern 2007). As part of the national energy efficiency strategy of Jordan is enforcing thermal insulation in residential and commercial building in certain places as well as the preparation of an "Energy Efficiency Code" is a part of such a strategy (Shahin 2005).

According to an analysis of GHG mitigation China curbed the use of low-grade coal, resulting in avoided emissions of some 366 $MtCO_2$ and in India, energy policy initiatives including demand side efficiency improvements are estimated to have reduced emissions by 66 $MtCO_2$ (Chandler et al. 2002). Technology has also moved to the forefront of a numerous international and national climate policy initiatives, such as Technology Strategy (GTSP 2001), the Japanese 'New Earth 21' Project (RITE 2003). Increased use of natural gas has recently occurred all over the Asian region (BP 2006), and a liquefied natural gas (LNG) market has recently emerged in the region (Sims et al. 2007). While renewable energy remains a small share of the global energy production, they have continued to grow rapidly. Government support has helped to increase global wind and solar generation capacity by 31 and 47% respectively. Wind energy growth was led by China and the US accounting a combined 62.4% of total growth (BP 2010).

18.2 International Collaboration

At the end of the 1960s, representatives of the developing countries considered that environmental protection was fundamentally a battle against pollution, produced principally by industry. It was therefore an "illness of the rich" that did not worry poor countries. But later, a better understanding of the many aspects confirmed that desertification, lack of clean water, erosion, deforestation and climate change affect Southern countries as much or more than they harm Northern ones.

International teamwork makes research and development efforts more productive but development of new climate-friendly technologies require local governments to set up a policy framework providing the right set of incentives. There are numerous, barriers, that may impede such technology transfers. To bring new technologies, research and development efforts remain necessary.

Some of the international collaboration where Asian are listed below:

Asia-Pacific Economic Cooperation forum (APEC) – The Asia-Pacific Economic Cooperation forum, founded in 1989, consists of 21 members (Australia; Brunei; Canada; Chile; the People's Republic of China; Chinese Taipei; Hong Kong, China; Indonesia; Japan; the Republic of Korea (ROK); Malaysia; Mexico; New Zealand; Papua New Guinea; Peru; the Republic of the Philippines; the Russian Federation; Singapore; Thailand; the United States; and Vietnam). APEC's Energy Working Group (EWG) has undertaken technical projects with respect to climate change adaptation.

Asia Pacific Partnership on Clean Development and Climate (APPCDC) – is a voluntary international public-private partnership among Australia, Canada, India, Japan, China, South Korea and Untied States. These countries account for nearly half of world's GHG emission engage themselves to accelerate the development and use of clean energy technologies with no compulsory enforcement mechanism.

Environmental Cooperation-Asia Clean Development and Climate Program (ECO-Asia CDCP) – promotes clean energy solutions for Asia that to mitigate climate impact.

East Asia Climate Partnership (EACP) – The East Asia Climate Partnership (EACP) aims to mitigate climate change, by identifying an East Asian low carbon development path.

South Asian Association for Regional Cooperation (SAARC) – SAARC established in 1985, is creating network among the environmental, and energy officials of member countries.

International Centre for Integrated Mountain Development (ICIMOD) – is a knowledge development and learning centre based in Kathmandu, Nepal. The centre is serving the regional member countries of the Hindu Kush-Himalayas – Afghanistan, Bangladesh, Bhutan, China, India, Myanmar, Nepal, and Pakistan.

In addition to the efforts of regional institutions the area has a few multilateral and bilateral activities like Mekong River Commission (MRC) and the Coral Triangle (CTI) Initiative and on the efforts to protect Himalayan glaciers. The MRC an agreement among the governments of Cambodia, Laos, Thailand, and Vietnam, founded in 1995 is jointly put efforts on management of their shared water resources. The CTI has brought together governments, NGOs, and multilateral agencies to protect the coral reefs in Indonesia, Malaysia, Papua New Guinea, the Philippines, Timor Leste, and the Solomon Islands (the CT6 countries) to protect 5.7 million square km of the Coral Triangle which is home to the highest diversity of marine life on Earth. In 2010, China, India, and Nepal began a project for the conservation of the greater Mount Kailash Region of the Himalayas.

18.3 Kyoto Protocol and Asia

The Kyoto Protocol an international agreement linked to the United Nations Framework Convention on Climate Change (UNFCCC) sets binding targets for 37 industrialized countries and the European community for reducing greenhouse gas (GHG) emissions amounting to an average of 5% against 1990 levels over the 5-year period 2008–2012.

The UNFCCC has two Annexes:

- Annex I – These countries shall adopt national policies and take corresponding measures on the mitigation of climate change and
- Annex II – These countries shall provide new and additional financial resources to meet the agreed full costs incurred by developing country

The Berlin Mandate of 1995 stipulated that 'quantified emission limitation and reduction objectives' should be set for developed countries and no new obligation should be introduced for developing countries. This formed the basis for the 1997 Kyoto Protocol. The countries most at risk from sea-level rise are united in the Alliance of Small Island States (AOSIS) which includes some of the Asian countries.

The Kyoto Protocol was adopted in Kyoto, Japan, on 11 December 1997 and entered into force on 16 February 2005, but detailed rules for the implementation of the Protocol were adopted at COP 7 (Conference of Parties) at Marrakesh in 2001, and called the "Marrakesh Accords."

Kyoto Protocol has two Annexes:

- Annex A – GHGs and sectors/source categories and
- Annex B – Party (Countries) and Quantified emission limitation or reduction commitment (percentage of base year or period)

Under the treaty, countries must meet their targets through national measures and offers them an additional means of meeting their targets by following mechanisms:

- **Emissions trading** – Countries that have spare emission units can sell excess capacity to countries that are over their target
- **Clean development mechanism (CDM)** – Allows a country with an emission-reduction commitment (Annex B Party) under the Kyoto Protocol to implement an emission-reduction project in developing countries and earn saleable certified emission reduction (CER) credits and can be counted towards meeting Kyoto targets
- **Joint implementation (JI)** – Allows a country with an emission reduction under the Kyoto Protocol (Annex B Party) to earn emission reduction units (ERUs) from an emission-reduction or emission removal project in another Annex B Party

The Kyoto Protocol requires industrialized countries (Listed in Annex B of protocol) to reduce their GHG emissions to 5.2% below that of 1990 otherwise they must buy emission credits from countries that are under these levels.

Hence developing countries have no requirements under the Protocol. Further, treaty provides that developed countries pay for costs of developing countries. Developing counties may sell emission credits and receive funds and technology from Annex II (of UNFCCC) countries for climate-related studies and projects. Many Annex I (of UNFCCC) and Annex II (of UNFCCC) countries overlap.

Some targets for some countries are higher than for others, depending on their emission status.

USA has not signed Kyoto Protocol even though it is a major emitter of GHG. Australia signed the Treaty on the 3rd December 2007.

Table 18.1 gives list of Asian countries and remarks with respect to Kyoto Protocol.

Six billion people live in developing countries, but have contributed only around one quarter of GHG in the atmosphere. Energy use and emissions per person in developing countries is around a quarter of that in industrialized countries. Resources for economic restructuring is limited in developing countries, as average per capita income is less than one quarter than those in the developing countries.

Table 18.1 Kyoto Protocol and Asian countries – 'Mixed Basket'

Country	Remarks with respect Kyoto Protocol
Japan	Enlisted in Annex I and II of UNFCCC. Enlisted in Annex B of Kyoto Protocol
Russian Federation	Largest oil producer in the non-OPEC countries, and second biggest in the world. World's third largest emitter of greenhouse gases. Have the world's second largest coal reserves. Annex I country of UNFCCC and Annex B country of Kyoto Protocol
Bahrain, Brunei Darussalam, China, India, Indonesia, Israel, Jordan, Kazakhstan, Korea Democratic Peoples Republic (North Korea), Korea Republic (South Korea), Kyrgyzstan Republic, Laos, Lebanon, Malaysia, Mongolia, Oman, Pakistan, Philippines, Sri Lanka, Syrian Arab Republic, Tajikistan, Thailand, Timor-Leste, Turkey, Turkmenistan, Uzbekistan, Viet Nam	Comprises mostly developing and Economics in Transition countries
Papua New Guinea, Singapore	AOSIS countries – face most at risk from sea-level rise
Iran, Iraq, Kuwait, Qatar, Saudi Arabia, United Arab Emirates	OPEC – economies are highly dependent on income generated from the fossil fuels
Afghanistan, Bangladesh, Bhutan, Cambodia, Myanmar, Nepal, Yemen	Least Developed Countries (LDC). Hardly contributed to the climate change but vulnerable for changing climate

18.4   Public Programs, Policies, and Innovation

Different countries in Asia have different action for common goal. India through its national action plan released in 2008 has concentrated promoting the development and use of solar energy, energy efficiency, Water use efficiency, conserving biodiversity in the Himalayan region, afforestation, adaptation of climate-resilient crops, establishment of Climate Science Research Fund (GoI 2008).

China established the National Coordination Committee on Climate Change through China's national Climate Change Programme. As per the programme it has laid emphasis on scientific approach of development; promote the construction of socialist harmonious society; advance the fundamental national policy of resources conservation and environmental protection; control GHG emission; secure economic development; conserve energy, to optimize energy; rely on the advancement of science and technology; enhance the capacity to address climate change (People's Republic of China 2007).

Singapore's government released the 'National Climate Change Strategy' (NCCS) in 2008. The policy stresses on Increasing energy efficiency; Using less carbon-intensive fuels; Increasing carbon 'sinks' such as forests.

Republic of Russia has adopted legal acts defining national climate policy by initiating actions to increase energy efficiency (Gershinkova and Dinara 2010).

The most impressive policy among Asian country is Japan's Kyoto Target Achievement Plan Approved by Cabinet April 2005 which has Policies for Systems to reduce its greenhouse gas emissions through energy efficiency and savings.

OPEC countries have begun various activities designed to protect the environment. Government of the Emirate of Abu Dhabi has created funding for infrastructure, manufacturing, and renewable energy projects such as solar power, hydrogen, wind power, carbon reduction and management technologies, and carbon capture and storage (CCS). It is also created International Renewable Energy Agency (IRENA) to promote sustainable use of renewable energy sources (ICAO 2009).

Policies are important but effective only if implemented. Since 1972 after The United Nations Conference on The Human Environment countries around the world have churned policies, legislation, programmes but the effectiveness of implementation is still questionable as these policies have created one more set of government institutions which are obedient and dictated by ruling parties. The implementation usually deviate from what is enacted and published due to difference in off the paper communication which is rather oral.

In spite of numerous environmental legislations supplied by countries over the world the quality of environment has deteriorated compared to 1972 due to week institutions without sufficient manpower, skill, commitment and funds. Corruption has not spared institutions created for environment protection and legislation is used as means to generate political and personal funds by people in the power.

18.5 Existing Climate-Finance Instruments

States may influence the decision-making process in three ways. The first is the command-and-control approach which may forbid or license, directly regulate by standards, bans, permits, zoning, quotas, use restrictions, etc. The second is economic measures as incentives to behavior. A third approach is through education, information, and training, as well as social pressure, negotiation and moral arguments.

Economic incentives can either (1) pressurize or apply disincentives; (2) provide incentives; or (3) allow participants to negotiate the level of benefits they receive (e.g., tradable emissions).

The usual instruments used in Environment protection is:

Taxation – Taxation in the context to environment protection is the "price" paid for damaging environment. Charges may have a revenue-raising impact that can finance environmental investments and provide incentives to reduce pollution and waste.

1. **Loans** – Loans in the context of environment protection is financial assistance and incentives for the construction and operation of more environmentally safe installations and recycling systems. International instruments usually refer to as "concessional" access to funds.
2. **Insurance** – Governments can oblige those whose activities likely to damage environment to be insured against damage to third parties or national property.
3. **Grants and subsidies** – Environmental funds, created to directly finance environmental protection. Subsidies can include fiscal measures such as reduced taxes for less polluting business, accelerated depreciation allowances, favourable interest rates for, etc.
4. **Negotiable Permits and joint implementation** – Negotiable permits, sometimes referred to as "bubbles", fixes the total amount of pollution permissible within an area. These are financial instruments that help to lower the costs of environment protection by incentives emissions trading through tradable renewable certificates.
5. **Deposits** – This instrument can be used to receive deposit from polluter to compel him to adopt pollution control system failing which the deposit can be forfeited.
6. **Labeling** – The "*green*" or "*ecolabel*" is an incentive to environmental protection where in eco-friendly products are labeled to advice encourage "learned costumers" so that these products will capture greater market share.

CDM, the main source of mitigation finance to date for developing countries, has shortcomings discussed below.

18.5.1 Uncertain Environmental Integrity

As per *Kyoto Protocol* CDM emission reductions have to be additional to the reductions that could have occurred otherwise. The amount of additionally provided by the CDM has been argued extensively (Michaelowa and Pallav 2007; Schneider 2007).

18.5.2 Inadequate Contribution to Sustainable Development

The CDM was created with two objectives: the global mitigation of climate impact; and the sustainable development of developing countries. But the CDM has been more effectual in bringing mitigation costs than in advancing sustainable development (Olsen 2007; Sutter and Parreno 2007; Olsen and Fenhann 2008; Nussbaumer 2009). Corruption within regulatory setup may lead to certification of all projects as environmentally sustainable.

18.5.3 Weak Governance and Inefficient Operation

The CDM regulating a market dominated by private players through an executive board – essentially a United Nations committee – that approves the calculation methods. The trustworthiness of the CDM depends largely on the strength of its regulatory framework and the private sector's assurance in the opportunities the mechanism provides (Streck and Chagas 2007; Meijer 2007; Streck and Lin 2008).

18.5.4 Limited Scope

CDM projects are not consistently distributed. Nearly half of total Carbon credit accrues to China and quarter of total Carbon credits accrue to Brazil and India.

18.5.5 Uncertainty About Market Continuity

The CDM has not moved to all developing countries onto of low-carbon development (Figueres et al. 2005; Wara 2007; Wara and Victor 2008). Hence there is high uncertainty about market continuity.

18.5.6 Cash and Carbon May Not Move in Same Direction

As shown in Fig. 18.1, carbon movement need not always follow the direction of cash. Movement of gaseous pollutants is always subject to movement of air, and other climatic conditions. It is quite possible that the pollutants along with CO_2 released from developed countries move in opposite direction of cash used in

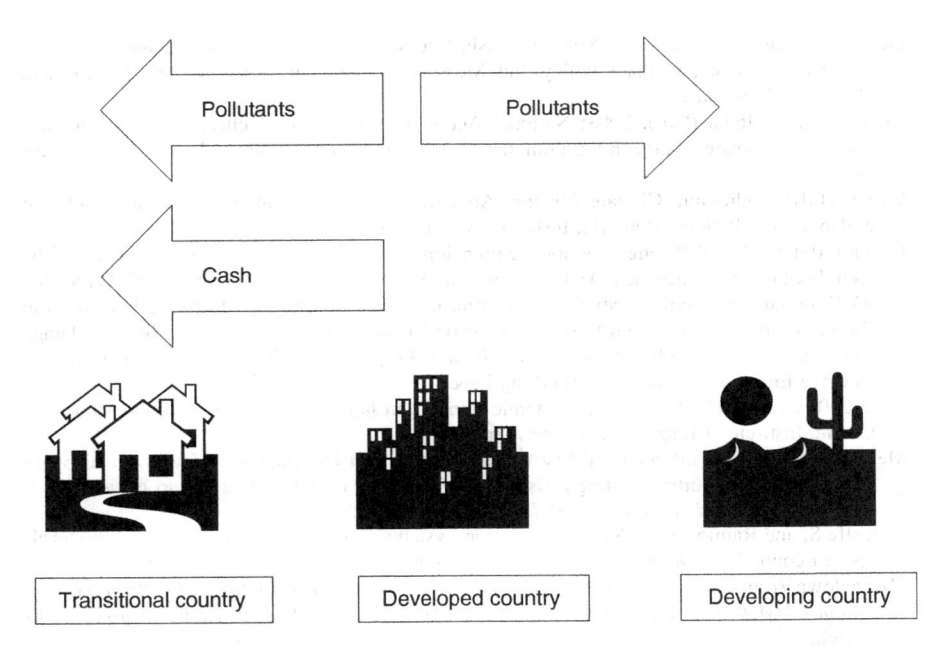

Fig. 18.1 Pollutants need not follow same route as cash in case of 'negotiable permit' instrument

carbon trading. In the process developed countries may just get pollution and GHG where as transitional countries get the monetary benefit.

Kyoto Protocol was appreciated as well as criticized. It is criticized many as it was invented to pay for the transition in the poorer world but used for attaining emission reduction by the cheapest options. CERs used in this transaction never reflected the cost of renewable and other high-technology choices (Narain 2010). Taking these aspects into consideration there is huge scope to evolve better financial instruments in the future after rectifying the flaws in current instruments.

References

BP, 2006: BP Statistical review of world energy, BP Oil Company Ltd., London.

BP, 2010: BP Statistical Review of World Energy, BP Oil Company Ltd., London

Chandler W, Schaeffer R, Dadi Z, Shukla P.R, Tudela F, Davidson O and Alpan-Atamer S, 2002: Climate change mitigation in developing countries: Brazil, China, India, Mexico, South Africa, and Turkey. Pew Center on Global Climate Change.

Edmonds J, Wise M.A, Dooley J.J, Kim S.H, Smith S.J, Runci P.J, Clarke L.E, Malone E.L, and Stokes G.M: 2007: *Global Energy Technology Strategy Addressing Climate Change: Phase 2 Findings from an International Public-Private Sponsored Research Program.* Washington, DC: Battelle Pacific Northwest Laboratories.

Figueres C, Haites E, and Hoyt E. 2005: *Programmatic CDM Project Activities: Eligibility, Methodological Requirements and Implementation.* Washington, DC: World Bank Carbon Finance Business Unit.

Gershinkova and Dinara, 2010: Northeast Asia International Conference for Economic Development in Nigata, Climate Policy and Measures in Russian Federation in "Kyoto" and "Post-Kyto" Periods.

Government of India (GoI), 2008: National Action Plan on Climate change, Prime ministers council on climate change. http://pmindia.nic.in/Pg01-52.pdf downloaded on 14th July 2010. PP49

GTSP, 2001: Addressing Climate Change: An initial report of an international public-private collaboration. Battelle Memorial Institute, Washington, D.C.

Ibrahim Abdel Gelil, 2009: Ghg Emissions - Mitigation Efforts In The Arab Countries, Report Of The Arab Forum For Environment And Development, Edited By Mostafa K. Tolba Najib W. Saab

ICAO (International Civil Aviation Organization), 2009: Emissions Reductions: United Arab Emirates Contribution, High-Level Meeting on International Aviation and Climate Change Montréal, 7 To 9 October 2009, Agenda Item 2: Proposals for Strategies And Measures To Achieve Emissions Reductions, Working Paper

IEEJ, 2005: Handbook of energy & economic statistics in Japan. The Energy Data and Modelling Centre, Institute of Energy Economics, Japan.

Meijer E. 2007: "The International Institutions of the Clean Development Mechanism Brought before National Courts: Limiting Jurisdictional Immunity to Achieve Access to Justice." *NYU Journal of International Law and Politics* 39 (4): 873–928.

Metcalfe S, and Ramlogan R, 2008: "Innovation Systems and the Competitive Process in Developing Economies." *Quarterly Review of Economics and Finance* 48 (2): 433–46.

Michaelowa A and Pallav P. 2007: *Additionality Determination of Indian CDM Projects. Can Indian CDM Project Developers Outwit the CDM Executive Board?* Zurich: University of Zurich.

Narain Sunita, 2010: A 'just' climate agreement: the framework for an effective global deal; Global Sustainability - A Nobel Cause , Edited By Hans Joachim Schellnhuber, Mario Molina, Nicholas Stern, Veronika Huber, Susanne Kadner, 108–121

Nussbaumer P, 2009: "On the Contribution of Labelled Certified Emission Reductions to Sustainable Development: A Multi-criteria Evaluation of CDM Projects." *Energy Policy* 37 (1): 91–101.

Olsen K.H, 2007: "The Clean Development Mechanism's Contribution to Sustainable Development: A Review of the Literature." *Climatic Change* 84 (1): 59–73.

Olsen K.H and Fenhann J, 2008: "Sustainable Development Benefits of Clean Development Mechanism Projects. A New Methodology for Sustainability Assessment Based on Text Analysis of the Project Design Documents Submitted for Validation." *Energy Policy* 36 (8): 2819–30.

People's Republic of China, China's National Climate Change Programme, 2007: Prepared under the Auspices of National Development and Reform Commission, PP66

RITE, 2003: New Earth 21 Project. Research Institute of Innovative Technologies for the Earth. Kyoto, Japan,

Schneider, L. 2007: *Is the CDM Fulfilling Its Environmental and Sustainable Development Objective? An Evaluation of the CDM and Options for Improvement.* Berlin: Institute for Applied Ecology.

Sutter C, and Parreno J.C. 2007: "Does the Current Clean Development Mechanism (CDM) Deliver Its Sustainable Development Claim? An Analysis of Officially Registered CDM Projects." *Climatic Change* 84 (1): 75–90.

Streck C, and Chagas T.B. 2007: "The Future of the CDM in a Post-Kyoto World." *Carbon & Climate Law Review* 1 (1): 53–63.

Streck C and Lin J, 2008: "Making Markets Work: A Review of CDM Performance and the Need for Reform." *European Journal of International Law* 19 (2): 409–42.

Sims R.E.H, Schock R.N, Adegbululgbe A, Fenhann J, Konstantinaviciute I, Moomaw W, Nimir H.B, Schlamadinger B, Torres-Martínez J, Turner C, Uchiyama Y, Vuori S.J.V, Wamukonya N, Zhang X, 2007: Energy supply. In Climate Change 2007: Mitigation. Contribution of

Working Group III to the Fourth Assessment Report of the Intergovernmental Panel on Climate Change [Metz B, Davidson O.R, Bosch P.R, Dave R, Meyer L.A (eds)], Cambridge University Press, Cambridge, United Kingdom and New York, NY, USA.

Shahin W.R, 2005: NERC Presentation. CLASP Online. At: http://www.clasponline.org/files/WkshpTunisiaNov05_Jordan.pdf (Accessed 27 September, 2009).

Singapore's National Climate Strategy, 2008, PP44

Stern N. 2007: *The Economics of Climate Change: The Stern Review*. Cambridge, UK: Cambridge University Press.

Wara M. 2007: "Is the Global Carbon Market Working?" *Nature* 445: 595–96.

Wara M, and Victor D, 2008: "A Realistic Policy on International Carbon Markets." Working Paper 74, Program on Energy and Sustainable Development, Stanford University, Stanford, CA.

Appendix A: Glossary

Absorption spectrum Absorption spectrum is the portion of electromagnetic wavelength that is absorbed by a given substance from a array of frequencies.

Acclimatization Acclimatization is the physiological *adaptation of living organism to change in* climate and environment.

Accountability Accountability is to answer or be judged for one's conduct; to be responsible or liable for wrongdoing.

Active layer Active layer is the top layer of soil or rock in *permafrost which* is subjected to seasonal freezing and thawing.

Adaptation Adaption is getting adjusted to changing scenario.

Adoption Adoption is the acceptance of changing scenario.

Aerodynamics Dynamics concerned with movement of air is called aerodynamics.

Aerosols A collection of particles, with size between 0.01 and 10 μm, that exist in the *atmosphere* for at least several hours.

Affected (in the context of disaster) People needing immediate basic survival need like food, water, shelter, medical assistance.

Affected (in the context of epidemics) Affected in the context of epidemics is the appearance of disease to a person/people free from that disease.

Afforestation Human-induced forest plantation.

Agreement The term "international agreement" reflects the widest range of international instruments, referring both to binding treaties and to political commitments. It is general tendency to apply the term "agreement" to bilateral or restricted multilateral treaties.

Aggregate impacts Overall *impacts* integrated across sectors and/or regions.

Albedo Albedo is the portion of solar radiation reflected by a surface or object.

R. Chandrappa et al., *Coping with Climate Change*,
DOI 10.1007/978-3-642-19674-4, © Springer-Verlag Berlin Heidelberg 2011

Algae Algae are a variety of plant without distinct functional plant tissue occurring in marine and freshwater ecosystems.

Algal bloom Increase in population of *algae* in marine and freshwater *ecosystems*.

Alpine The biogeographic region made up of slopes above the *tree line* distinguished by the existence of rosette-forming *herbaceous* plants and low, shrubby, slow-growing woody plants.

Alpine climate Alpine is the climate of region above the tree line. This climate is also referred to as mountain climate or highland climate.

Amendment Amendment is a subsequent change or modification of the terms of a legal instrument.

Anabatic winds An anabatic wind is a wind which blows up a steep slope due to heating of the slope.

Anaerobic process Process that occurs in the absence of air is called anaerobic process. Such process are often used in treatment of degradable waste.

Anatolia Anatolia is a term used for the westernmost protrusion of Asia, comprising the majority of the Republic of Turkey.

Annex I countries/parties Countries/parties listed in Annex I of the UNFCCC are referred as Annex I countries/parties.

Annex II countries/parties Countries/parties listed in Annex II of the UNFCCC are referred as Annex II countries/parties.

Annex B countries/parties Countries/parties listed in Annex B of Kyoto Protocol is referred as Annex B countries/parties.

Anthropogenic Anything resulting from human activity such as transportation, agriculture, industrial activity is termed as anthropogenic. Human activity is called anthropogenic activity.

Anthropogenic emissions Emissions from human activity is called anthropogenic emissions.

APEC (Asia-Pacific Economic Cooperation) Asia-Pacific Economic Cooperation forum is forum founded in 1989 with 21 member economies: Australia, Brunei, Canada, Chile, the People's Republic of China, Chinese Taipei, Hong Kong, China, Indonesia, Japan, the Republic of Korea (ROK), Malaysia, Mexico, New Zealand, Papua New Guinea, Peru, the Republic of the Philippines, the Russian Federation, Singapore, Thailand, the United States, and Vietnam.

APP (Asia-Pacific Partnership for Clean Development and Climate) APP is international non-treaty agreement among Australia, Canada, India, Japan, the People's Republic of China, South Korea and United States of America to cooperate, development and transfer of technology with respect to reduction of GHG.

Aquaculture   The cultivation of aquatic plants or animals held in captivity for the purpose of harvesting.

Aquifer   Layer of permeable rock which holds water below the ground is called aquifer.

Archipelago   An archipelago cluster of islands formed tectonically.

Arid region   Area with low rainfall ('low' is usually considered to be less than 250 mm precipitation per year).

Atmosphere   The gaseous layer surrounding the Earth is termed as atmosphere.

Avalanche   Snow or ice that slides down a mountainside by gravity due to natural or anthropogenic causes is termed as avalanche. It occurs when the load on the upper snow layers surpasses the bonding of the mass of snow above sliding layer.

Axial tilt (in the context of earth)   Inclination of the Earth's axis to plane of orbit around the Sun is termed as axial tilt.

Baseline/reference   The baseline (or reference) is the situation against which change is measured.

Basin   The drainage area which contributes water to water body (stream, river or lake).

Before the Current Era (BCE)   Also known as Before Christian Era or Before Common Era is a secular alternative to BC (Before Christ).

Before Present (BP)   Before Present (BP) is a time scale used to specify when events in the past occurred. Since time changes the standard practice is to use 1st January 1950 as the arbitrary origin of time scale.

Benthic community   Benthic community is the community of organism living on or near the base of a water body.

Biodiversity   Biodiversity is the overall diversity of organisms and *ecosystems.*

Bioenergy   Energy generated using biomass is referred as Bioenergy.

Biofuel   Biofuel is the fuel produced from organic matter.

Biogeochemical cycles   The cycling of chemical elements through the biotic and abiotic components of an ecosystem is called Biogeochemical cycle.

Biological disasters   *Disaster* due to exposure of living organisms, to germs is called Biological disaster.

Biomass   The total mass of living organisms located in area or volume of interest.

Biome   Biome is the major and distinct regional element of the *biosphere.*

Biosphere   The part of the Earth system encompassing all *ecosystems* and living organisms in the *atmosphere*, on land, or in the oceans.

Biota   All living organisms of an area is referred as biota.

Blizzard/snowstorm   A snowstorm refers to a storm, where a large amounts of snow fall. If it is accompanied by strong winds and blowing snow is called blizzard.

Boreal forest   Forests characterized by pine, spruce, fir and larch is termed as Boreal forest. The climate is continental, with long, very cold winters (temperatures below freezing), and short, cool summers (50–100 frost-free days).

Bottom-up approach   Study carried out progressing from smaller subordinate units to higher units is called bottom-up approach.

Breakwater   Human made structure built in the sea for the purpose of breaking waves.

C3 plants   Plants that produce a three-carbon compound during *photosynthesis* (e.g. rice, wheat, soybeans, potatoes and vegetables) are categorized as C3 plants.

C4 plants   Plants that produce a four-carbon compound during *photosynthesis* (e.g. maize, sugar cane, millet and sorghum) are categorized as C4 plants.

Calcareous organisms   Organisms that use *calcite* or *aragonite* to form shells or skeletons are categorized as calcareous organisms.

Calcite   A mineral characterized by calcium carbonate.

Capacity building   Developing the knowledge and institutional capabilities

Carbon credit   Carbon credit is the credit for a reduction of GHGs in Kyoto protocol. Each carbon credit is equal to the decrease of 1 metric ton of CO_2 or CO_2 equivalent.

Carbon cycle   Carbon cycle is the circulation of carbon atoms between biotic and abiotic components of the Earth.

Carbon dioxide (CO_2)   Chemical compound composed of one carbon and two oxygen atoms.

Carbon dioxide equivalent   Carbon dioxide equivalent of a GHG is the concentration of GHG that has the same CO_2 global warming potential.

Carbon footprint   The amount of GHG emissions produced by an individual, organization, event, or product is called carbon foot print.

Carbon intensity   The quantity of carbon emitted per unit of energy generated is termed as carbon intensity.

Carbon market   Carbon market is the market where in right to emit is traded like equities.

Carbon leakage   Carbon leakage is the term use the context of climate policy referred to rise in CO_2 in one nation as a result of reduction in CO_2 emission by another.

Carbon offset Compensation for carbon emissions those are impossible or too costly to be avoided is termed as carbon off set.

Carbon price Carbon price is the cost of carbon per unit (usually metric ton) of avoided CO_2 emission from the atmosphere.

Carbon sequestration Carbon sequestration is the process of enhancing the carbon content of a *reservoir*/pool.

Carbon sink A medium which removes carbon from the atmosphere and stores it for a long period of time is called carbon sink.

Catchment Catchment area at any given point is the area that collects and drains rainwater through that point.

CCS (Carbon Capture and Storage) CCS is the process wherein the CO_2 is captured and stored to reduce changing climate.

CDM (Clean Development Mechanism) Mechanism under the Kyoto Protocol that allows GHG gas emission reduction projects to take place in countries with no emission targets.

CDM-EB (Clean Development Mechanism – Executive Board) CDM-EB is the executive board which overseas execution of CDM.

Chemical spill Chemical spill is the accident release of chemicals during the production, transportation or handling of hazardous chemical materials.

Cholera Cholera is a water-borne intestinal infection that results in frequent watery stools, cramping abdominal pain, and dehydration. This disease is caused by a bacterium – *Vibrio cholerae*.

Climatological disasters Climatological disasters are events caused by meso to macro scale processes in the spectrum of intra-seasonal to multi-decadal climate variation.

Climate In short climate is defined as the 'average weather'. In wider sense climate is the state, including a statistical description, of the *climate system*. The classical period of time is 30 years, as per World Meteorological Organization (WMO).

Climate change Change in climate over time.

Climate model Climate model is any mathematical representation of the climate system.

Climate prediction (climate forecast) Estimate of climate in the future.

Climate risk management An approach to methodically manage climate-related risks is called climate risk management.

Climate sensitivity   Climate sensitivity is the equilibrium temperature increase that would happen for a doubling of CO_2 concentration beyond *pre-industrial* levels.

Climate system   Dynamics and interactions of five major components: atmosphere, hydrosphere, cryosphere, land surface, and biosphere.

Climate threshold   The point at which external forcing, initiates a significant climatic or environmental event that is considered unalterable or recoverable after long time (e.g. widespread bleaching of corals or a collapse of oceanic circulation systems).

Climate variability   Climate variability is the natural variation in climate that lies within the normal range of extremes in a given region.

Coastal erosion   Landward progress of the shoreline due to the forces of waves and currents is called coastal erosion.

Coastal squeeze   Coastal squeeze is the squeeze of coastal *ecosystems* between rising sea levels and shorelines.

Cold wave   Lengthy period of extremely cold weather and the abrupt invasion of very cold air over a large area are termed as cold wave.

Committed to extinction   Committed to extinction in the context of living organism is process in which species becomes extinct in the absence of human intervention.

Communicable disease   Disease caused due to transmission of an infective biological agent.

Continental climate   Climate characterized by hot summers, cold winters, and little rainfall, representative of the interior of a continent.

Customary law   Customary law is the binding rules emerging from the practices and usages of states and recognized as legally binding by them.

Coral   Corals are organisms belonging to order *Scleractinia* which have hard limestone skeletons.

Coral bleaching   The paling of coral color due to loss of symbiotic, energy-providing, organisms is called coral bleaching.

Coral reefs   Structures built by corals are referred as coral reefs.

Coriolis effect   Coriolis effect is an inertial force explained by the nineteenth-century French engineer-mathematician Gustave-Gaspard Coriolis in 1835. The effect of the Coriolis force is a noticeable deviation of the path of an object that moves within a rotating coordinate system. The object in reality does not deviate from its path, but it appears to do so.

Coriolis force   Coriolis force a fictitious force exerted on a body when it moves in a rotating reference frame.

CO_2-equivalent emission CO_2-equivalent emission of a GHG is the amount of CO_2 emission that would end up in the same time-integrated radiative forcing as that of GHG. The equivalent CO_2 emission is calculated by multiplying the emission of a GHG by its Global Warming Potential (GWP) for the given time horizon.

CO_2-equivalent concentration CO_2-equivalent concentration of a GHG is the concentration of CO_2 that would cause the same amount of radiative forcing as that of GHG.

Cirrus clouds   Cirrus colds are atmospheric clouds that are characterized by thin strands.

Cretaceous period   The cretaceous period refers to the period from around 145.5 ± 4 to 65.5 ± 0.3 million years ago.

Crop failure   Abnormal reduction in crop yield which is not sufficient to meet the nutritional or economic needs of the community is termed as crop failure.

Cryosphere   The component of the *climate system* comprising of all snow along with ice located on and below the surface of the Earth and ocean.

CSLF (Carbon Sequestration Leadership Forum)   A ministerial-level international forum for development and improving cost based CCS technologies.

Ctenophores   Ctenophores, commonly known as comb jellies, are type of animals that live in marine waters.

Cyclone   Cyclone (sometimes referred as tropical cyclones) is a large-scale closed circulation system in the atmosphere above the Indian Ocean and South Pacific. It is characterised by low barometric pressure and strong rain and winds of 64 knots or more. The most severe versions are called hurricanes (in the North Atlantic, the Northeast Pacific east of the dateline, or the South Pacific east of 160E) or typhoon (in the Northwest Pacific west of the dateline). Cyclones only form and intensify above warm water, and are becoming more intense due to the warming of the ocean surface due to global warming.

Deforestation   Deforestation is the conversion of forest land to non-forest land.

Dengue fever   An infectious viral disease spread by mosquitoes, characterized by severe pain in the joints and back.

Desert   A region of very low rainfall ('very low' is widely considered to be <100 mm/year).

Disaster   Occurrence of an extreme hazard event that impacts on vulnerable communities causing significant damage, disruption and casualties is called disaster.

Disaster management Disaster management is a systematic process of implementing policies, strategies, and measures to reduce the impacts of disasters.

Disaster preparedness Disaster preparedness is the range of activities that contribute to the pre-planned effective response to reduce the impact of disaster and deal with the consequences of future disasters.

Disaster recovery Disaster recovery is the process and actions taken after a disaster with a view to restoring or improving the pre-disaster conditions of the stricken community.

Disaster rehabilitation Disaster rehabilitation is the set of actions taken after a disaster to bring back to disaster affected community to normal life by enabling basic services to resume functioning, repairing physical damage to community facilities, to revive economic activities and well-being of the survivors.

Disaster relief/response Disaster relief is the coordinated activities aimed at meeting the needs of people who are affected by a disaster.

Disaster risk reduction Disaster risk reduction is the measures at all levels to curb disaster losses, by reducing exposure to different hazards.

District heating District heating is a system where in the heat generated in centralized location is distributed for use by end users.

Drought Drought is the phenomenon that arises when precipitation is significantly below normal levels.

Dyke Human-made wall along a shore to avoid flooding of low-lying land is called dyke.

Early warning Providing timely and effective information regarding an forthcoming hazard that allows people to take action to avoid a disaster or prepare for effective response is called early warning.

Earthquake Earthquake is shaking of ground due to seismic waves.

Ecological community Ecological community is a community of living organism characterized by a distinctive assemblage of species and their population.

Ecological corridor Ecological corridor is a thin strip of vegetation that allows movement of organisms between two areas.

Economies in Transition (EIT) Countries undergoing the process of transition to market economy are called economies in transition.

Ecosystem Ecosystem is the interactive system of living things and their abiotic (physical and chemical) environment.

Ecosystem services Ecosystem services are the ecological processes having monetary or non-monetary value.

Ecotone   Ecotone is changeover area between adjacent *ecological communities* (e.g., between forests and grasslands).

El Niño   El Niño is a warm-water current that occasionally flows along the coast of Ecuador and Peru. The event is usually accompanied by heavy rainfall in the coastal region of Peru and Chile, and reduction of rainfall in equatorial Africa and Australia.

El Niño-Southern Oscillation (ENSO)   Variation of the inter-tropical surface pressure pattern and movement in the Indian and Pacific Oceans is called the *southern oscillation*. Collective atmosphere-ocean phenomenon of *El Niño* and *southern oscillation* is known as *El Niño-Southern Oscillation*. The opposite of an El Niño event is *La Niña*.

Emergency   Emergency is the unforeseen event that requires immediate action to minimise its adverse consequences.

Emission registry   Emission registry is the database of issuance, allocation, holding, transfer and cancellation of emission allowances.

Emission scenario   Emission scenario is the future scenario about carbon emissions based on set of assumption about driving forces.

Emission trading   Emission trading in the context of carbon market is the market based approach where in carbon credits are treaded like equity.

Enactment   Enactment is a duly adopted law.

Endangered species   Endangered species are the species threatened with extinction due to declining population levels.

Endemic   Endemic means restricted to a region. The term endemic species is often used to show the importance of a region with respect to species.

End-user sectors   Major energy-consuming sectors are referred as end-user sectors.

Energy balance   Difference between energy entering and leaving the system is called energy balance.

Enforcement   Executing or making effective or compelling obedience is termed as enforcement.

Eocene epoch   The eocene epoch refers to period from about 56 to 34 million years ago

Eustatic sea-level rise   Change in global average sea level due to rise in the volume of the world ocean.

Eutrophication   The phenomenon wherein nutrient deficient water body becomes rich in dissolved nutrients.

Evapotranspiration   Evapotranspiration is the loss of water due to evaporation from the surface of earth and *transpiration* from vegetation.

Ex-situ conservation   Protection of biological resources outside their normal habitat is called ex-situ conservation.

Extinction   Extinction in the context of a species is the disappearance of a species from earth.

Extreme weather event   Weather that is extreme in a particular place (such as extreme intense rainfall, extreme heat, a very strong windstorm) is called extreme weather event.

Expiration   Expiration in the context of species is the loss of population of species in a given location

Famine   Famine is the food shortage affecting large numbers of people as a result of climatic, environmental and socio-economic reasons.

Feedback   Feed back is the process through which a system is controlled or modulated in response to its own output.

Feed-in tariffs   Feed-in tariffs is a mechanism to encourage the adaptation of renewable energy through legislation.

Flash flood   Sudden flooding with short duration is called flash flood.

Flood   Significant rise of water level in a surface water body stream, lake, reservoir or coastal region is called flood.

Food chain   The chain formed in respect of feeding relationships when several species feed on each other.

Food web   Several interconnected food chains together is called food web.

Forest fire   Burning of forest is called forest fire.

Fossil fuel   Fuel derived from fossils such as coal, natural gas and petroleum products are called fossil fuel.

Frozen ground   Ground in which the entire pore water in soil and rocks are frozen.

Fuel cell   Fuel cell is the cell in which energy of a fuel is directly converted in to energy.

Gondwana   Gondwana is the southernmost of two precursor supercontinents formed due to split of the huge Pangaea supercontinent

Geophysical disasters   Events originating from solid earth are termed as geophysical disasters.

Geologic epoch The geologic epoch of geologic time scale is a system to describe the timing and relationships among events that have happened during the history of earth.

Geologic storage Storage in the ground is called geologic storage.

Genetic resources Genetic material of actual or potential value is called genetic resources.

Glacier A mass of land ice movement due to internal deformation and sliding is called glacier.

GLOF (Glacier Lake Outburst Flood) Flooding due to the outburst of a glacier lake is called glacier lake outburst.

Green building Green building is the practice of creating structures that are environment-friendly and resource-efficient throughout a building's life-cycle.

Green certification Green certification is the certification process with respect to activity, product and services for its eco-friendliness.

Greenhouse effect The phenomenon in which the absorption of infrared radiation by the atmosphere rises temperature of the Earth is called greenhouse effect.

GHG (Greenhouse Gas) GHG are the gases that absorb and emit radiation at particular wavelengths within the range of infrared radiation.

GHG registry GHG registry is the database of emitters with respect to GHG emission.

Gross primary production The total carbon fixed by plant through photosynthesis.

Groundwater recharge Groundwater recharge is the process of adding external water to the zone of saturation of an aquifer.

G8 Group of eight countries namely Canada, France, Germany, Italy, Japan, Russia, United Kingdom and United States of America.

GWP (Global Warming Potential) GWP of a substance is the ratio of warming of atmosphere of the substance to that caused by same mass of CO_2.

Habitat Habitat is the locality in which a particular living organisms lives. It is the place or type of site where an organism or population naturally occurs.

Hail storm A hail storm is a type of storm characterised by hail as the prevailing part of its precipitation. The size of the hailstones can vary between pea size and softball size and therefore cause considerable damage.

Halocarbons Halocarbons are group of halogenated organic compounds.

Hazard Hazard is the threatening event within a given time period and area.

Heat island   An urban area distinguished by ambient temperatures greater than that of the surrounding non-urban area.

Heat wave   A heat wave is a prolonged period of excessively hot weather relative to normal climate patterns in a given region.

Heating value   Heating value is the measure of useful energy in a fuel.

Hibernation   Hibernation is state of inactivity due to lower metabolic rate. Some species will hibernate continuously for several days in winter to conserve energy in the body when availability of food is short.

High-latitude region   Part of earth near the North Pole and South Pole is termed as high-latitude region.

Hirnantian   The Hirnantian is the seventh and final stage of Ordovician period of the Paleozoic Era occurred from 445.6 ± 1.5 to 443.7 ± 1.5 million years ago.

Holocene   Holocene is a geological time period which began about 12,000 years ago.

Hybrid vehicles   Vehicles that use two sources for energy are called hybrid vehicles.

Hydrographic events   Hydrographic events are events that change the condition or current of waters in oceans, rivers or lakes.

Hydrologic cycle   Hydrologic cycle or water cycle is the term used to define the continuous circulation of water throughout the earth and atmosphere.

Hydrological disasters   Events caused by deviations in the normal water cycle and/or overflow of surface water bodies.

Hypolimnetic   Part of a lake below the *thermocline* comprising of stagnant water with uniform temperature except during the period of overturn is termed as hypolimnetic.

Hypoxic events   Events that lead to a deficiency of oxygen are called hypoxic events.

Hurricane   Large-scale closed clockwise circulation system in the southern hemisphere and counter-clockwise circulation system in the northern hemisphere in the atmosphere above the western Atlantic with low barometric pressure and strong winds with speed of 64 knots or more.

Ice cap   Ice mass covering a highland area that is smaller than an *ice sheet* is called ice cap.

Ice flow   A stream of ice flowing faster than surrounding sheet of ice is called ice flow.

Ice sheet   A mass of land ice that covers most of the underlying bedrock topography is called Ice sheet.

Ice shelf Ice shelf is a floating *ice sheet* attached to a coast often a seaward.

Industrial revolution A period of rapid industrial growth is termed as *industrial revolution*.

Infectious disease Any disease caused by microbial agents.

In situ conservation The conservation of ecosystems and natural habitats as well as the maintenance along with recovery of viable populations of species in their natural surroundings is called in situ conservation.

Integrated assessment A method of analysis, which combines results and models of physical, biological, economic and social sciences is called integrated assessment.

Intergovernmental Panel on Climate Change (IPCC) IPCC is an international organization of scientists established by the United Nations and the WMO whose task is to publish the most recent scientific findings on climate change.

Invasive species and invasive alien species (IAS) A species assertively expanding its range and population in a region in which it is not native.

Isohyet A line on a map joining locations that receive the same quantity of rainfall is called Isohyet.

Jurisdiction Jurisdiction is an area where legal authority enacts and enforces laws. Jurisdiction of is an area where they take and for courts to take cognizance of law suits.

Katabatic winds Katabatic winds are winds that carry high density air from a higher altitude down a slope under the force of gravity.

Katian age The Katian age of the Late Ordovician Epoch refers to the period from 456 to 446 million years ago

Kyoto protocol Protocol adapted to UNFCCC during third conference of parties in Kyoto, Japan.

Landfill A landfill is an engineered solid waste disposal site where waste is deposited in trenches or pits taking all precaution to control leachate and gases emitted.

Landslide Landslide is the rapid movement of a mass of soil, rock or debris down a slope. It includes lahar, mudslide, debris flow.

Land-use Various types of uses for which land is utilized (e.g., forest, agriculture, urban settlement, mining).

La Niña It is the opposite of *El Niño* in which the ocean becomes much cooler than normal. It is thought to occur due to rise in the strength of the trade winds which increases the amount of cooler water toward the West Coast of South America resulting in fall in water temperatures.

Leaching The elimination of soil elements or applied chemicals by movement of water through the soil is termed as leaching.

Leapfrogging Leapfrogging is the ability of developing countries to bypass intermediate technologies and jump straight to advanced clean technologies.

Levee Levee is a natural or artificial bund or wall to regulate water levels.

Liability Liability is all forms of obligation imposed by law; assignment of responsibility for injury or harm.

Lightening Lightning is an atmospheric discharge of electricity, usually occurring during thunderstorms. It sometime occurs during volcanic eruptions or dust storms.

Littoral zone Littoral zone is the zone between high and low watermarks in a coastal region.

Local windstorm Local windstorm refers to strong winds caused due to regional atmospheric phenomena that are typical for a certain area.

Long-lived GHGs GHGs that have long residence time in the atmosphere is called long-lived GHGs.

Low latitude region Part of the earth near to equator is called low latitude region.

Mandate Mandate is a command, order, or direction, indicating action to be taken.

Marrakech accords Accords during seventh session of UNFCCC Conference of Parties held at Marrakech from 20 October to 10 November 2001.

Mediterranean region Countries adjacent to Mediterranean sea is called Mediterranean region.

Meridional overturning circulation Meridional overturning circulation in the ocean measured by zonal sums of mass transport in depth or density layers.

Methane to market partnership Methane to market partnership is the international initiative that advances cost-effective, methane recovery and use.

Meteorological disasters Meteorological disasters are events caused by short-lived/small to mesoscale atmospheric processes (occurs in range from minutes to days).

Microclimate Local climate at or near the Earth's surface is referred as microclimate.

Mitigation (in the context of climate change) Human involvement to reduce the sources of GHGs or increase the sinks that remove them from the atmosphere is called mitigation of climate change.

Mixing height Top of layer within the atmosphere where rigorous mixing of gases take place is called mixing height.

Monitoring   Monitoring is the repeated or continuous supervision of performance.

Monsoon   A seasonal prevailing wind in tropical and sub-tropical regions is called monsoon. The phenomenon lasts for several weeks and leads to substantial changes in rainfall.

Montane   The biogeographic region made up of relatively moist, cool upland slopes below the *sub-alpine* zone characterized by mixed deciduous at lower region and coniferous evergreen forests at higher places.

Moraine   An accumulation of boulders, stones and other mass carried and deposited by glacier is called moraine.

Negotiate   Negotiate in the context of international agreements is to bargain or arrange a settlement or draft a text.

No regrets GHG-reduction   GHG-reduction actions that have negative net costs is termed as no regrets GHG-reduction costs.

Non-permanence   Non-permanence refers to risks involved in emission removals by sinks (as sinks like forest is reduced due to deforestation and desertification).

Nutrient loads   The total amount of nitrogen and phosphorous entering a water body during a given time is referred as nutrient loads.

Miocene era   The Miocene era spans from about 23.03 to 5.332 million years ago.

Montreal protocol   The Montreal protocol is a protocol on production and consumption of chlorine and bromine-containing chemicals that destroy stratospheric ozone. It was adopted in Montreal in 1987, and subsequently amended in London (1990), Copenhagen (1992), Vienna (1995), Montreal (1997) and Beijing (1999).

Morbidity   Morbidity is the rate of occurrence of health disorders within a population.

Mortality   Mortality is the rate of occurrence of death within a population.

Natural hazard   Natural hazard is natural event that has the potential to cause harm or loss.

Neolithic period   The Neolithic period, or New Stone Age, was last part of the Stone Age. It is divided into (1) Neolithic 1 – Pre-Pottery Neolithic A (PPNA) around 10,700–9400 BC; (2) Neolithic 2 – Pre-Pottery Neolithic B (PPNB) began around 8500 BC (3) Neolithic 3 – Pottery Neolithic (PN) began around 6500 BC.

Nitrogen oxides (NOx)   Any of oxides of nitrogen is referred as Nitrogen oxides.

Nosocomial Infection   Infection transferred from health care facilities to healthy individuals.

Ocean acidification   Ocean acidification is the increase in acidity in sea water.

OECD (Organization for Economic Co-operation and Development) OECD is an organization of 30 countries that accept the principles of representative democracy and free-market economy.

Ocean heat uptake efficiency Ocean heat uptake efficiency is a measure of the rate at which heat storage by global ocean augments as global surface temperature increases.

Ordovician period The Ordovician refers to time between 488.3 ± 1.7 and 443.7 ± 1.5 million years ago.

Orogeny Orogeny refers to forces and events that lead to structural deformation of the Earth's crust due to the movement of tectonic plates.

Paleocene – Ecocen boundary The most extreme change in Earth surface started 55.8 million years ago which is termed as Paleocene – Ecocen boundary.

Paleoclimatic records Reconstruction of climate data by studying tree rings, corals and ice cores.

Paleolithic Era The Paleolithic (or Palaeolithic) Era is an era during which development of the first stone tools occurred. It spans from about 2.5 million years ago, to around 12,000 BP.

Paleozoic Era The Paleozoic Era refers to the period from about 542 to 251 million years ago. It is further subdivided into following eras: Cambrian, Ordovician, Silurian, Devonian, Carboniferous, and Permian.

Pangea Supercontinent that existed during about 250 million years ago in the Paleozoic and Mesozoic eras is termed as Pangea.

Particulates Particulates are particles with size more than molecule and less than 500 μm.

Pastoral Pastoral refers to the lifestyle of pastoralists, such as shepherds herding livestock moving according to changing availability of water and pasturage.

Peat Partially decomposed dead plants typically *Sphagnum* mosses.

Peatland A wetland of slowly accumulating *peat* is called Peatland.

Permafrost Perennially frozen ground where the temperature remains below 0°C for several years.

Permian period Period from 299.0 ± 0.8 to 251.0 ± 0.4 million years before the present is termed as Permian period.

Penal law Penal law is a law which prohibits acts and imposes penalties for the commission of those acts.

Penalty Penalty means any punishment imposed as the consequence for committing an offense.

Phenology Periodic plant and animal life cycle events and how these are influenced by seasonal and interannual variations in climate.

Phytoplankton The plant forms of *plankton*.

Plankton Plankton are microscopic aquatic organisms that drift or swim weakly.

ppm (parts per million) Measure used for chemicals of very low concentration. It is one part of any substance out of million parts of solvent or media in which it is dissolved or suspended.

Primary energy Primary energy is all energy consumed by end-users. This comprises the energy used to generate electricity, but does not comprise only the electricity.

Primary production Production of organic matter through photosynthesis is termed as primary production.

Protected area Protected area is a geographically defined area which is designated or regulated and managed to achieve specific conservation objectives.

Protocol Protocol, is the agreement less formal than those entitled "treaty" or "convention". It is usually used to designate an instrument subsidiary or supplementary to a treaty, and drawn up by the same parties. A protocol based on a framework or treaty is an instrument with specific obligations that implements the general objectives of the framework or umbrella convention.

Proterozoic era The Proterozoic era is a period that occurred before the first abundant complex life on Earth.

Quota Quota is a proportional part or share; quantitative restrictions.

Radiative forcing Radiative forcing is a measure of the influence a factor has in changing the balance of incoming and outgoing energy in the Earth-atmosphere system and is expressed in watts per square metre (W/m^2).

Radiolarians Radiolarians are amoeboid protozoa that produce mineral skeletons dividing the cell into inner and outer parts, called endoplasm and ectoplasm.

Rangeland Rangelands are unmanaged grasslands, shrub lands, savannas and tundra.

Reforestation Converting deforested area in to forest.

Regenerative braking Regenerative braking is a mechanism in which kinetic energy generated during breaking in the vehicle is stored for future use.

Resilience Resilience is the ability of a social or ecological system to cope with change without disturbing structure and functioning.

Snow line   Snow line is the lower limit of permafrost snow cover below which snow does not accumulate.

Social cost of carbon   Social cost of carbon is the quantification of the damage due to climate change in monetary terms.

Soil carbon stocks   Carbon stored in soil is termed as soil carbon stocks.

Solar radiation   Electromagnetic radiation emitted from sun is called solar radiation.

Spillover effect   The economic effect of domestic or sectoral mitigation measures on other nations of sectors is called spillover effect.

Stabilization scenarios   Stabilization scenarios in the context of GHG, is the scenarios wherein they match to levels that would achieve certain temperature goal or certain goal of atmospheric CO_2 concentration.

Risk   Risk is the expected losses (like loss of lives, persons injured, and property damaged) due to a particular hazard for a given area and period.

Rock fall   Rock fall is the rock or stone falling freely from a cliff face caused by undercutting, weathering or permafrost degradation.

Ruminant   A mammal that digests food by softening initially in one of the four chambers, then regurgitating the semi-digested mass called cud, and chewing it again.

SAARC (South Asian Association of Regional Cooperation)   Association of South Asian countries, namely Afghanistan, Bangladesh, Bhutan, India, Maldives, Nepal, Pakistan and Sri Lanka.

Sahel   Sahel is ecoclimatic and biogeographic zone of transition between the Sahara desert and the Sudanian savannas. It covers parts of the countries of Senegal, Mauritania, Mali, Burkina Faso, Algeria, Niger, Nigeria, Chad, Sudan, Somalia, Ethiopia and Eritrea.

Salinisation   The accumulation of salts in soils is termed as salinisation.

Salt-water intrusion/encroachment   Displacement of fresh surface water or ground water by the movement of salt water due to its greater density is called slat-water intrusion/encroachment.

Sandstorm/dust storm   A sandstorm/dust storm occurs in arid or semi-arid regions when high wind speeds transports fine particles like sand.

Savanna   Tropical or sub-tropical grassland or woodland *biomes* with strewn shrubs, individual trees or a very open canopy of trees is termed as savanna.

Sea-ice biome   The *biome* created by all marine organisms living within or on the floating sea ice (frozen sea water) of the polar oceans.

Sea-level rise   An increase in the mean level of the ocean often occurring due to thermal expansion of sea water and global melt down of ice is called sea-level rise.

Sea wall   A human-made wall or embankment beside a shore to prevent wave erosion is termed as sea wall.

Semi-arid regions   Regions of moderately low rainfall ('moderately low' is usually accepted as between 100 and 250 mm precipitation per year).

Sequestration   See *carbon* sequestration.

Sericulture   Rearing of silkworms for the production of raw silk is called sericulture.

Siberia   Region, comprising of Northern Asia

Silviculture   Development and care of forests is termed as silviculture.

Snowpack   A seasonal accumulation of slow-melting snow is refereed as snow pack.

Soil Organic Carbon (SOC)   Carbon in the organic form in the soil is termed as soil organic carbon.

Stratosphere   Stratosphere is the layer of atmosphere above the troposphere extending from about 10 km to about 50 km.

Sub-alpine   Sub-alpine region is the biogeographic region below the tree line and above the montane zone that is distinguished by the occurrence of coniferous forest and trees.

Surface runoff   Surface runoff is the water that flows over the land surface to the nearest surface stream.

Taiga   The northernmost belt of boreal forest adjacent to the Arctic tundra is referred as Taiga.

Thermocline   Thermocline is the area in the world's ocean, usually at a depth of 1 km, where temperature reduces rapidly with depth and marks the boundary between the surface and the ocean.

Thermohaline Circulation (THC)   Large-scale, density-driven water movement in the ocean, due to differences in temperature and salinity is termed as Thermohaline circulation. It is also called *meridional overturning circulation (MOC)*.

Thermokarst   Landscape full of depressions often filled with water (ponds) as a result of thawing of ground ice or permafrost.

Thrace   Thrace is region in southeast of Europe bounded by the Balkan Mountains, Rhodope Mountains, Aegean Sea, Black Sea, and Sea of Marmara.

Transition economies   An economy of countries which is changing to market economy is called as transition economies.

Transpiration Transpiration is loss of water vapor from the aerial parts of plant which helps movement of water and nutrients from soil to different parts of plant.

Triassic time The Triassic era is the period that spanned from about 250 to 200 million years ago.

Tropical region Global region between latitudes 20°N and 20°S.

Troposphere The lower most layer of the atmosphere from the surface of the earth to about 10 km in altitude in mid-latitudes.

Sand casting Deposition of large amounts of sand by flood water is called sand casting.

Scrub fire Fires in scrub or bush that cover extensive damage is termed as scrub fire. They can start by natural causes or anthropogenic causes.

Severe storm A severe storm is the storm that comes along with high winds, heavy precipitation, thunder and lightning.

Storm surge Storm surge is the coastal flood on coasts and lake shores caused by wind.

Subsidence Downward movement of the Earth's surface relative to a datum (e.g. the sea level) is termed as subsidence.

Subsidy Subsidy is direct payment from the government or a tax reduction to a private party for implementing a practice the government wishes to encourage.

Tornado Tornado is a rotating column of air that arises out of the base of a cumulonimbus cloud and has touching base to the Earth's surface.

Tree line The tree line is the edge of the habitat at which trees can grow.

Tropical cyclone *See cyclone.*

Tsunami A tsunami is a series of waves caused by a quick movement of a body of water (ocean, lake).

Turbulence (in the context of atmosphere) Turbulence in air caused due to mixing of cold and warm air in atmosphere

Tundra A level, or gently undulating plain devoid of trees in the Arctic and sub-Arctic regions distinguished by low temperatures and short growing seasons.

Typhoon Typhoon is large-scale closed circulation system in the atmosphere above the western pacific with maximum wind speed of 64 knots or more.

Uncertainty Uncertainty is an expression of the degree to which a value is unknown.

Upwelling region Upwelling region is an area in the ocean where cold, typically nutrient-rich waters from the bottom of the ocean rises to surface.

Urbanization   Urbanization is the process of the transformation of land to cities.

Urban heat island   A warmer urban area compared to adjacent nonurban area.

Urolithiasis   It is the condition wherein kidney stone (urinary calculi) are formed in urinary track.

Vector   A blood-sucking organism that transfers a pathogen from one host to another is termed as vector.

Vector-borne diseases   Disease that is transmitted by vectors (e.g., *malaria*, *dengue fever* and leishmaniasis) is termed as vector-borne diseases.

Vulnerability   The potential to suffer harm or loss is termed as vulnerability.

Weather   The exact condition of the atmosphere at a particular place and time is called weather.

Westerlies   Prevailing wind blowing from west to east in the latitudes between 35 and 65° latitude is called westerlies.

Wildfire   *Wildfire* describes an uncontrolled burning fire, in wild lands.

Winter storm   Extra-tropical cyclones in occurring in spring, autumn or winter. A winter storm appears from an extra-tropical cyclone. It comes along with high wind speeds, breezes, thunderstorms, and rain and storm surges.

Zoonosis   Zoonosis are the diseases which occur normally in animals that are transmitted from animals to people (e.g. Swine flu).

Zooplankton   The animal forms of plankton are termed as zooplankton.

Appendix B: Abbreviation Used in This Book

ABC	Atmospheric brown cloud
ACDR	Asian Conference on Disaster Reduction
ADB	Asian Development Bank
ADPC	Asian Disaster Preparedness Centre
ADRC	Asian Disaster Reduction Centre
AIJ	Activities implemented jointly
AFN	Asian Flood Network
AFOLU	Agriculture, forestry and other land use
ALT	Active layer thickness
AOSIS	Alliance of Small Island States
APPCDC	Asia Pacific Partnership on Clean Development and Climate
APEC	Asia-Pacific Economic Cooperation
ASEAN	Association of South East Asian Nations
BAPPENA	State Ministry of National Development Planning (in Indonesia)
BC	Before Christ
BCE	Before the Current Era or Before the Christian Era, or Before the Common Era
BP	Before present
Cal BP	Calibrated years before the present
CBFEWS	Community based flood early warning system
CCCD	Commission on Climate Change and Development
CCS	Carbon capture and storage
CDM	Clean development mechanism
CER	Certified emission reduction
CFC	Chlorofluoro carbon
CFC-11	Trichlorofluoromethane
CFC-12	Dichlorodifluoromethane
CH_4	Methane
CITES	Convention on international trade in endangered species of wild fauna and flora
CO_2	Carbon dioxide
CRED	Centre for Research and Epidemiological Research

CSIS	Centre for Strategy and International Studies
CTEC	Community Tsunami Early Warning Centre
CTI	Coral Triangle Initiative
EACP	East Asia Climate Partnership
EIA	Environmental Impact Assessment
EIT	Economies in transition
EM-DAT	Emergency Events Database prepared and hosted by CRED
ERU	Emission reduction units
ESI	Energy Security Initiative
FAO	Food and Agriculture Organization
GBM	Ganaga Bramhaputra Meghna
GDP	Gross domestic product
GHG	Green house gases
GLOF	Glacial lake outburst floods
GoI	Government of India
GoS	Government of Sri Lanka
Gt	Giga ton
GWP	Global warming potential
HEI	Health Effect Institute
IAEM	International Association of Environmental Managers
IAS	Invasive alien species
ICAO	International Civil Aviation Organisation
ICIMOD	International Centre for Integrated Mountain Development
ICT	Information communication technology
IOTWS	Indian Ocean Tsunami Warning and Mitigation System
INCOIS	Indian National Centre for Ocean Information Services
IPCC	Intergovernmental Panel for Climate Change
IPPU	Industrial processes and product use
IRADe	Integrated Research and Action for Development
IRENA	International Renewable Energy Agency
Jaxa	Japan Aerospace Exploration Agency
JI	Joint implementation
ka	Kilo years ago
kyr	Kilo years
LDC	Least developed country
LGM	Last glacial maximum
LLDC	Land locked developed countries
LNG	Liquefied natural gas
LUCF	Land-use change and forestry
LULCF	Land use, land-use change and forestry
Ma	Million years ago
MDG	Millennium Development Goal
MEA	Millennium Ecosystem Assessment
MNPRA	Ministry of Nature Protection of the Republic of Armenia

MRC	Mekong River Commission
MRTS	Mass rapid transport system
$MtCO_2$-eq	Metric ton carbon dioxide equivalent
Myr	Million Years
NCCS	National Climate Change Strategy
NDRCPRC	National Development and Reform Commission People's Republic of China
NGO	Non-governmental organisations
N_2O	Nitrous oxide
NO_2	Nitrogen dioxin
NO_x	Oxides of nitrogen
O_3	Ozone
OECD	Organisation for Economic Co-operation and Development
OECD-DAC	Organisation for Economic Co-operation and Development – Development Assistance Committee
OFDA	Office of U.S. Foreign Disaster Assistance
OPEC	Organization of the Petroleum Exporting Countries
PPNA	Pre-pottery Neolithic A
PPP	Per capita and per $1 GDP
RGOB	Royal Government of Bhutan
RHAP	Regional haze action plan
RTS	Reservoir triggered seismicity
SAARC	South Asian Association for Regional Cooperation
SCC	Social cost of carbon
SCE	Snow cover extent
SCZMC	SAARC Coastal Zone Management Centre
SDMC	SAARC Disaster Management Centre
SF_6	Sulphur hexafluoride
SIDS	Small island developing states
SMRC	SAARC Meteorology Research Centre
SMS	Small message service
SO_x	Oxides of sulphur
SO_2	Sulphur dioxide
SOC	Soil organic carbon
SST	Sea surface temperature
UHI	Urban heat island
UN	United Nations
UNDRO	UN Disaster Relief Organization
UNFCCC	United Nations Framework Convention on Climate
UN-OCHA	UN Office for the Coordination of Humanitarian Affairs
UN-OHRLLS	UN Office of the High Representative for the Least Developed Countries, Landlocked Developing Countries and Small Island Developing States
USAID	United State Agency for International Development

VOCs	Volatile organic compounds
WCED	World Commission on Environment and Development
WEC	World Energy Council
WHO	World Health Organization
WTO	World Trade Organization
WWF	World Wild Fund

Appendix C: Impact of Climate Change on Individual Asian Countries

Afghanistan

Afghanistan is a land locked country bordered by China, Iran (Islamic Republic of), Pakistan, Tajikistan, Turkmenistan, Uzbekistan. Afghanistan has an area of 652,230 km². Its terrain of this land locked country is mostly mountains and desert. The population as estimated in 2009 is 28.396 million. Natural resources of country contains natural gas, oil, coal, petroleum, copper, chromite, talc, barites, sulphur, lead, zinc, iron ore, salt, precious and semiprecious stones. Agriculture contributes to about 31% of GDP with main crops being wheat, corn, barley, rice, cotton, fruit, nuts, karakul pelts, wool, and mutton. Industrial activity comprises small-scale production of textiles, soap, furniture, shoes, fertilizer, cement, carpets, natural gas, coal, and copper.

Services sector include transport, retail, and telecommunications. Main export items are fruits and nuts, hand-woven carpets, wool, cotton, hides and pelts, precious and semiprecious gems. About 85% of Afghans are dependent on agriculture for their livelihoods.

Climate is dry with cold winters, low rainfall and hot summers. Climate system is characterized by dust storms with northerly winds in warm months. The snow fall varies considerably with elevation. Average summer temperatures do not exceed 15°C, and winter temperatures go below zero in some parts.

Summarized table of natural disasters in Afghanistan from 1900 to 2010 is given in Table C.1. The vulnerability of the agricultural sector due to rise in temperatures and changes in precipitation will be due to increased soil evaporation, reduced river flow, and water shortages. But as evident from past events, occurrence of flood can not be over ruled. The country will be affected by increase in snow melt and subsequent consequences resulting in increases in glacial runoff and GLOFs, and an increased frequency of events such as floods, mudflows, and avalanches affecting human settlements. GLOF events can have impacts on socioeconomic systems, hydrology, and ecosystems. The climate change can bring in reduction of glacier volume and eventual disappearance of many glaciers may cause water stress to millions of people. The area will also get affected due to increase in number of mass

Table C.1 Summarized table of natural disasters in Afghanistan from 1900 to 2010

		No. of events	Killed	Total affected	Damage (000 US$)
Drought	Drought	5	37	4,808,000	250
	Ave. per event		7.4	961,600	50
Earthquake (seismic activity)	Earthquake (ground shaking)	28	11,325	624,778	54,060
	Ave. per event		404.5	22,313.5	1,930.7
Epidemic	Unspecified	7	2,775	7,372	–
	Ave. per event		396.4	1,053.1	–
	Bacterial infectious diseases	7	340	40,627	–
	Ave. per event		48.6	5,803.9	–
	Parasitic infectious diseases	1	–	200,000	–
	Ave. per event		–	200,000	–
	Viral infectious diseases	5	730	6,318	–
	Ave. per event		146	1,263.6	–
Extreme temperature	Cold wave	5	572	200,200	10
	Ave. per event		114.4	40,040	2
Flood	Unspecified	12	681	793,304	312,000
	Ave. per event		56.8	66,108.7	26,000
	Flash flood	15	838	37,865	4,000
	Ave. per event		55.9	2,524.3	266.7
	General flood	34	2,195	342,862	80,000
	Ave. per event		64.6	10,084.2	2,352.9
Insect infestation	Locust	1	–	–	–
	Ave. per event		–	–	–
Mass movement dry	Avalanche	1	100	–	–
	Ave. per event		100	–	–
Mass movement wet	Avalanche	7	305	432	–
	Ave. per event		43.6	61.7	–
	Landslide	4	582	301,174	–
	Ave. per event		145.5	75,293.5	–
Storm	Unspecified	3	299	22,656	5,000
	Ave. per event		99.7	7,552	1,666.7
	Local storm	2	1,359	170,684	–
	Ave. per event		679.5	85,342	–
Wildfire	Forest fire	1	–	–	–
	Ave. per event		–	–	–

Created on: Oct-4-2010 – Data version: v12.07
Source: "EM-DAT: The OFDA/CRED International Disaster Database http://www.em-dat.net – Université Catholique de Louvain – Brussels – Belgium"

movements (like rock falls and landslides). The reduced volume of runoff due to disappearance of glaciers will have serious consequences for downstream hydro-power and agriculture.

Armenia

Armenia is a land locked country. Armenia borders with Georgia to the North, Azerbaijan to East, Turkey to the West and Iran in South-East. It has an area of 29,800 km^2 with population of about three million. Armenia is a mountainous country with about 90% of its territory is over 1,000 m above sea level. In the lowlands the average air temperature in July and August varies from 24 to 26°C, and in the Alpine belt the temperature does not exceed 10°C. Temperature varies from 1 to 13°C in January. Maximum and minimum temperatures are 41 and −42°C. The average annual precipitation is about 570 mm. The territory of Armenia includes 11 climatic zones. The country has about 3,500 plant species and more than 500 species of vertebrate, about 300 birds species, about 50 species of reptiles, 8 amphibian species, more than 20 fish species and more than 80 mammal species. Important crops include grains, potatoes, almonds, grapes, and cotton. Natural resources include copper, zinc, aluminium, molybdenum, marble and granite. Manufacturing activity include cement, chemicals, textile, food industry and carpet-weaving.

As can be seen from Table C.2 the country has been affected by series of climate induced disasters. According to the first national communication of the republic of Armenia under the UNFCCC, climate change and internal micro-climatic changes in Armenia may result in (1) change in the borders of natural climatic zones; (2) change of the biota; (3) change in water resources and (4) change in amount of precipitation and soil moisture content.

The country has undergone significant changes in the ecosystems in the last three millenniums resulting in reduction in forest, expansion of the semi-desert and steppe vegetation. Further change in climate is likely to affect the current ecosystems. The climate change will impact agriculture and animal husbandry as well as health of people.

Table C.2 Summarized table of natural disasters in Armenia from 1900 to 2010

		No. of events	Killed	Total affected	Damage (000 US$)
Drought	Drought	1	–	297,000	100,000
	Ave. per event		–	297,000	100,000
Earthquake (seismic activity)	Earthquake (ground shaking)	1	–	15,000	33,333
	Ave. per event		–	15,000	33,333
Flood	Unspecified	2	4	7,144	8,120
	Ave. per event		2	3,572	4,060
	General flood	1	1	–	–
	Ave. per event		1	–	–

Created on: Nov-13-2010 – Data version: v12.07
Source: "EM-DAT: The OFDA/CRED International Disaster Database http://www.em-dat.net – Université Catholique de Louvain – Brussels – Belgium"

Azerbaijan

Azerbaijan is located at the between Europe and Asia besides Caspian Sea. It borders with the Russian Federation in the north; Iran and Turkey in south; Georgia in the west. Total area of Azerbaijan is 86.6 thousand km^2 and has population of about eight million. Agricultural products include grain, cotton, fruits, vegetables and tobacco. Natural resources include oil, iron, aluminium, copper, lead, zinc, precious metals, lime stone, and salt. Industrial activity includes oil, copper, chemicals, building material, food, timber and fishing.

As can be seen from Table C.3 the country has been affected by series of climate induced disasters. According to the first national communication under UNFCCC, Caspian sea level has raised due to climate change. Increase in temperature will increase evaporation to about 30–35%, resulting in degradation of natural humidification and moisture deficiency as well as reduction of river water resources. Recurrence of drought and dry winds will augment, especially in the Kura-Araz lowland. Climate change will result in significant increase of upper climatic boundary of forest cover on the Greater and Lesser Caucasus. In the Talish Mountains, upper boundary is expected to be decreasing about 50–200 m, with up to 300 m in some places. Lower boundary of forest will go up to 100–200 m. Draught resistant tree and bush species will increase. Density and productivity of the forests will diminish resulting in decrease of wood stock.

Bahrain

Bahrain is a generally flat and arid archipelago comprising of 36 islands in the Arabian Gulf. Area of the country is about 727 km^2. Desert constitutes 92% of

Table C.3 Summarized table of natural disasters in Azerbaijan from 1900 to 2010

		No. of events	Killed	Total affected	Damage (000 US$)
Drought	Drought	1	–	–	100,000
	Ave. per event		–	–	100,000
Earthquake (seismic activity)	Earthquake (ground shaking)	3	33	712,474	15,000
	Ave. per event		11	237,491.3	5,000
Flood	Unspecified	2	16	81,000	29,000
	Ave. per event		8	40,500	14,500
	General flood	4	3	1,756,500	61,700
	Ave. per event		0.8	439,125	15,425
	Storm surge/coastal flood	1	–	2,800	5,500
	Ave. per event		–	2,800	5,500
Mass movement wet	Avalanche	1	11	–	–
	Ave. per event		11	–	–

Created on: Nov-13-2010 – Data version: v12.07

Source: "EM-DAT: The OFDA/CRED International Disaster Database http://www.em-dat.net – Université Catholique de Louvain – Brussels – Belgium"

Table C.4 Summarized table of natural disasters in Bahrain from 1900 to 2010

		No. of events	Killed	Total affected	Damage (000 US$)
Epidemic	Bacterial infectious diseases	1	111	–	–
	Ave. per event		111	–	–

Created on: Oct-13-2010 – Data version: v12.07
Source: "EM-DAT: The OFDA/CRED International Disaster Database http://www.em-dat.net – Université Catholique de Louvain – Brussels – Belgium"

Bahrain, and periodic droughts and dust storms are the main natural hazards for Bahrainis. Climate is hot and humid from May-September; with average highs ranging from 30 to 40°C. Maximum temperatures average 20–30°C the remainder of the year. Population as on 2008 is estimated to be around one million. The summer is very hot since and summer temperatures may reach more than 40°C with minimal and irregular rains reaching maximum of about 70 mm. Terrain is mainly flat alluvial plain, with hills in the northeast and southeast. Climate of the country is semitropical, monsoonal. Natural resources include Oil, aluminium, textiles, natural gas, fish, and pearls. Agriculture contributes to less than 1% of GDP. The main agricultural Products are fruit, vegetables, poultry, dairy products, shrimp, and fish. Major industrial activity is oil and gas, manufacturing, aluminium. Services in the country include finance, transport and communications, and real estate. Bahrain is a regional financial and business centre with international financial institutions operating in Bahrain. The financial sector is currently the largest contributor to GDP at 30%.

As could be seen from Table C.4 the country is safe from climate induced disasters. The country does not depend on its own agriculture for sustenance. With less than 1% depending on agriculture country does not have to bother much of agricultural loss in its own territory. But agricultural, damage at other parts of world may affect food supply to the country. The possible effect due to climate change is sea level rise, health impact, and water scarcity. The region could be affected due to inundation as well as increasing salinity of soil and available freshwater resources. Increase in temperature will also affect number of tourist visiting the country.

Bangladesh

Bangladesh is located in southern Asia with India and Myanmar as neighbouring country. The country has coastal stretch on Bay of Bengal. It has an area of 147,570 km^2 and population of about 156 million. Terrain is mainly flat alluvial plain, with hills in the northeast and southeast. Climate is semitropical and monsoonal characterized by heavy seasonal rainfall, moderately warm temperatures, and high humidity. Main natural resources are natural gas, fertile soil, and water. Agriculture contributes to 19.1% of GDP with major products being rice, jute, tea, sugar, wheat. Industry contribute 28.6% of GDP comprising garments and knitwear, jute goods, frozen fish and seafood, textiles, fertilizer, sugar, tea, leather, ship-breaking, pharmaceuticals, ceramic tableware and newsprint. Half of GDP is generated through service sector. Two thirds of people are employed in agriculture. Rice is single most important crop. Bangladesh is a low-lying, country in South Asia with a largely marshy jungle coastline of 710 km on the Bay of Bengal. Formed by a deltaic plain at the confluence of the Ganges, Brahmaputra, and Meghna Rivers and their tributaries, Bangladesh's alluvial soil is highly fertile but at the same time vulnerable to flood and drought. Hills rise above the plain only in the Chittagong hill tracts.

As can be seen from Table C.5 the country has been affected by series of climate induced disasters. Rise in temperature and subsequent losses could reduce agricultural land reducing production of food crops. Natural calamities, such as floods, tropical cyclones, tornadoes, and tidal bores affect the country almost every year. Bangladesh also is affected by major cyclones on average 16 times a decade. The flooding has been and would affect economic activity. The country is constantly affected by storms. Increase in intensity and frequency of extreme weather events will affect infrastructure, economy, health and agriculture.

Given its vulnerability to extreme climate events, numerous adaptation measures are already in place in Bangladesh. This includes both hard infrastructure as well as soft, policy measures combined with communal practice. Hard infrastructure includes coastal embankments, foreshore afforestation, cyclone shelters, early warning systems, and relief operations. Soft measures adopted by Bangladesh include design standards for roads which make them lie above the highest flood levels with a return period of 50 years while feeder roads are designed to lie above the normal flood level. The country has also introduced high yielding varieties of *aman* and *boro* rice crops.

Bangladesh is one of the most vulnerable countries to climate change and most vulnerable to tropical cyclones. Between 1877 and 1995 Bangladesh was affected by 154 cyclones which included 43 severe cyclonic storms and 68 tropical depressions. Cyclone disaster in Bangladesh on 29–30 April 1991, had a maximum wind speed of more than 225 km/h. and a maximum surge height of about 8-m that killed about 140,000 people. Bangladesh is affected by about 40% of the impact of total storm surges in the world. Bangladesh also experiences harsh monsoon flooding, resulting in significant damage to crops and properties. The performance of the agriculture sector depends on the type of the annual floods. Nearly two-thirds of the country is less than 5 m above sea levels making it is susceptible to flooding and sea level rise. In a typical year approximately one quarter of the country is inundated. Low frequency high magnitude floods contribute to adverse impacts on poor.

Bhutan

Bhutan is a small landlocked country in South Asia, located at the eastern end of the Himalayas with a land area of about 40,077 km^2 with a population of 0.56 million. It is located between India and China. The northern region of the country consist Eastern Himalayan alpine shrub and meadows reaching up to glaciated mountain.

Table C.5 Summarized table of natural disasters in Bangladesh from 1900 to 2010

		No. of events	Killed	Total affected	Damage (000 US$)
Drought	Drought	7	1,900,018	25,002,000	–
	Ave. per event		271,431.1	3,571,714.3	–
Earthquake (seismic activity)	Earthquake (ground shaking)	6	34	19,125	–
	Ave. per event		5.7	3,187.5	–
	Tsunami	1	2	–	500,000
	Ave. per event		2	–	500,000
Epidemic	Unspecified	17	5,068	2,503,118	–
	Ave. per event		298.1	147,242.2	–
	Bacterial infectious diseases	5	3,639	420,479	–
	Ave. per event		727.8	84,095.8	–
	Parasitic infectious diseases	3	1,396	69,904	–
	Ave. per event		465.3	23,301.3	–
	Viral infectious diseases	5	393,085	48,928	–
	Ave. per event		78,617	9,785.6	–
Extreme temperature	Cold wave	15	2,014	136,200	–
	Ave. per event		134.3	9,080	–
	Extreme winter conditions	2	230	101,000	–
	Ave. per event		115	50,500	–
	Heat wave	2	62	–	–
	Ave. per event		31	–	–
Flood	Unspecified	31	44,847	177,076,392	4,024,100
	Ave. per event		1,446.7	5,712,141.7	129,809.7
	Flash flood	10	251	6,064,018	729,000
	Ave. per event		25.1	606,401.8	72,900
	General flood	39	7,005	126,605,442	7,285,300
	Ave. per event		179.6	3,246,293.4	186,802.6
	Storm surge/coastal flood	2	51	473,335	–
	Ave. per event		25.5	236,667.5	–
Mass movement wet	Landslide	1	13	50	–
	Ave. per event		13	50	–
Storm	Unspecified	50	5,756	2,376,857	850,000
	Ave. per event		115.1	47,537.1	17,000
	Local storm	24	1,732	1,170,666	16,401
	Ave. per event		72.2	48,777.8	683.4
	Tropical cyclone	85	626,846	73,566,523	4,765,979
	Ave. per event		7,374.7	865,488.5	56,070.3

Created on: Oct-13-2010 – Data version: v12.07
Source: "EM-DAT: The OFDA/CRED International Disaster Database http://www.em-dat.net – Université Catholique de Louvain – Brussels – Belgium"

The country is characterized by cold climate. Southern part of the country is covered with dense Himalayan subtropical broadleaf forests, alluvial lowland river valleys, and mountains. The topography is highly uneven. Rising from a height of 100 m above sea level in the south, terrain reaches over 7,550 m in the north. Geographically, Bhutan can be divided into southern foothills, inner Himalayas and higher Himalayas. The southern foothills rise from the plains to a elevation of 1,500 m. The inner Himalayas gradually rise to about 3,000 m. The northern area comprises the major Himalayan range of the high mountains.

The climate in Bhutan varies with altitude. It varies from subtropical in the south to temperate north. Bhutan experiences five distinct seasons: summer, monsoon, autumn, winter and spring. Western Bhutan has the heavier monsoon rains whereas southern Bhutan has hot humid summers and cool winters. Northern highlands has polar-type climate, with year-round snow. Central and eastern Bhutan is temperate and drier than the west with warm summers and cool winters. The country can be divided into three climatic zones corresponding to the three major geographical divisions. The southern belt has a hot, humid climate, with temperatures varying between 15 and 30°C with rainfall ranging between 2,500 and 5,000 mm. The central inner Himalayas is characterized by cool, temperate climate, with annual average rainfall of around 1,000 mm. Higher and more northern region has an alpine climate, with annual rainfall of about 400 mm.

The flora of Bhutan is diverse. Forests of Bhutan are divided into broadleaf forest, conifer forest and scrub forest. Bhutan is part of one of the ten global biodiversity "hotspots". Forests of alpine scrub, fir, mixed coniferous species, temperate scrub and broadleaf species cover about 72% of the country covering about 47 indigenous species. More than 160 species of mammals have been reported which includes tiger, leopard, snow leopard, red panda, gaur, Himalayan black bear, wild boar, musk. Phobjikha valley is one of the global wintering grounds for the rare black-necked crane. One-quarter of the country is has been declared as protected reserves/sanctuaries/nature parks.

Bhutan's economy is based on agriculture, forestry, manufacturing, mining, tourism and hydroelectric power.

Table C.6 shows the natural disasters occurred since the beginning of last century. Climate change is expected to cause an increase in frequency and intensity of natural disasters such as floods, landslides. Bhutan is vulnerable to climate due to its ecological fragility and economic marginality. Bhutan is important due to its location in Eastern Himalayas a biodiversity hotspot. Several species are already endangered by climate change and extreme events. The combination of climate change, land use changes, habitat degradation and fragmentation presents a significant threat to biodiversity. Climate change can affect biodiversity by changing the physiological responses of species, or by changing the relationships between species. The region is vulnerable to climate change due to its ecological fragility and economic marginality. Himalayas is biodiversity hotspot and the source of the headwaters of four major Asian river systems. Hence change in climate will have huge impact on biodiversity as well water resources originating from the area. The region has witnessed extraordinary melting of permanent glaciers during the past three decades. Himalayan glaciers are showing the fast rate of retreat, resulting in increases in glacial runoff and GLOFs, and an increased frequency of events such as floods, mudflows, and avalanches affecting human settlements. GLOF events can

Table C.6 Summarized table of natural disasters in Bhutan from 1900 to 2010

		No. of events	Killed	Total affected	Damage (000 US$)
Earthquake (seismic activity)	Earthquake (ground shaking)	1	11	12	–
	Ave. per event		11	12	–
Epidemic	Bacterial infectious diseases	2	41	741	–
	Ave. per event		20.5	370.5	–
Flood	Flash flood	1	22	600	–
	Ave. per event		22	600	–
	General flood	2	200	1,000	–
	Ave. per event		100	500	–
Storm	Tropical cyclone	2	29	65,000	–
	Ave. per event		14.5	32,500	–
Wildfire	Forest fire	1	–	–	3,500
	Ave. per event		–	–	3,500

Created on: Oct-13-2010 – Data version: v12.07

Source: "EM-DAT: The OFDA/CRED International Disaster Database http://www.em-dat.net – Université Catholique de Louvain – Brussels – Belgium"

have impacts on socioeconomic systems, hydrology, and ecosystems. The 1994 glacier lake outburst in Lunana, one of the northern highlands, seriously damaged the lower valleys of Punakha. The climate change can bring in reduction of glacier volume, and eventual disappearance of many glaciers, may cause water stress to millions of people. A rare dry-spell with no snowfall was experienced in the winter of 1998. Mid-summer snowfall occurred in some places in the north in July 1999. Flash floods in August 2000 caused by torrential rains have claimed significant damage to infrastructures and natural resources. The area will also get affected due to increase in number of mass movements. The reduced volume of runoff due to disappearance of glaciers will have serious consequences for downstream hydropower and agriculture. According to the first national communication under UNFCCC warming may have positive impacts on crop yields if moisture is not a restriction, but rise in the frequency of extreme events or pests may offset any potential benefits. Global warming will also result in increase in the number of generations of harmful insects per year, as well as to a spread in their distribution areas.

Brunei Darussalam

Brunei Darussalam is located on the north coast of the island of Borneo, in Southeast Asia surrounded by the state of Sarawak, Malaysia. Brunei Darussalam has a tropical rainforest climate with average annual temperature about 27°C. The country is influenced by south-western and north-eastern monsoons; diurnal wind system (land and sea breezes); and anabatic and katabatic winds. The country is affected by ENSO in 1972/1973, 1982/1983, 1997/1998. The country receives mean annual rainfall: 2,300 mm to over 4,000 mm.

Table C.7 Summarized table of natural disasters in Brunei Darussalam from 1900 to 2010

		No. of events	Killed	Total affected	Damage (000 US$)
Wildfire	Forest fire	1	–	–	2,000
	Ave. per event		–	–	2,000

Created on: Oct-4-2010 – Data version: v12.07

Source: "EM-DAT: The OFDA/CRED International Disaster Database http://www.em-dat.net – Université Catholique de Louvain – Brussels – Belgium"

The country occupies 5,765 km^2 with population of 0.4 million as on 2005. Crude oil and natural gas production is major economic activity. Rice is the chief food crop. Other crops include banana, vegetables, cassava, pepper, coconut, sago and rubber.

Summarised table of natural Disasters in Brunei Darusalam from 1900 to 2010 is given in Table C.7. Apart from a forest fire the country had not recorded any major natural calamity including climate induced disaster. Projected sea-level rise may impact coastal ecosystems and economy.

Cambodia

Cambodia is a country in Southeast Asia that borders Thailand, Laos, Vietnam, and the gulf of Thailand. Cambodia has an area of 181,035 km^2 and the geography of Cambodia is dominated by the Mekong River.

Cambodia's is dominated by monsoons with temperature range from 21 to 35°C. Cambodia has two distinct seasons. The rainy season, which runs from May to October, and the dry season lasts from November to April.

Cambodia is an underdeveloped country with 50% of the land covered with virgin forests. Rice is the chief crop occupying 90% of the arable land followed by maize, bens, black pepper, and rubber.

As can be seen from Table C.8 country has witnessed drought, flood and storm suffering considerable damage considering size of the country. The change in climate is likely to affect the country with similar disaster but with changed frequency and magnitude. Global warming may result in increase in the number of generations of harmful insects per year, as well as to a spread in their distribution areas.

China

China is the most populous state in the world with over 1.3 billion people located in East Asia. The country has an area of 9.6 million km^2. The country is bordered by Vietnam, Laos, Burma, India, Bhutan, Nepal, Pakistan, Afghanistan, Tajikistan, Kyrgyzstan, Kazakhstan, Russia, Mongolia and North Korea. The territory of China

Table C.8 Summarized table of natural disasters in Cambodia from 1900 to 2010

		No. of events	Killed	Total affected	Damage (000 US$)
Drought	Drought	5	–	6,550,000	138,000
	Ave. per event		–	1,310,000	27,600
Epidemic	Bacterial infectious diseases	3	121	1,343	–
	Ave. per event		40.3	447.7	–
	Parasitic Infectious Diseases	1	–	380,000	–
	Ave. per event		–	380,000	–
	Viral infectious diseases	5	667	36,595	–
	Ave. per event		133.4	7,319	–
Flood	Unspecified	1	506	29,000	–
	Ave. per event		506	29,000	–
	Flash flood	1	7	535,904	500
	Ave. per event		7	535,904	500
	General flood	10	614	8,844,235	327,600
	Ave. per event		61.4	884,423.5	32,760
	Storm surge/coastal flood	1	–	124,475	–
	Ave. per event		–	124,475	–
Storm	Tropical cyclone	3	44	178,091	10
	Ave. per event		14.7	59,363.7	3.3

Created on: Oct-13-2010 – Data version: v12.07
Source: "EM-DAT: The OFDA/CRED International Disaster Database http://www.em-dat.net – Université Catholique de Louvain – Brussels – Belgium"

contains a large variety of landscapes: sea shores, alluvial plains, mountain ranges, deltas, plateaus and desert. The climate in China differs from region to region.

China is 1 of 17 mega diverse countries. The country lies in two of the world's major eco-zones, the Palearctic and the Indomalaya. The country has animals such as the horse, camel and tapir, Leopard Cat, bamboo rat, tree shrew, monkeys and apes. The giant panda is found only in a limited area of the country.

China is essentially agricultural country with main crops being rice, tea, tobacco, sugarcane, jute, soya, groundnut and hemp. Since 1978, the economy has boomed and stands second largest economy measured on a purchasing power parity basis. The tremendous industrial activity has already shown its effect in terms of degradation of environment. Major agriculture products are rice, wheat, potatoes, corn, peanuts, tea, millet, barley, cotton, other fibers, apples, oilseeds, pork, fish and other livestock products. Industrial activity include mining and ore processing, metals, coal, machine building, armaments, textiles, petroleum, cement, chemicals, fertilizers, consumer products, toys, electronics, food processing, transportation equipment and telecommunications equipment.

China is one of the top ten natural climate induced disaster affected country. As could be seen in Table C.9 the country has suffered lot both in terms of economy and human loss.

The Himalayan region of China is vulnerable to climate change due to its ecological fragility and economic marginality. Himalayas is biodiversity hotspot and the source of the headwaters of four major Asian river systems. Hence change

Table C.9 Summarized table of natural disasters in China from 1900 to 2010

		No. of events	Killed	Total affected	Damage (000 US$)
Drought	Drought	32	3,503,534	441,274,000	20,110,714
	Ave. per event		109,485.4	13,789,812.5	628,459.8
Earthquake (seismic activity)	Unspecified	1	4	13,529	–
	Ave. per event		4	13,529	–
	Earthquake (ground shaking)	119	874,606	68,807,683	94,213,307
	Ave. per event		7,349.6	578,215.8	791,708.5
	Tsunami	1	47	–	–
	Ave. per event		47	–	–
Epidemic	Unspecified	1	–	1,000	–
	Ave. per event		–	1,000	–
	Bacterial infectious diseases	5	1,561,133	842	–
	Ave. per event		312,226.6	168.4	–
	Viral infectious diseases	4	365	7,987	–
	Ave. per event		91.3	1,996.8	–
Extreme temperature	Cold wave	3	26	132,000	29,000
	Ave. per event		8.7	44,000	9,666.7
	Extreme winter conditions	2	145	77,000,000	21,100,000
	Ave. per event		72.5	38,500,000	10,550,000
	Heat wave	5	166	3,880	–
	Ave. per event		33.2	776	–
Flood	Unspecified	50	2,254,492	165,018,799	16,865,744
	Ave. per event		45,089.8	3,300,376	337,314.9
	Flash flood	20	2,099	89,073,073	4,493,090
	Ave. per event		105	4,453,653.7	224,654.5
	General flood	128	4,338,632	1,474,001,300	126,093,219
	Ave. per event		33,895.6	11,515,635.2	985,103.3
	Storm surge/coastal flood	5	391	1,000,015	–
	Ave. per event		78.2	200,003	–
Insect infestation	Locust	1	–	–	–
	Ave. per event		–	–	–
Mass movement dry	Landslide	6	454	5,473	–
	Ave. per event		75.7	912.2	–
Mass movement wet	Avalanche	2	68	554	–
	Ave. per event		34	277	–
	Landslide	49	3,205	2,193,353	1,091,400
	Ave. per event		65.4	44,762.3	22,273.5
Storm	Unspecified	41	2,029	39,331,901	1,769,963
	Ave. per event		49.5	959,314.7	43,169.8
	Local storm	58	1,654	161,850,956	4,388,863
	Ave. per event		28.5	2,790,533.7	75,670.1
	Tropical cyclone	108	169,790	227,939,914	41,281,249
	Ave. per event		1,572.1	2,110,554.8	382,233.8
Wildfire	Forest fire	5	243	56,613	110,000
	Ave. per event		48.6	11,322.6	22,000

Created on: Oct-13-2010 – Data version: v12.07

Source: "EM-DAT: The OFDA/CRED International Disaster Database http://www.em-dat.net – Université Catholique de Louvain – Brussels – Belgium"

in climate will have huge impact on biodiversity as well water resources originating form the area. The region has witnessed extraordinary melting of permanent glaciers during the past three decades. Himalayan glaciers are showing the fast rate of retreat, resulting in increases in glacial runoff and GLOFs, and an increased frequency of events such as floods, mudflows, and avalanches affecting human settlements. GLOF events can have impacts on socioeconomic systems, hydrology, and ecosystems. The climate change can bring in reduction of glacier volume, and eventual disappearance of many glaciers may cause water stress to millions of people. The area will also get affected due to increase in number of mass movements. The reduced volume of runoff due to disappearance of glaciers will have serious consequences for downstream hydropower and agriculture. Water resources in north China are vulnerable to climate change due to floods on the Asian monsoon and the ENSO phenomenon. Agriculture in this region appears to be especially sensitive to climate change due to potential increases in the soil moisture deficit. Warming and augmented evapotranspiration with declines in precipitation would make it hard to maintain the current crop pattern. Regular water logging in the south and spring droughts in the north may inhibit the growth of subtropical crops. Climate change is already serious problem with respect to sea-level rise due to seawater intrusion. Further sea level may result in increase in impact. As per China's national action plan, the country experienced temperature rise in annual average by 0.5–0.8°C during the past 100 years, which was higher than the average global temperature. Warming is more significant in western, eastern and northern China compared to the south of the Yangtze River. The decrease in annual precipitation was significant in most of northern China, eastern part of the northwest, and northeastern China, averaging 20–40 mm/10 years while precipitation increased in southern China and southwestern China, averaging 20–60 mm/10 years; Sea level increased along China's coasts during the past 50 years by 2.5 mm/year, slightly higher than the global average; the glaciers in the Qinghai-Tibetan Plateau and the Tianshan mountains would retreat at speedy rate, and some smaller glaciers would vanish.

East Timor

East Timor, also known as Timor-Leste, is a state in Southeast Asia. Timor-Leste has an area of approximately 16,000 km² and a population of about 1.1 million. East Timor is a lower-middle-income economy. The agricultural products include coffee, coconuts, cinnamon, areca nuts, paddy and other crops. It comprises the eastern half of the island of Timor, and islands of Atauro and Jaco. It also contains Oecusse, an exclave on the northwestern side of the island, within Indonesian West Timor. The country is surrounded by Wetar Strait on the north, the Timor Sea on the south and the Maluku Sea on the east. Timor-Leste is mountainous in about one-third of its area, mainly in the west. The central and eastern parts of Timor-Leste contain several low plateaus and coastal lowlands. Climate is tropical and generally hot,

Table C.10 Summarized table of natural disasters in East Timor from 1900 to 2010

		No. of events	Killed	Total affected	Damage (000 US$)
Drought	Drought	1	–	–	–
	Ave. per event		–	–	–
Epidemic	Viral infectious diseases	1	22	336	–
	Ave. per event		22	336	–
Flood	Unspecified	1	–	450	–
	Ave. per event		–	450	–
	General flood	2	4	3,108	–
	Ave. per event		2	1,554	–
Storm	Unspecified	1	–	8,730	–
	Ave. per event		–	8,730	–

Created on: Oct-13-2010 – Data version: v12.07
Source: "EM-DAT: The OFDA/CRED International Disaster Database http://www.em-dat.net –
Université Catholique de Louvain – Brussels – Belgium"

humid and dry in the most part of the year. The wet season is from December to March in the north and from December to July in the south. The north and mid-central parts of the country have a hot and dry climate with rainfall ranging between 50 and 110 cm and a mean temperature of 26°C. In mountainous areas the climate is cool with rainfall above 320 cm. Northern coast, receives rainfall in the range of 500–1,000 mm/year. The southern coastal plain, receives over 2,000 mm with two wet seasons. The country hosts sparsely populated unique plant and animal species and northern coast is characterized by a number of coral reef systems that have been determined to be at risk due to climate change.

Table C.10 shows summarized table of natural disasters in East Timor from 1900 to 2010. Timor-Leste is a small island country, with depleting natural resources and growing population. The majority of the population depends on subsistence agri-culture with unstable agricultural practices like 'slash-and-burn'. The country is already facing food insecurity that may be aggravated by an increase in extreme climate events. The country will be affected by global mean sea-level rise and El Niño events. Because country is located in tectonically active region, local land movements are important part of relative sea level rise. Low lying coastal areas, may be adversely affected if the relative rate of rise is high. The infrastructure, agriculture and freshwater resources may all be at risk. Global warming may result in increase in the number of generations of harmful insects per year, as well as to a spread in their distribution areas.

India

India is the seventh-largest country by geographical area located in south Asia with an area of 3.29 million km^2. This second-most populous country has a population of over 1.18 billion people (as per 2010 estimates).

Agriculture contributes to 17% of GDP with major agriculture products being wheat, rice, coarse grains, oilseeds, sugar cane, cotton, jute, tea, coffee, spices, and potato. Industry contributes to 28.2% of GDP. Major Industrial products are textiles, jute, processed food, steel, machinery, transport equipment, cement, aluminium, fertilizers, mining, petroleum, chemicals, and computer software. Services and transportation contributes to 54.9% of GDP.

The benefits of high growth economy are unequally shared. Twenty-eight percent still live under poverty line with three quarter of population below poverty line. The country which has suffered disaster historically has few pockets which are less vulnerable to disasters. The country which has suffered variety of disasters historically will be affected by drop in agriculture production due to climate change. Economy may suffer due to increase in climate shocks.

The north-eastern coast adjacent to Bangladesh is often affected by storms where as Andaman and Nicobar Islands are affected by volcano. Droughts are common and are witnessed every year. But floods often occur in area least expected and in the time unexpected. Tsunami event in 2004 has been very much new to country and people of the country for the first time learnt the drastic effect of the disaster. The event changed coping mechanism and country changed for ever with regard to disaster management. The country now has disaster management institutions both country and state level.

Since the country has long coastal line it is likely to get affected by rising sea water. The country is likely to witness flood and drought due to uneven precipitation. The melting of ice in Himalayas will increase flow for some time and reduce there after as there would be no snow to melt. While increase in melting causes flood, the loss of snow will cause drought in the area that depend on rivers arising from Himalayas.

The northern part of country has always witnessed hot summer due to heat wave from Thar Desert in Rajasthan and cold winter due to cold waves form Himalayas. The severity and duration in coming days would depend on severity of climate change.

India's topography, geo-climatic conditions and the dominance of socioeconomic vulnerability among the weaker sections of the population make it one of the most disaster prone countries in the world. Summarised table of natural disasters in India from 1900 to 2010 is given in Table C.11. The country will have impact on agriculture and economy due to climate triggered disasters. About 40 million hectares of land in India is identified as flood-prone and on an average 8.6 million hectares of land gets flooded annually. Nearly 75% of coastline is prone to cyclones arising from the Bay of Bengal and the Arabian Sea. Cyclonic storms and storm surges affect the lives, property and livelihoods of the coastal poor. The hilly regions of India are vulnerable to forest fires, landslides and snow avalanche hazards. The emerging concerns of climate change pose solemn challenges, impacting the lives and livelihoods of people in the coastal areas and the mountainous regions. The most vulnerable to landslides are the Himalayan Mountains and North-Eastern hill ranges. Over million houses are getting destroyed annually due to disasters. Agricultural land getting silted during floods,

Table C.11 Summarized table of natural disasters in India from 1900 to 2010

		No. of events	Killed	Total affected	Damage (000 US$)
Drought	Drought	14	4,250,320	1,061,841,000	2,441,122
	Ave. per event		303,594.3	75,845,785.7	174,365.9
Earthquake (seismic activity)	Earthquake (ground shaking)	25	61,705	27,265,183	4,079,900
	Ave. per event		2,468.2	1,090,607.3	163,196
	Tsunami	1	16,389	654,512	1,022,800
	Ave. per event		16,389	654,512	1,022,800
Epidemic	Unspecified	6	293	95,997	–
	Ave. per event		48.8	15,999.5	–
	Bacterial infectious diseases	24	4,103,948	70,856	–
	Ave. per event		170,997.8	2,952.3	–
	Parasitic infectious diseases	5	3,411	57,135	–
	Ave. per event		682.2	11,427	–
	Viral infectious diseases	33	436,222	197,485	–
	Ave. per event		13,218.8	5,984.4	–
Extreme temperature	Cold wave	23	4,752	25	144,000
	Ave. per event		206.6	1.1	6,260.9
	Extreme winter conditions	1	180	–	–
	Ave. per event		180	–	–
	Heat wave	23	8,869	225	400,000
	Ave. per event		385.6	9.8	17,391.3
Flood	Unspecified	95	30,860	455,269,954	11,673,059
	Ave. per event		324.8	4,792,315.3	122,874.3
	Flash flood	21	7,223	23,430,801	416,200
	Ave. per event		344	1,115,752.4	19,819
	General flood	109	21,170	304,693,377	19,631,929
	Ave. per event		194.2	2,795,352.1	180,109.4
	Storm surge/coastal flood	4	569	11,500,000	275,000
	Ave. per event		142.3	2,875,000	68,750
Insect infestation	Locust	1	–	–	–
	Ave. per event		–	–	–
Mass movement dry	Landslide	1	45	–	–
	Ave. per event		45	–	–
Mass movement wet	Avalanche	7	878	10,456	50,000
	Ave. per event		125.4	1,493.7	7,142.9
	Landslide	35	3,934	3,828,660	4,500
	Ave. per event		112.4	109,390.3	128.6
Storm	Unspecified	32	2,702	5,337,261	225,000
	Ave. per event		84.4	166,789.4	7,031.3
	Local storm	22	2,280	558,095	2,226,000
	Ave. per event		103.6	25,368	101,181.8
	Tropical cyclone	97	159,130	87,392,126	8,600,900
	Ave. per event		1,640.5	900,949.8	88,669.1
Wildfire	Forest fire	2	6	–	2,000
	Ave. per event		3	–	1,000

Created on: Oct-8-2010 – Data version: v12.07
Source: "EM-DAT: The OFDA/CRED International Disaster Database http://www.em-dat.net – Université Catholique de Louvain – Brussels – Belgium"

results in widespread malnutrition, homelessness, disruption of education, and loss of livelihoods.

Himalayan region of India will have impact due to melting ice and its impact on biomes and agriculture is major concern to the country and its people. The region is vulnerable to climate change due to its ecological fragility and economic marginality. Himalayas is biodiversity hotspot and the source of the headwaters of four major Asian river systems. Hence change in climate will have huge impact on biodiversity as well water resources originating form the area. The region has witnessed extraordinary melting of permanent glaciers during the past three decades. Himalayan glaciers are showing the fast rate of retreat, resulting in increases in glacial runoff and GLOFs, and an increased frequency of events such as floods, mudflows, and avalanches affecting human settlements. GLOF events can have impacts on socioeconomic systems, hydrology, and ecosystems. The climate change can bring in reduction of glacier volume, and eventual disappearance of many glaciers, may cause water stress to millions of people. The area will also get affected due to increase in number of mass movements. The reduced volume of runoff due to disappearance of glaciers will have serious consequences for downstream hydropower and agriculture.

Indonesia

Indonesia is a country in Southeast Asia and Oceania comprising 17,508 islands and population of around 240 million people. It has an area of 2 million km^2 with average population density of 134 people per km^2. The country has 81,000 km of coastline. Indonesia has a tropical climate, with two distinct monsoonal wet (June–September) and dry (December–March) seasons. Temperatures remain high throughout the year with very little difference from month to month. The temperature is cooler at higher altitudes. Average annual rainfall in the lowlands varies from 1,780 to 3,175 mm and up to 6,100 mm in mountainous regions. Indonesia's size, tropical climate, and archipelagic geography, support the world's second highest level of biodiversity (after Brazil). Indonesia is second only to Australia in terms of total endemic species. Agriculture form 15.3% of GDP in 2009 with major products being timber, rubber, rice, palm oil and coffee. About 17% of land is cultivated. Manufacturing contributed to 26.4% of GDP in 2009 with major products being garments, footwear, electronic goods, furniture, and paper products.

Indonesia is one of the world's most disaster-prone countries, regularly overwhelmed by droughts, epidemics, floods, earthquakes, landslides, volcanic eruptions, tsunami waves and wildfires. Summarised table of natural Disasters in Indonesia from 1900 to 2010 is given in Table C.12. There have been over 100 major floods in the last century, 97 earthquakes and 48 volcanic eruptions. These disasters have serious economic consequences. Floods that affected area of West Java and Banten in February 2007 left about half a million people homeless or

Table C.12 Summarized table of natural disasters in Indonesia from 1900 to 2010

		No. of events	Killed	Total affected	Damage (000 US$)
Drought	Drought	9	9,329	4,804,220	160,200
	Ave. per event		1,036.6	533,802.2	17,800
Earthquake (seismic activity)	Earthquake (ground shaking)	97	29,964	8,468,140	7,053,476
	Ave. per event		308.9	87,300.4	72,716.2
	Tsunami	7	167,841	568,561	4,506,600
	Ave. per event		23,977.3	81,223	643,800
Epidemic	Unspecified	4	819	9,984	–
	Ave. per event		204.8	2,496	–
	Bacterial infectious diseases	15	744	38,030	–
	Ave. per event		49.6	2,535.3	–
	Parasitic infectious diseases	3	225	504,000	–
	Ave. per event		75	168,000	–
	Viral infectious diseases	13	2,178	137,015	–
	Ave. per event		167.5	10,539.6	–
Flood	Unspecified	52	1,811	2,562,100	91,144
	Ave. per event		34.8	49,271.2	1,752.8
	Flash flood	26	1,718	1,216,802	169,500
	Ave. per event		66.1	46,800.1	6,519.2
	General flood	56	2,362	4,950,207	2,157,909
	Ave. per event		42.2	88,396.6	38,534.1
	Storm surge/coastal flood	1	11	2,000	–
	Ave. per event		11	2,000	–
Mass movement dry	Landslide	1	131	701	1,000
	Ave. per event		131	701	1,000
Mass movement wet	Landslide	41	2,119	392,951	120,745
	Ave. per event		51.7	9,584.2	2,945
Storm	Unspecified	3	35	12,000	–
	Ave. per event		11.7	4,000	–
	Local storm	1	4	2,400	–
	Ave. per event		4	2,400	–
	Tropical cyclone	6	1,953	5,298	–
	Ave. per event		325.5	883	–
Volcano	Volcanic eruption	48	17,945	1,015,453	344,390
	Ave. per event		373.9	21,155.3	7,174.8
Wildfire	Forest fire	9	300	3,034,478	9,329,000
	Ave. per event		33.3	337,164.2	1,036,555.6

Created on: Oct-13-2010 – Data version: v12.07

Source: "EM-DAT: The OFDA/CRED International Disaster Database http://www.em-dat.net – Université Catholique de Louvain – Brussels – Belgium"

displaced. The flood resulted in disruption of commerce and telecommunications systems for several weeks. Rivers and streams burst their banks resulting in inundation of heavily polluted water. Flood of coastal region during spring tide, has affected coastal regions of Indonesia. Climate change is likely to affect agriculture in the country. Agriculture in coastal area is likely to affect due to salt water intrusion. Increase in forest fires cannot be over ruled. Peat lands are significant for the hydrology of an area as they buffer flooding as well as for conservation of biodiversity. Peat lands cover more than 10% of Indonesia's land area. When peat lands are drained, upper layers dry-out and turn out to be prone to fire. Indonesia is vulnerable to the impact of climate change as raise in sea levels and flood affect coastal farming areas. Disruption in climate resulting in prolonged droughts and floods could bring damage to food security, health, habitats and livelihoods of coastal communities.

Iran

Iran is a country in Central Eurasia and Western Asia. The country has an area of 1.6 million km^2 with a population of over 74 million. Iran consists of the Iranian Plateau and coasts of the Caspian Sea and Khuzestan. It is one of the world's most mountainous countries, and northern part of Iran is covered by dense rain forests. The eastern part consists mostly of desert. Iran's climate ranges from arid or semiarid. Subtropical summer temperatures rarely exceed 29°C and annual precipitation vary from is 680 to 1,700 mm. Western part of the country will receive heavy snowfall during winter. The eastern and central basins are arid, with less than 200 mm of rain. The coastal plains have mild winters, and very humid and hot summers with annual precipitation ranging from 135 to 355 mm. Major natural resources are petroleum, natural gas, coal, chromium, copper, iron ore, lead, manganese, zinc and sulphur. Major crops of the country are wheat, rice, other grains, sugar beets, fruits, nuts, cotton, dairy products and wool. Major industries in the country are petroleum, petrochemicals, textiles, cement and building materials, food processing and metal fabricating. Iran's wildlife is composed of several animal species including bears, wild pigs, wolves, jackals, panthers, Eurasian lynx, and foxes.

Summarised table of natural disasters in Iran from 1900 to 2010 is given in Table C.13. The possible effect due to climate change is sea level rise, health impact and water scarcity leading to subsequent consequences on agriculture. The region could be affected due to inundation as well as increasing salinity of soil and available freshwater resources. The country will also be affected de to El Niño-Southern Oscillation. Global warming will also result in increase in the number of generations of harmful insects per year, as well as to a spread in their distribution areas.

Iraq

Iraq is bordered by Jordan, Syria, Turkey, Iran, Kuwait and Saudi Arabia. Iraq consists of desert, and alluvial plains of Euphrates and Tigris the local climate is

Table C.13 Summarized table of natural disasters in Iran from 1900 to 2010

		No. of events	Killed	Total affected	Damage (000 US$)
Drought	Drought	2	–	37,625,000	3,300,000
	Ave. per event		–	18,812,500	1,650,000
Earthquake (seismic activity)	Earthquake (ground shaking)	92	147,106	2,600,008	10,518,628
	Ave. per event		1,599	28,261	114,332.9
Epidemic	Unspecified	1	76	–	–
	Ave. per event		76	–	–
	Bacterial infectious diseases	2	296	2,500	–
	Ave. per event		148	1,250	–
Extreme temperature	Heat wave	1	158	–	–
	Ave. per event		158	–	–
Flood	Unspecified	27	3,816	1,285,520	408,300
	Ave. per event		141.3	47,611.9	15,122.2
	Flash flood	14	2,689	1,291,066	253,700
	Ave. per event		192.1	92,219	18,121.4
	General flood	31	1,262	1,075,948	6,990,528
	Ave. per event		40.7	34,708	225,500.9
Mass movement wet	Avalanche	3	73	44	–
	Ave. per event		24.3	14.7	–
	Landslide	1	43	100	–
	Ave. per event		43	100	–
Storm	Unspecified	8	248	19,785	13,540
	Ave. per event		31	2,473.1	1,692.5
	Local storm	3	88	–	15,000
	Ave. per event		29.3	–	5,000
	Tropical cyclone	1	12	160,009	–
	Ave. per event		12	160,009	–
Wildfire	Scrub/grassland fire	1	–	–	–
	Ave. per event		–	–	–

Created on: Oct-12-2010 – Data version: v12.07
Source: "EM-DAT: The OFDA/CRED International Disaster Database http://www.em-dat.net – Université Catholique de Louvain – Brussels – Belgium"

mostly desert, with hot arid summer and with mild to cool winters. Summer temperatures average above 40°C and frequently exceed 48°C. Winter temperatures vary between 15 and 21°C. The country receives low precipitation less than 250 mm annually.

As can be seen in Table C.14, the country has witnessed flood, and drought. The change in climate will have these disasters to reoccur with change in magnitude and frequency. The possible effect due to climate change is sea level rise, health impact and water scarcity leading to subsequent consequences on agriculture. The region could be affected due to inundation as well as increasing salinity of soil and available freshwater resources. Increase in temperature will also affect number of tourist visiting the country. Global warming will also result in increase in the number of generations of harmful insects per year, as well as to a spread in their distribution areas.

Table C.14 Summarized table of natural disasters in Iraq from 1900 to 2010

		No. of events	Killed	Total affected	Damage (000 US$)
Drought	Drought	2	–	500,000	2,000
	Ave. per event		–	250,000	1,000
Earthquake (seismic activity)	Earthquake (ground shaking)	1	20	500	–
	Ave. per event		20	500	–
Epidemic	Bacterial infectious diseases	5	36	5,824	–
	Ave. per event		7.2	1,164.8	–
	Viral infectious diseases	1	2	–	–
	Ave. per event		2	–	–
Flood	Unspecified	3	–	410,000	58,000
	Ave. per event		–	136,666.7	19,333.3
	Flash flood	3	22	29,020	–
	Ave. per event		7.3	9,673.3	–
	General flood	2	4	42,490	1,300
	Ave. per event		2	21,245	650

Created on: Oct-12-2010 – Data version: v12.07
Source: "EM-DAT: The OFDA/CRED International Disaster Database http://www.em-dat.net –
Université Catholique de Louvain – Brussels – Belgium"

Israel

Israel is bordered by Lebanon, Syria, Jordan, West Bank, Egypt and Gaza. Temperatures in Israel vary widely. It has an area of 24,000 km^2 with a population of 6.2 million. Coastal line runs parallel to Mediterranean Sea. Mountain belt runs the length of the country with hills Galilee ranging from 500 to 1,200 m above sea level. The area of Beersheba and the Northern Negev has a semi-arid with hot summers. The Southern Negev and the Arava areas have Desert. Four different phyto-geographic regions exist in Israel, with extremely diverse the flora and fauna.

As can be seen in Table C.15, the country has witnessed extreme temperature, flood, drought and storm. The change in climate will result in change in magnitude and frequency of these disasters. The possible effect due to climate change is sea level rise, health impact, and water scarcity leading to subsequent consequences on agriculture. Flat sandy Mediterranean coastline is highly vulnerable to sea level rise where a 30 cm rise in sea level can flood as much as 60 m of land from seashore. The region could be affected due to inundation as well as increasing salinity of soil and available freshwater resources. Increase in temperature will also affect number of tourist visiting the country. Increase in surface run off due to increase in rain intensity to drop in overall precipitation may result in desertification, reduced ground water recharge, loss of water and top soil erosion. Both anthropogenic activity and climate change will bring in changes in population of wild life. According to the first national communication under UNFCCC, degradation of coral reef fin Gulf of Eilat will adversely affect coastal ecosystem and tourism. Warming may have positive impacts on crop yields if moisture is not a restriction, but rise in the frequency of extreme events or pests may offset any potential

Table C.15 Summarized table of natural disasters in Israel from 1900 to 2010

		No. of events	Killed	Total affected	Damage (000 US$)
Drought	Drought	1	–	–	75,000
	Ave. per event		–	–	75,000
Epidemic	Viral infectious diseases	1	12	139	–
	Ave. per event		12	139	–
Extreme temperature	Cold wave	1	–	–	550,000
	Ave. per event		–	–	550,000
	Heat wave	1	–	–	–
	Ave. per event		–	–	–
Flood	Unspecified	1	15	–	40,000
	Ave. per event		15	–	40,000
	Flash flood	1	–	1,000	–
	Ave. per event		–	1,000	–
	General Flood	1	2	–	–
	Ave. per event		2	–	–
Mass movement wet	Landslide	1	20	13	–
	Ave. per event		20	13	–
Storm	Unspecified	2	8	410	2,750
	Ave. per event		4	205	1,375
Wildfire	Forest fire	2	–	240	45,000
	Ave. per event		–	120	22,500

Created on: Oct-12-2010 – Data version: v12.07
Source: "EM-DAT: The OFDA/CRED International Disaster Database http://www.em-dat.net – Université Catholique de Louvain – Brussels – Belgium"

benefits. Both crops and livestock would be affected by means of increased pestilence of alien/invasive pests and diseases. Fisheries in Sea of Galilee may decline due to decline in oxygen concentration associated with rise in temperature. Eutrophication due to increased surface runoff and pollution will increase algal bloom which will further affect fish population. Sea level rise may erode coastal structures and collapse of coastal beach cliff. Increase in storm frequency and changes in wind direction may result in coastal erosion. Sea level rise may lead to loss of valuable lands, buildings and tourist facilities. Eilat's narrow recreational beaches on Read sea coast may be affected due to rising water levels and wave activity. Increase in evaporation along with decrease in precipitation will further reduce flow to Dead Sea coast resulting in effects on ecosystem as well as formation of hollows along coast line endangering human life and property. Underwater archaeological sites now covered with sand may be lost due to coastal erosion and disappearance of sand exposing site to wave attack and oxidation. Sea level rise will further decrease hydraulic slope between water outlets and seawater table reducing efficiency of coastal power stations.

Japan

Japan is an archipelago of more than 6,800 islands with an area of 37,790,000 ha in the Pacific Ocean. Eighty percent of the land is either forested or agricultural land.

The country has a population of 127 million with a population density of 340 inhabitants per hectare.

The climate of Japan is predominantly temperate, with rainy season beginning in early May. Japan has variety of forest that include subtropical moist broadleaf forests, temperate broadleaf and mixed forests and temperate coniferous forests. Rice is cultivated in about half the area of arable land. Other crops include wheat, barley, potatoes and tobacco. Japan is one of the most industrially advanced country. The principal industries are automobile, iron and steel, chemical, textile, fishing, ceramics, precision instruments, electronic goods, machinery, fertilisers and ship building.

As can be seen in Table C.16, the country has witnessed extreme temperature, flood, drought and storm. The change in climate will make these disasters to re-occur with change in magnitude and frequency. Climate change will have a significant effect on Japan's agriculture, forestry, fisheries, water resources, coastal management, natural ecosystems, and human health. Country may face risk of malaria and other tropical infectious diseases. Sea levels rise will result in submergence of coastal area and the damage from storms will grow. As per second national communication under UNFCCC sea level rise by just 65 cm, will result in loss of over 80% of Japan's sandy beaches. Both crops and livestock would be affected by means of increased pestilence of alien/invasive pests and diseases. An augment in temperature can reduce the ability of farmers to work. Japan is an island nation with long coastlines and changing precipitation, typhoon patterns and sea level rise will affect economic activities concentrated in coastal area. Lake, marsh, and coastal ecosystems which are affected due to anthropogenic activities will further affected by global warming. Rise in sea levels, increase in precipitation, and global warming changes will increase riverbed erosion. The climate change may also increase mass movement disasters due to effect on stability of slopes in mountainous areas. Fresh water species with low resistance to rise in temperature will be affected. As per third national communication under UNFCCC increase in global temperature and sun shine may cause dehydration and death of trees in forest due to dehydration. Phytoplankton species that used to live near lower latitude will appear near seas of Japan due to increase in temperature. Production capability of sea of Okhotsh will deteriorate as the ice algae attached to ice in the sea is expected to decline due to reduction in ice. Numbers of jelly fish will increase with increase in global temperature. Rise in temperature will increase in population of rivals of fish in food chain. Changing climate will bring in variation in composition and spread of species.

Jordan

Jordan is located on bank of the river Jordan in west-central Eurasia. It borders Saudi Arabia, Iraq, Syria, and Israel. It has an area if 0.08 million km^2 with a population of around six million. Climate is characterized by hot, dry summers

Table C.16 Summarized table of natural disasters in Japan from 1900 to 2010

		No. of events	Killed	Total affected	Damage (000 US$)
Drought	Drought	1	–	–	–
	Ave. per event		–	–	–
Earthquake	Earthquake (ground shaking)	47	166,160	994,127	148,122,400
(seismic	Ave. per event		3,535.3	21,151.6	3,151,540.4
activity)	Tsunami	8	6,477	25,319	450,000
	Ave. per event		809.6	3,164.9	56,250
Epidemic	Bacterial infectious diseases	2	1	534	–
	Ave. per event		0.5	267	–
	Viral infectious diseases	1	–	2,000,000	–
	Ave. per event		–	2,000,000	–
Extreme	Heat wave	2	72	3,300	–
temperature	Ave. per event		36	1,650	–
Flood	Unspecified	31	12,814	7,015,269	268,300
	Ave. per event		413.4	226,299	8,654.8
	Flash flood	1	21	25,807	1,950,000
	Ave. per event		21	25,807	1,950,000
	General flood	11	187	99,266	1,814,000
	Ave. per event		17	9,024.2	164,909.1
	Storm surge/coastal flood	2	34	384,143	7,440,000
	Ave. per event		17	192,071.5	3,720,000
Mass movement	Avalanche	1	13	–	–
wet	Ave. per event		13	–	–
	Landslide	20	989	25,706	210,000
	Ave. per event		49.5	1,285.3	10,500
Storm	Unspecified	24	1,890	192,814	453,500
	Ave. per event		78.8	8,033.9	18,895.8
	Local storm	6	27	100,499	363,000
	Ave. per event		4.5	16,749.8	60,500
	Tropical cyclone	109	32,500	7,512,095	53,055,500
	Ave. per event		298.2	68,918.3	486,747.7
Volcano	Volcanic eruption	15	515	99,979	132,000
	Ave. per event		34.3	6,665.3	8,800
Wildfire	Forest fire	1	–	222	–
	Ave. per event		–	222	–

Created on: Oct-12-2010 – Data version: v12.07
Source: "EM-DAT: The OFDA/CRED International Disaster Database http://www.em-dat.net – Université Catholique de Louvain – Brussels – Belgium"

with average temperature of about 30°C and cool winters with average temperature around 13°C during which the precipitation occurs. The country has a Mediterranean. Most of the land receives less than 620 mm of rain a year. Jordan is largely deserted but the western portion is fertile. Main agricultural products are citrus fruits, wheat, barley, lentils and water melons. Phosphate and potash is important export item. Tourism is the main activity which earns foreign exchange.

As can be seen in Table C.17, the country has witnessed extreme temperature, flood, drought and storm. The change in climate will have these disasters to re occur

Table C.17 Summarized table of natural disasters in Jordan from 1900 to 2010

		No. of events	Killed	Total affected	Damage (000 US$)
Drought	Drought	2	–	330,000	–
	Ave. per event		–	165,000	–
Earthquake (seismic	Earthquake (ground shaking)	1	242	–	–
activity)	Ave. per event		242	–	–
Epidemic	Bacterial infectious diseases	1	4	715	–
	Ave. per event		4	715	–
Extreme temperature	Cold wave	1	15	–	400,000
	Ave. per event		15	–	400,000
	Heat wave	1	–	12	–
	Ave. per event		–	12	–
Flood	Unspecified	4	44	529	2,000
	Ave. per event		11	132.3	500
	Flash flood	2	267	23,792	1,400
	Ave. per event		133.5	11,896	700
Insect infestation	Locust	1	–	–	–
	Ave. per event		–	–	–
Storm	Unspecified	3	16	225	–
	Ave. per event		5.3	75	–

Created on: Oct-12-2010 – Data version: v12.07
Source: "EM-DAT: The OFDA/CRED International Disaster Database http://www.em-dat.net – Université Catholique de Louvain – Brussels – Belgium"

with change in magnitude and frequency. But the country will be affected by fall in agricultural productivity and tourist activity. The warming of the earth can drastically reduce flow in Jordan River. Climate change could bring threat to some of endangered and endemic species of the area. Both crops and livestock would be affected by means of increased pestilence of alien/invasive pests and diseases. An augment in temperature can reduce the ability of farmers to work.

Kazakhstan

Kazakhstan is located in Central Asia. It is bordered by Russia, China, Kyrgyzstan, Uzbekistan, Turkmenistan. The country has an area of 2.7 million km^2. with a population of about 15 million. The climate is characterized by warm summers and colder winters. Precipitation varies between arid and semi-arid conditions. Major agricultural products of the country are grain, sugar beet, potatoes, vegetables, meat, milk, eggs, and cotton. Major natural resources are ores of copper and lead. It has second largest oil field in the world in Kasagan.

As can be seen in Table C.18, the country has witnessed extreme temperature, flood, mass movement, storm and wild fire. The change in climate will have these disasters to reoccur with change in magnitude and frequency. The most part of Kazakhstan is deserts and semi-deserts that are slightly vulnerable to climate change. Redistribution of precipitation, increase of frequency and intensity of

Table C.18 Summarized table of natural disasters in Kazakhstan from 1900 to 2010

		No. of events	Killed	Total affected	Damage (000 US$)
Earthquake (seismic activity)	Earthquake (ground shaking)	1	3	36,626	–
	Ave. per event		3	36,626	–
Epidemic	Unspecified	1	–	166	–
	Ave. per event		–	166	–
	Bacterial infectious diseases	1	7	593	–
	Ave. per event		7	593	–
	Viral infectious diseases	1	–	114	–
	Ave. per event		–	114	–
Extreme temperature	Cold wave	2	3	600,012	–
	Ave. per event		1.5	300,006	–
Flood	Unspecified	2	–	6,168	1,500
	Ave. per event		–	3,084	750
	Flash flood	2	44	41,200	40,202
	Ave. per event		22	20,600	20,101
	General flood	3	11	56,000	166,532
	Ave. per event		3.7	18,666.7	55,510.7
Mass movement wet	Landslide	1	48	–	–
	Ave. per event		48	–	–
Storm	Unspecified	1	112	–	3,000
	Ave. per event		112	–	3,000
Wildfire	Forest fire	1	–	8,000	–
	Ave. per event		–	8,000	–

Created on: Oct-12-2010 – Data version: v12.07

Source: "EM-DAT: The OFDA/CRED International Disaster Database http://www.em-dat.net – Université Catholique de Louvain – Brussels – Belgium"

droughts will affect agriculture, forestry, and water resources of the country. Human health will be affected due to strengthening of heat stress, and distribution of many kinds of diseases. As per first communication under UNFCCC the yield of wheat one of the main grain crops, can be especially vulnerable to climate change. Sheep-breeding productivity will be affected due to decrease in grass land yield and impact of hot weather periods on sheep. Both crops and livestock would be affected by means of increased pestilence of alien/invasive pests and diseases. An augment in temperature can reduce the ability of farmers to work.

Korea, Dem. People's Republic

North Korea, officially the Democratic People's Republic of Korea occupies the northern portion of the Korean Peninsula, covering an area of 120,540 km² neighboured by China, Russia, South Korea, Yellow Sea and Korea Bay. The country is characterised by mountain ranges that crisscross the peninsula. North Korea's climate is relatively temperate with four distinct seasons. Long winters bring bitter cold and with average snowfall of 37 days during the winter. Summer tends to be

Table C.19 Summarized table of natural disasters in Korea Demographic People's Republic of Korea from 1900 to 2010

		No. of events	Killed	Total affected	Damage (000 US$)
Earthquake (seismic activity)	Earthquake (ground shaking)	1	–	–	–
	Ave. per event		–	–	–
Epidemic	Bacterial infectious diseases	1	–	200	–
	Ave. per event		–	200	–
	Viral infectious diseases	1	4	3,000	–
	Ave. per event		4	3,000	–
Flood	Unspecified	4	431	3,790,071	2,200,000
	Ave. per event		107.8	947,517.8	550,000
	Flash flood	4	627	320,517	11,400
	Ave. per event		156.8	80,129.3	2,850
	General flood	7	762	7,148,838	15,320,000
	Ave. per event		108.9	1,021,262.6	2,188,571.4
	Storm surge/coastal flood	2	–	36,324	–
	Ave. per event		–	18,162	–
Storm	Tropical cyclone	5	55	638,730	6,110,510
	Ave. per event		11	127,746	1,222,102

Created on: Oct-12-2010 – Data version: v12.07
Source: "EM-DAT: The OFDA/CRED International Disaster Database http://www.em-dat.net – Université Catholique de Louvain – Brussels – Belgium"

short, hot, humid, and rainy. Chief crops of the nation are rice, wheat, barley, potatoes and vegetables. Major natural resources include coal, iron ore, tungsten ore and graphite. Industrial activity includes textiles, steel, petrochemicals and automobiles.

Natural hazards include late spring floods and storms. As could be seen in Table C.19 drought has not affected the area so as to qualify as a disaster. Temperate zone fruit like apples, grapes, pears and peaches will be affected because of change in climate. Both crops and livestock would be affected by means of increased pestilence of alien/invasive pests and diseases. An augment in temperature can reduce the ability of farmers to work.

Korea, Republic of

South Korea, officially the Republic of Korea is neighbored by China, Japan and North Korea. The country has a total area of 100,032.00 km² and has a population of 50 million. The country has humid continental climate and a humid subtropical climate, and is affected by the East Asian monsoon. The average January temperature range is −7 to 1°C, and the average August temperature range is 22–30°C. The average annual precipitation varies from 1,370 to 1,470 mm with occasional typhoons that bring high winds and floods. Chief crops

Table C.20 Summarized table of natural disasters in Republic of Korea from 1900 to 2010

		No. of events	Killed	Total affected	Damage (000 US$)
Drought	Drought	2	–	2,800,000	–
	Ave. per event		–	1,400,000	–
Epidemic	Bacterial infectious diseases	2	137	1,888	–
	Ave. per event		68.5	944	–
	Viral infectious diseases	2	6	39,534	–
	Ave. per event		3	19,767	–
Extreme temperature	Heat wave	1	40	–	–
	Ave. per event		40	–	–
Flood	Unspecified	18	2,402	1,271,823	402,834
	Ave. per event		133.4	70,656.8	22,379.7
	Flash flood	7	608	480,848	1,915,000
	Ave. per event		86.9	68,692.6	273,571.4
	General flood	8	938	1,480,279	591,045
	Ave. per event		117.3	185,034.9	73,880.6
	Storm surge/coastal flood	1	10	13	–
	Ave. per event		10	13	–
Mass movement wet	Landslide	8	346	5,688	248,700
	Ave. per event		43.3	711	31,087.5
Storm	Unspecified	14	626	220,704	530,393
	Ave. per event		44.7	15,764.6	37,885.2
	Local storm	2	26	2,000	572,854
	Ave. per event		13	1,000	286,427
	Tropical cyclone	31	3,621	417,127	10,944,431
	Ave. per event		116.8	13,455.7	353,046.2
Wildfire	Forest fire	3	2	5,150	–
	Ave. per event		0.7	1,716.7	–

Created on: Oct-12-2010 – Data version: v12.07
Source: "EM-DAT: The OFDA/CRED International Disaster Database http://www.em-dat.net –
Université Catholique de Louvain – Brussels – Belgium"

of the nation are rice, wheat, barley, potatoes and vegetables. Fish is both source of food and export item. Major natural resources include coal, iron ore, tungsten ore, graphite and fluorite. Industrial activity includes textiles, steel, petrochemicals and automobiles.

As can be seen in Table C.20, the country has witnessed drought, extreme temperature, flood, mass movement, storm and wild fire. The change in climate will have these disasters to re occur with change in magnitude and frequency. The expected climate change can affect the health of country due to increase in infectious diseases and heat stress. As per National Communication of the Republic of Korea under UNFCCC, climate variation cause changes in soil moisture and water resources. The southern coastal areas and the lower southern regions would be experiencing almost no winter. Temperate zone fruit like apples, grapes, pears and peaches will be affected because of change in climate. Habitat of cold-water fish like salmon and herring is likely move northward where as middle-depth cold-water fish like the Alaska pollack and codfish would disappear from Korea's waters if no "cold-water masses" flowed from the north. Warm-water fish from the East China Sea will move to seas of the Korean Peninsula. Both crops and livestock would be affected by means of increased pestilence of alien/invasive pests and diseases. An augment in temperature can reduce the ability of farmers to work.

Kuwait

Kuwait is located in the northeast of the Arabian Peninsula in Western Asia bordered by Saudi Arabia to the south, and Iraq to the north. The flat, sandy Arabian desert covers most of Kuwait. The spring season in March is warm and pleasant with occasional thunderstorms. The frequent winds from the northwest are cold in winter and spring and hot in summer. Southeasterly winds, usually hot and damp, spring up between July and October; hot and dry south winds prevail in spring and early summer. Northwesterly wind common during June and July, causes dramatic sandstorms.

Major industries include petroleum, shipping, construction, cement, water desalination, construction materials and financial services.

As can be seen in Table C.21, flood is the only climate induced disaster since 1900 AD. The country does not have any major agricultural activity. Since the country is land locked, rise in sea level will not affect the area. The expected climate change can affect the health of country due to increase in infectious diseases and heat stress.

Kyrgyzstan

Kyrgyzstan is bordered by Kazakhstan to the north, Uzbekistan to the west, Tajikistan to the southwest and China to the east. The mountainous region of the Tian Shan covers over 80% of the country. The climate varies regionally. The southwestern Fergana Valley is subtropical and extremely hot in summer, with temperatures reaching 40°C The northern foothills are temperate and the Tian Shan varies from dry continental to polar climate. Agriculture is an important sector and crops

Table C.21 Summarized table of natural disasters in Kuwait from 1900 to 2010

		No. of events	Killed	Total affected	Damage (000 US$)
Epidemic	Viral infectious diseases	1	–	1	–
	Ave. per event		–	1	–
Flood	Flash flood	1	2	200	–
	Ave. per event		2	200	–

Created on: Oct-12-2010 – Data version: v12.07
Source: "EM-DAT: The OFDA/CRED International Disaster Database http://www.em-dat.net – Université Catholique de Louvain – Brussels – Belgium"

Table C.22 Summarized table of natural disasters in Kyrgyzstan from 1900 to 2010

		No. of events	Killed	Total affected	Damage (000 US$)
Earthquake (seismic activity)	Unspecified	1	74	1,197	–
	Ave. per event		74	1,197	–
	Earthquake (ground shaking)	5	58	153,086	163,000
	Ave. per event		11.6	30,617.2	32,600
Epidemic	Bacterial infectious diseases	1	22	336	–
	Ave. per event		22	336	–
	Viral infectious diseases	1	–	458	–
	Ave. per event		–	458	–
Extreme temperature	Cold wave	1	11	–	–
	Ave. per event		11	–	–
Flood	Flash flood	1	1	7,728	2,400
	Ave. per event		1	7,728	2,400
	General flood	2	3	2,895	2,860
	Ave. per event		1.5	1,447.5	1,430
Mass movement wet	Avalanche	1	11	2	–
	Ave. per event		11	2	–
	Landslide	7	238	68,159	37,500
	Ave. per event		34	9,737	5,357.1
Storm	Unspecified	1	4	9,075	–
	Ave. per event		4	9,075	–

Created on: Oct-12-2010 – Data version: v12.07
Source: "EM-DAT: The OFDA/CRED International Disaster Database http://www.em-dat.net – Université Catholique de Louvain – Brussels – Belgium"

include wheat, sugar beets, potatoes, cotton, tobacco, vegetables, fruit and livestock raising.

As seen by Table C.22 flood, mass movement and storm major climate induced risks in the area. As the country is land locked hence it will not have any threat due to rise in sea level. Most of the rivers in the country are snow and glacier fed rivers and hence flow be affected due rise in temperature. As per the first national communication under UNFCCC change in climate can increase urolithiasis in the country. The expected climate change can also affect the health of country due to increase in infectious diseases and heat stress. Change in climate will lead to increase in desert and steppe area. Both crops and livestock would be affected by means of increased pestilence of alien/invasive pests and diseases. An augment in temperature can reduce the ability of farmers to work.

Lao

The Lao is a landlocked country covering an area of 236,800 km². The country shares its borders with the Vietnam, Thailand, Cambodia, China and Myanmar. The topography of the nation is mainly mountainous ranging from 200 to 28,820 m with mountains covering nearly two-thirds of the land area. The chief agricultural

Table C.23 Summarized table of natural disasters in Lao from 1900 to 2010

		No. of events	Killed	Total affected	Damage (000 US$)
Drought	Drought	5	–	4,250,000	1,000
	Ave. per event		–	850,000	200
Epidemic	Unspecified	3	44	9,685	–
	Ave. per event		14.7	3,228.3	–
	Bacterial infectious diseases	2	534	8,244	–
	Ave. per event		267	4,122	–
	Viral infectious diseases	3	208	2,000	–
	Ave. per event		69.3	666.7	–
Flood	Unspecified	10	76	1,878,600	2,480
	Ave. per event		7.6	187,860	248
	General flood	8	358	1,569,740	37,128
	Ave. per event		44.8	196,217.5	4,641
Storm	Unspecified	2	8	38,435	302,301
	Ave. per event		4	19,217.5	151,150.5
	Tropical cyclone	3	64	1,397,764	103,650
	Ave. per event		21.3	465,921.3	34,550

Created on: Nov-15-2010 – Data version: v12.07

Source: "EM-DAT: The OFDA/CRED International Disaster Database http://www.em-dat.net – Université Catholique de Louvain – Brussels – Belgium"

products are rice, maize, tobacco and cotton. The major industries include tin, timber and textiles.

Table C.23 gives the summary of natural disasters in Lao from 1900 to 2010. As the country is land locked it will not have any threat due to rise in sea level. Most of the rivers in the country are snow and glacier fed rivers and hence flow be affected due rise in temperature. The expected climate change can affect the health of country due to increase in infectious diseases and heat stress. Redistribution of precipitation, increase of frequency and intensity of droughts will affect agriculture, forestry, and water resources of the country. Both crops and livestock would be affected by means of increased pestilence of alien/invasive pests and diseases. An augment in temperature can reduce the ability of farmers to work.

Lebanon

Lebanon is located on the seashore of Mediterranean sea. The country shares its borders with Syria and Israel. Lebanese climate is typically Mediterranean with six Bioclimatic levels: arid, semi-arid, sub-humid, humid, pre-humid and oromediterranean, all the more remarkable by their variability over short distances. The country produces olive oil, grains and fruits. The major industries are oil refining, food processing, textiles, chemicals and cement.

Table C.24 gives the summary of natural disasters in Lebanon from 1900 to 2010. The expected climate change can affect the health of country due to increase in infectious diseases and heat stress. Redistribution of precipitation as well as

Table C.24 Summarized table of natural disasters in Lebanon from 1900 to 2010

		No. of Events	Killed	Total Affected	Damage (000 US$)
Earthquake (seismic activity)	Earthquake (ground shaking)	1	136	200	–
	Ave. per event		136	200	–
Flood	Unspecified	2	440	1,500	10,000
	Ave. per event		220	750	5,000
	General flood	1	–	17,000	–
	Ave. per event		–	17,000	–
Mass movement dry	Landslide	1	20	–	–
	Ave. per event		20	–	–
Storm	Unspecified	1	–	500	–
	Ave. per event		–	500	–
	Local storm	1	25	104,075	155,000
	Ave. per event		25	104,075	155,000
Wildfire	Forest fire	1	1	15	–
	Ave. per event		1	15	–

Created on: Nov-15-2010 – Data version: v12.07
Source: "EM-DAT: The OFDA/CRED International Disaster Database http://www.em-dat.net – Université Catholique de Louvain – Brussels – Belgium"

increase of frequency and intensity of droughts will affect agriculture, forestry, and water resources of the country. Both crops and livestock would be affected by means of increased alien/invasive pests and diseases. An augment in temperature can reduce the ability of farmers to work.

Malaysia

Malaysia is a coastal nation with an area of 329,733 km^2 and is divided into two landmasses separated by the South China Sea. The country has a population of about 25.7 million. Malaysia depicts relatively uniform temperatures throughout the year. Mean temperature in the lowlands ranges between 26 and 28°C. Annual variation of the daily mean temperature is about 2–3°C, but the diurnal variation is as large as 12°C. The Northeast monsoon dominates from November to March, with wind speeds of 15–50 km/h. Southwest monsoon winds blow between June and September, with wind speeds rarely exceeding 25 km/h. Rainfall distribution closely follows the topography of the land with highlands being usually wetter. More than 3,550 mm of rainfall a year is recorded in the lowlands. Pockets of rain-shadow areas located in-between highland ridges receive less than 1,780 mm of rainfall a year. Malaysia is one of the largest producer of rubber, tin and palm oil. It is also leading exporter of pepper and timber. The other agriculture products include rice, coconut, vegetables, fruits, coffee, tea, cocoa etc., Major mineral resources are ores of iron, gold, limonite and bauxite. Major industrial activities include food products, tobacco, wood products, electrical goods, textiles and chemicals.

Table C.25 Summarized table of natural disasters in Malaysia from 1900 to 2010

		No. of events	Killed	Total affected	Damage (000 US$)
Drought	Drought	1	–	5,000	–
	Ave. per event		–	5,000	–
Earthquake (seismic activity)	Tsunami	1	80	5,063	500,000
	Ave. per event		80	5,063	500,000
Epidemic	Unspecified	2	30	2,620	–
	Ave. per event		15	1,310	–
	Bacterial infectious diseases	4	19	662	–
	Ave. per event		4.8	165.5	–
	Viral infectious diseases	7	491	28,765	–
	Ave. per event		70.1	4,109.3	–
Flood	Unspecified	10	141	736,276	75,100
	Ave. per event		14.1	73,627.6	7,510
	Flash flood	7	36	157,600	23,000
	Ave. per event		5.1	22,514.3	3,285.7
	General flood	18	134	338,182	978,000
	Ave. per event		7.4	18,787.9	54,333.3
Mass movement dry	Landslide	1	72	–	–
	Ave. per event		72	–	–
Mass movement wet	Landslide	3	80	285	–
	Ave. per event		26.7	95	–
Storm	Unspecified	4	22	51,500	–
	Ave. per event		5.5	12,875	–
	Local storm	1	2	155	–
	Ave. per event		2	155	–
	Tropical cyclone	2	272	6,291	53,000
	Ave. per event		136	3,145.5	26,500
Wildfire	Forest fire	4	–	3,000	302,000
	Ave. per event		–	750	75,500

Created on: Nov-15-2010 – Data version: v12.07
Source: "EM-DAT: The OFDA/CRED International Disaster Database http://www.em-dat.net –
Université Catholique de Louvain – Brussels – Belgium"

Table C.25 gives summary of natural disasters in Malaysia from 1900 to 2010. Climate change may increase in the frequency and intensity of extreme weather events, such as, droughts, storms and floods. Within a climatic zone, change in rainfall, sunshine hours, temperature, relative humidity and length of the drought period, results in year-to-year variability of crop production. As per initial communication under UNFCCC, drought-prone areas in parts of Kelantan, Terengganu, Pahang, Johore, Kedah, Perak, Negeri Sembilan and Melaka are, most vulnerable with respect to cultivation of oil palm. The country will also get affected with respect other agricultural activity. Expansion of upland forest by 5–8% will be nullified by a loss of between 15 and 20% of mangrove forests located along the coastline due to sea level rise. Changing climate will have effects on species composition of the forest. The country will be affected by a higher risk of slope failures of riverbanks and hills; rapid rate of sedimentation of reservoirs; channels; and loss of soil nutrients due to soil erosion. Mangrove belt will be lost due to sea level rise because the rise in the elevation of existing mudflats due to deposition of organic and inorganic deposition. The country will also be affected by shore line erosion and impact due to increased wave action. The country may face economic loss associated with increased flooding. The expected climate change can affect the health of country due to increase in infectious diseases and heat stress. Both crops and livestock would be affected by means of increased alien/invasive pests and diseases. An augment in temperature can reduce the ability of farmers to work.

Mongolia

Mongolia is a landlocked country in Northeast Asia neighboured by Russia and China with an average altitude of 1,580 m above sea level, 81.2% of the province is greater than 1,000 m. Mongolia has a total area of 1,566,600 km^2 and population of about three million. The country has high mountains, wide steppe, desert and semi-desert zones. The country hosts the great Siberian taiga forest, the Central Asian steppe, the high Altai Mountains and the Gobi desert converge. Mongolia has harsh continental climate with four distinctive seasons – high annual and diurnal temperature fluctuations, and low rainfall. Average annual temperatures range between 8.5 and 7.8°C. The extreme minimum temperature is between ranges −31.1 and −52.9°C. The extreme maximum temperature varies from 28.5 to 42.2°C in July. The rain fall intensity varies between 38.4 and 389 mm with average annual precipitation between 200 and 220 mm/year. The majority of precipitation happens in the months of June, July and August. November and March are the driest months. Animal husbandry is the principal occupation and comprises of horses, oxen, sheep, goats and camels. Minerals include coal; fluorspar; ores of tungsten, tin and copper. The industrial sector includes wool and cashmere processing, leather goods production, food processing, construction, and garment manufacturing. Traditionally, crop production is not a significant agricultural activity and only about 1.3 million hectares is used for cultivation of crops. The major crops are wheat, potatoes and other vegetables.

Table C.26 gives summary of natural disasters in Mongolia from 1900 to 2010. Country's major economic sectors, such as animal husbandry, rain fed arable farming, and the mining industry, are very sensitive to climate change. The expected climate change can affect the health of country due to increase in infectious diseases and heat stress. As per initial national communication to the Conference of the Parties to the UNFCCC, about 200,000 animals are lost every year due to heavy snow and frost. Due to its location in the transition zone between the Siberian taiga and the Central Asia desert, the country is very sensitive to a shift of geo-climate zones that can be caused due to climate change. Frequency of extreme events such as droughts, floods, dust storms, thunderstorms, heavy snowfall, flash floods etc., has increased during the past 30 years. Livestock is also being affected due to extreme winters. The frequency of wild fires is increasing because of the extremely dry springs. Rivers provide the main source of water in mountainous regions, but in

Table C.26 Summarized table of natural disasters in Mongolia from 1900 to 2010

		No. of events	Killed	Total affected	Damage (000 US$)
Drought	Drought	1	–	450,000	–
	Ave. per event		–	450,000	–
Earthquake (seismic activity)	Earthquake (ground shaking)	1	1,200	–	–
	Ave. per event		1,200	–	–
Epidemic	Bacterial infectious diseases	1	8	108	–
	Ave. per event		8	108	–
	Viral infectious diseases	2	–	3,160	–
	Ave. per event		–	1,580	–
Extreme temperature	Cold wave	1	5	769,113	62,000
	Ave. per event		5	769,113	62,000
Flood	Unspecified	3	62	270,000	25,000
	Ave. per event		20.7	90,000	8,333.3
	Flash flood	2	56	1,650	270
	Ave. per event		28	825	135
	General flood	2	26	14,000	94
	Ave. per event		13	7,000	47
Storm	Unspecified	4	14	1,265,000	90,000
	Ave. per event		3.5	316,250	22,500
	Local storm	5	129	746,000	–
	Ave. per event		25.8	149,200	–
Wildfire	Forest fire	2	25	5,061	1,822,800
	Ave. per event		12.5	2,530.5	911,400
	Scrub/grassland fire	1	–	–	–
	Ave. per event		–	–	–

Created on: Nov-15-2010 – Data version: v12.07
Source: "EM-DAT: The OFDA/CRED International Disaster Database http://www.em-dat.net – Université Catholique de Louvain – Brussels – Belgium"

some areas, such as the steppe and the Gobi, ground water is the only source for daily use. Water resources will be affected with increase in temperature. Forest area is being reduced at an accelerated rate and climate variation will have significant effects on the re-growth and productivity of forests. Permafrost covers about 63% of Mongolia's territory and is classified into seven categories: continuous, discontinuous, wide spread, rare spread, sporadic, pereletka, and seasonal. Change in climate will affect both permafrost and snow resulting in subsequent impacts on water resources and ecology of the region. The country will also get affected by soil erosion. Both crops and livestock would be affected by means of increased alien/invasive pests and diseases. An augment in temperature can reduce the ability of farmers to work.

Myanmar

Myanmar, is located in Southeast Asia bordered by China, Laos, Thailand, Bangladesh, India, Bay of Bengal and Andaman Sea. It has a total area of 678,500 km^2 with

Table C.27 Summarized table of natural disasters in Myanmar from 1900 to 2010

		No. of events	Killed	Total affected	Damage (000 US$)
Earthquake (seismic activity)	Earthquake (ground shaking)	5	551	160	–
	Ave. per event		110.2	32	–
	Tsunami	1	71	15,700	500,000
	Ave. per event		71	15,700	500,000
Epidemic	Bacterial infectious diseases	2	10	800	–
	Ave. per event		5	400	–
	Viral infectious diseases	1	30	–	–
	Ave. per event		30	–	–
Flood	Unspecified	7	161	386,988	55,115
	Ave. per event		23	55,284	7,873.6
	Flash flood	2	112	50,000	–
	Ave. per event		56	25,000	–
	General flood	10	193	2,135,390	79,840
	Ave. per event		19,3	213,539	7,984
Mass movement wet	Landslide	2	41	1,367	–
	Ave. per event		20,5	683.5	–
Storm	Tropical cyclone	16	144,618	3,675,795	4,022,388
	Ave. per event		9,038.6	229,737.2	251,399.3
Wildfire	Forest fire	2	8	78,588	11,000
	Ave. per event		4	39,294	5,500

Created on: Oct-12-2010 – Data version: v12.07
Source: "EM-DAT: The OFDA/CRED International Disaster Database http://www.em-dat.net –
Université Catholique de Louvain – Brussels – Belgium"

a population of 48 million. It lies in the monsoon region of Asia, receiving rain of
1,000 mm to over 5,000 mm depending on the region. Temperature is warm with
average temperature ranging from 21°C to of 32°C.

Typical jungle animals are tigers, leopards, rhinoceros, wild buffalo, wild boars,
deer, antelope, elephants, gibbons, monkeys, flying foxes, tapirs crocodiles, geckos,
cobras, Burmese pythons, and turtles. The common birds found in the region are
parrots, peafowl, pheasants, crows, herons, and paddy birds.

As can be seen form the Table C.27 country has been affected from flood, mass
movement, storm and wildfire. With change in climate the country will be affected
by these disasters with varied frequency and magnitude. Both crops and livestock
would be affected by means of increased alien/invasive pests and diseases. An
augment in temperature can reduce the ability of farmers to work.

Nepal

Nepal is a landlocked country in South Asia located in the Himalayas and bordered to
the north by China and India. The country has an area of 147,181 km^2 with a
population of approximately 30 million. The **Hill Region** varies from 800 to
4,000 m in altitude and the Mountain Region contains the highest elevations in the
world. Nepal experiences five seasons: summer, monsoon, autumn, winter and spring.

The dramatic differences in elevation has resulted in a variety of biomes: tropical savannas, subtropical broadleaf, coniferous forests, temperate broadleaf, coniferous, mountain grasslands and shrub lands.

The region play significant role in global atmospheric circulation, biodiversity, agriculture, and hydropower. The water resources of this region are currently facing threats from a large number of driving forces. Nepal's economy depends on agriculture. Out of total area of Nepal is 147,181 km^2 about 3,091,000 ha area is cultivated for agriculture. Changes in precipitation are uncertain with both increasing and decreasing trends due to climate change in different parts of the country. As could be seen in Table C.28 this small land locked country has witnessed almost all

Table C.28 Summarized table of natural disasters in Nepal from 1900 to 2010

		No. of events	Killed	Total affected	Damage (000 US$)
Drought	Drought	5	–	4,600,000	10,000
	Ave. per event		–	920,000	2,000
Earthquake (seismic activity)	Earthquake (ground shaking)	5	9,929	562,001	306,000
	Ave. per event		1,985.8	112,400.2	61,200
Epidemic	Unspecified	3	685	50,242	–
	Ave. per event		228.3	16,747.3	–
	Bacterial infectious diseases	6	1,834	109,883	–
	Ave. per event		305.7	18,313.8	–
	Viral infectious diseases	10	1,995	9,669	–
	Ave. per event		199.5	966.9	–
Extreme temperature	Cold wave	3	126	200	–
	Ave. per event		42	66.7	–
	Heat wave	1	–	10	–
	Ave. per event		–	10	–
Flood	Unspecified	13	1,729	781,907	766,313
	Ave. per event		133	60,146.7	58,947.2
	Flash flood	4	2,566	714,650	200,000
	Ave. per event		641.5	178,662.5	50,000
	General flood	17	1,625	1,992,173	70,929
	Ave. per event		95.6	117,186.6	4,172.3
Mass movement dry	Landslide	1	150	–	–
	Ave. per event		150	–	–
Mass movement wet	Avalanche	1	95	–	–
	Ave. per event		95	–	–
	Debris flow	1	106	–	–
	Ave. per event		106	–	–
	Landslide	14	1,387	442,618	–
	Ave. per event		99.1	31,615.6	–
Storm	Unspecified	4	27	184	3,600
	Ave. per event		6.8	46	900
	Local storm	2	70	–	–
	Ave. per event		35	–	–
Wildfire	Forest fire	2	88	54,000	6,200
	Ave. per event		44	27,000	3,100

Created on: Oct-12-2010 – Data version: v12.07
Source: "EM-DAT: The OFDA/CRED International Disaster Database http://www.em-dat.net – Université Catholique de Louvain – Brussels – Belgium"

climate induced disasters except sea level rise. Climate Change is likely to result in frequency and magnitude of extreme weather events, such as high intense rainfalls leading to flash floods, landslides and debris flows. The greater Himalayan region often referred as "the roof of the world" – contains the most widespread and uneven high altitude areas on Earth, and the largest areas covered by glaciers and permafrost outside the Polar Regions. The melting ice and its impact on biomes and agriculture is major concern to the country and its people. Climate change induced hazards such as floods, landslides, and droughts will affect livelihoods of people in mountain and downstream populations. The region is vulnerable to climate change due to its ecological fragility and economic marginality. Himalayas is biodiversity hotspot and the source of the headwaters of four major Asian river systems by snow and glacial melting, in the basins of which more than 1.3 billion people inhabit Hence change in climate will have huge impact on biodiversity as well water resources originating form the area. The warming in the greater Himalayas has been much greater than the global average. Nepal witnessed 0.6°C per decade, compared with a global average of 0.74°C over the last 100 years. The region has witnessed extraordinary melting of permanent glaciers during the past three decades. Himalayan glaciers are showing the fast rate of retreat, resulting in increases in glacial runoff and GLOFs, and an increased frequency of events such as floods, mudflows, and avalanches affecting human settlements. GLOF events can have impacts on socioeconomic systems, hydrology, and ecosystems. Global warming is having a harsh impact on the amount of snow and ice, which will influence downstream water availability in both short and long term. The climate change can bring in reduction of glacier volume, and eventual disappearance of many glaciers. This may cause water stress to millions of people. The area will also get affected due to increase in number of mass movements. The reduced volume of runoff due to disappearance of glaciers will have serious consequences for downstream hydropower and agriculture. The expected climate change can affect the health of country due to increase in infectious diseases and heat stress.

Oman

Oman is an Arab country in southwest Asia on the southeast coast of the Arabian Peninsula bordered by United Arab Emirates, Saudi Arabia and Yemen. The coast is formed by the Arabian Sea and the Gulf of Oman. The country is characterized by gravel desert plain and mountain ranges. Oman's climate is hot and dry in the interior and humid along the coast.

Annual rainfall in Muscat averages to 100 mm/year. Vegetation is sparse in the interior plateau, but vegetation on mountains is luxuriant during rains. Indigenous mammals include the Leopard, Hyena, Fox, Wolf, and Hare, Oryx and Ibex. Indigenous birds include the Vulture, Eagle, Stork, Bustard, Arabian Partridge, Bee Eater, Falcon and Sunbird. Country has limited renewable water resources, due to limited rainfall.

Table C.29 Summarized table of natural disasters in Oman from 1900 to 2010

		No. of events	Killed	Total affected	Damage (000 US$)
Storm	Unspecified	3	58	1,548	1,000
	Ave. per event		19.3	516	333.3
	Tropical cyclone	4	204	25,131	3,950,000
	Ave. per event		51	6,282.8	987,500

Created on: Oct-12-2010 – Data version: v12.07
Source: "EM-DAT: The OFDA/CRED International Disaster Database http://www.em-dat.net –
Université Catholique de Louvain – Brussels – Belgium"

As could be seen in Table C.29 in spite of limited rain fall the country has adopted to shortage of water without getting affected due to drought. The absence of drought in disaster list can be contributed to minimal impact or no impact of lower rainfall on the people due to their adaptation. The expected climate change can affect the health of country due to increase in infectious diseases and heat stress.

Pakistan

Pakistan is bordered by Afghanistan and Iran in the west, India in the east and the China in the northeast. The country has a population of 170.6 million as per 2010 estimates with geographical area of 796,095 km^2.

Pakistan is divided into three major geographic areas: the northern highlands; the Indus River plain; and the Balochistan Plateau. Pakistan's climate varies from tropical to temperate with arid conditions existing in the coastal south, characterized by a monsoon season with adequate rainfall and a dry season with lesser rainfall. The forests range from coniferous alpine and subalpine with trees such as spruce, pine, and deodar cedar in the northern mountains to deciduous trees such as the mulberry-type Shisham in the Sulaiman range in the south.

The southern plains are home to crocodiles, boars, porcupines, and small rodents. Sandy scrublands of central Pakistan are home to a jackals, hyenas, wild cats, panthers, and leopards. In the north, animals seen are Marco Polo sheep, Urial sheep, Markhor and Ibex goats, black and brown Himalayan bears, and the rare Snow Leopard. Rare species in Pakistan include Indus River Dolphin and Asiatic cheetahs.

Pakistan has a semi-industrialized economy.

Changing climate is causing damage to Pakistan. The Table C.30 shows past natural disaster in Pakistan. Major climate induced impacts are biodiversity loss, rise in the sea level, increased drought, shifts in the weather patterns, increased flooding, changes in freshwater supply and an increase in extreme weather events. Changing climate could also lead to alterations in forests, crop yields, human health and animals. The Indus irrigation scheme in Pakistan depends on runoff originating from snowmelt and glacial melt from Himalayas. Depending on the change in climate and subsequent runoff in Himalayas, the country will either face flood or

Table C.30 Summarized table of natural disasters in Pakistan from 1900 to 2010

		No. of events	Killed	Total affected	Damage (000 US$)
Drought	Drought	1	143	2,200,000	247,000
	Ave. per event		143	2,200,000	247,000
Earthquake (seismic activity)	Earthquake (ground shaking)	23	142,978	6,570,288	5,229,755
	Ave. per event		6,216.4	285,664.7	227,380.7
Epidemic	Unspecified	5	131	371	–
	Ave. per event		26.2	74.2	–
	Bacterial infectious diseases	3	142	11,103	–
	Ave. per event		47.3	3,701	–
	Parasitic infectious diseases	1	–	5,000	–
	Ave. per event		–	5,000	–
	Viral infectious diseases	1	10	12	–
	Ave. per event		10	12	–
Extreme temperature	Cold wave	3	18	–	–
	Ave. per event		6	–	–
	Heat wave	12	1,388	574	–
	Ave. per event		115.7	47.8	–
Flood	Unspecified	26	7,284	20,671,883	1,170,030
	Ave. per event		280.2	795,072.4	45,001.2
	Flash flood	11	949	1,734,229	573,118
	Ave. per event		86.3	157,657.2	52,101.6
	General flood	30	4,520	19,883,171	1,225,030
	Ave. per event		150.7	662,772.4	40,834.3
Insect infestation	Unspecified	1	–	–	–
	Ave. per event		–	–	–
Mass movement dry	Avalanche	1	50	–	–
	Ave. per event		50	–	–
Mass movement wet	Avalanche	10	423	4,322	–
	Ave. per event		42.3	432.2	–
	Landslide	9	204	16,019	–
	Ave. per event		22.7	1,779.9	–
Storm	Unspecified	7	184	2,988	–
	Ave. per event		26.3	426.9	–
	Local storm	9	180	1,385	–
	Ave. per event		20	153.9	–
	Tropical cyclone	7	11,555	2,599,940	1,635,036
	Ave. per event		1,650.7	371,420	233,576.6

Created on: Oct-12-2010 – Data version: v12.07
Source: "EM-DAT: The OFDA/CRED International Disaster Database http://www.em-dat.net – Université Catholique de Louvain – Brussels – Belgium"

shortage of water. The expected climate change can affect the health of country due to increase in infectious diseases and heat stress. Both crops and livestock would be affected by means of increased alien/invasive pests and diseases. An augment in temperature can reduce the ability of farmers to work.

Papua New Guinea

Papua New Guinea occupies the eastern half of the island of New Guinea sharing a border with the Indonesian province. It has four large islands – Manus, New Ireland, New Britain and Bougainville as well as about 600 smaller islands. The total land area is 465,000 km^2 and population is about four million. Weather and climate is determined by the oceans and land masses surrounding the country. December – April correspond to the southern summer influenced by northwest monsoons originating in Asia. Dry conditions prevail during June to October. The country is rich in natural resources with ores of gold, copper; oil and natural gas. The major crops are maize, cotton, tobacco and citrus fruits. Industries include food processing, wood products, textile and cement.

Table C.31 gives summary of natural disasters in Papua New Guinea from 1900 to 2010. The country is vulnerable to sea level rise and salt water intrusion. The rise temperature will result in coral bleach and impacts due to variation in food chain. The country will also be affected by loss of wetlands, freshwater sources due to seawater intrusion, and lands. Loss of mangrove ecosystems will impact fishing activity and coastal ecology. Water resources have been depleted due to anthropogenic activity will be further affected by changing climate. The expected climate change can affect the health of country due to increase in infectious diseases and heat stress. Yield of crops may increase due to increase in temperature, but increase in population of pest as well as crop diseases will affect the crops negatively.

Philippines

The Philippines in Southeast Asia, is composed of 7,107 islands, with a land area of 299,764 km^2. The Philippine coastline is 36,289 km long. Three prominent bodies of water surround the archipelago: the Pacific Ocean, the South China Sea, and the Celebes Sea. This location makes variations in geographic, climatic and vegetation conditions in the country. Except for the Cordilleras in Luzon and the mountainous regions of Mindanao, preferred sites for settlements in the country are the coastal plains. Eighty percent of the settlements are located along coastline. These sites are the fastest growing in terms of population. The mega metropolitan areas, in the country are in the coastal zone. The Philippines has a humid equatorial climate with high temperatures and heavy annual rainfall. Annual rainfall varies from 1,000 to 5,000 mm. The mean annual temperature is about 27°C. April, May and June are hottest month while December, January and February being the coldest month. The country is habitat for about 14,500 plant species, representing about 5% of the world's flora. The country has about 185 species of mammals; 558 species of birds; 252 species of reptiles; 95 species of amphibians; 54 species of millipedes; 44 species of centipedes; 341 species of spiders; 2,782 species of molluscs; 20,000 species of insects; 488 species of corals; 2,400 species of protozoa; and 6 species of sea grasses. As of 1991, 89 species of birds, 44 species of mammals, and

Table C.31 Summarized table of natural disasters in Papua New Guinea from 1900 to 2010

		No. of events	Killed	Total affected	Damage (000 US$)
Drought	Drought	2	60	540,000	–
	Ave. per event		30	270,000	–
Earthquake (seismic activity)	Earthquake (ground shaking)	12	86	36,071	10,875
	Ave. per event		7.2	3,005.9	906.3
	Tsunami	2	2,193	9,867	–
	Ave. per event		1,096.5	4,933.5	–
Epidemic	Bacterial infectious diseases	2	130	2,474	–
	Ave. per event		65	1,237	–
	Viral infectious diseases	5	318	8,527	–
	Ave. per event		63.6	1,705.4	–
Flood	Unspecified	4	56	157,000	14,400
	Ave. per event		14	39,250	3,600
	General flood	5	4	62,893	43,228
	Ave. per event		0.8	12,578.6	8,645.6
	Storm surge/coastal Flood	1	–	75,300	–
	Ave. per event		–	75,300	–
Mass movement dry	Unspecified	1	10	–	–
	Ave. per event		10	–	–
	Landslide	1	76	1,000	–
	Ave. per event		76	1,000	–
Mass movement wet	Landslide	10	434	16,103	–
	Ave. per event		43.4	1,610.3	–
Storm	Tropical cyclone	4	219	209,680	1,500
	Ave. per event		54.8	52,420	375
Volcano	Volcanic eruption	14	3,515	232,430	110,000
	Ave. per event		251.1	16,602.1	7,857.1
Wildfire	Scrub/grassland fire	1	–	8,000	–
	Ave. per event		–	8,000	–

Created on: Nov-16-2010 – Data version: v12.07
Source; "EM-DAT: The OFDA/CRED International Disaster Database http://www.em-dat.net – Université Catholique de Louvain – Brussels – Belgium"

8 species of reptiles are threatened. Important crops include rice corn, coconut, sugar cane, cassava, and banana. Natural resource includes ores of iron, gold, chromite, manganese and copper. Industrial activity includes food; beverages; tobacco; rubber products; textiles; clothing and footwear; pharmaceuticals; paints; plywood and veneer; paper and paper products; small appliances; and electronic goods.

As could be seen in Table C.32 the country is vulnerable to variety of disasters. National estimates show that 82.5% of the entire population are at risk to tropical cyclones, flooding and storm surge. By means of information from the 2003 official statistics, about 14.9 million homes are vulnerable to the impacts of climate change. The threat is mainly because the houses are either make-shift or made of substandard materials. The agriculture in Philippines will be affected due to the augmented occurrences of ENSO and La Nina events, bringing drought and

Table C.32 Summarized table of natural disasters in Philippines from 1900 to 2010

		No. of events	Killed	Total affected	Damage (000 US$)
Drought	Drought	8	8	6,553,207	64,453
	Ave. per event		1	819,150.9	8,056.6
Earthquake (seismic activity)	Earthquake (ground shaking)	22	9,580	2,223,269	519,575
	Ave. per event		435.5	101,057.7	23,617
	Tsunami	1	32	–	–
	Ave. per event		32	–	–
Epidemic	Unspecified	1	1	664	–
	Ave. per event		1	664	–
	Bacterial infectious diseases	3	43	327	–
	Ave. per event		14.3	109	–
	Parasitic infectious diseases	1	50	666	–
	Ave. per event		50	666	–
	Viral infectious diseases	9	594	42,436	–
	Ave. per event		66	4,715.1	–
Flood	Unspecified	33	1,440	7,680,373	351,857
	Ave. per event		43.6	232,738.6	10,662.3
	Flash flood	27	997	3,506,259	782,929
	Ave. per event		36.9	129,861.4	28,997.4
	General flood	34	419	3,432,308	92,868
	Ave. per event		12.3	100,950.2	2,731.4
	Storm surge/coastal flood	11	149	125,931	2,617
	Ave. per event		13.5	11,448.3	237.9
Insect infestation	Unspecified	2	–	200	925
	Ave. per event		–	100	462.5
Mass movement dry	Landslide	2	311	–	–
	Ave. per event		155.5	–	–
	Rockfall	1	50	–	–
	Ave. per event		50	–	–
Mass movement wet	Avalanche	1	6	1,200	–
	Ave. per event		6	1,200	–
	Landslide	24	2,044	312,596	33,281
	Ave. per event		85.2	13,024.8	1,386.7
	Subsidence	1	287	2,838	–
	Ave. per event		287	2,838	–
Storm	Unspecified	27	902	5,388,887	122,666
	Ave. per event		33.4	199,588.4	4,543.2
	Local storm	4	9	24,704	5
	Ave. per event		2.3	6,176	1.3
	Tropical cyclone	254	36,444	102,361,876	6,293,711
	Ave. per event		143.5	402,999.5	24,778.4
Volcano	Volcanic eruption	22	2,996	1,686,815	231,961
	Ave. per event		136.2	76,673.4	10,543.7
Wildfire	Forest fire	1	2	300	–
	Ave. per event		2	300	–

Created on: Oct-20-2010 – Data version: v12.07

Source: "EM-DAT: The OFDA/CRED International Disaster Database http://www.em-dat.net – Université Catholique de Louvain – Brussels – Belgium"

extreme rainfalls respectively. Sea level rise will increase risk of flooding. Coastal erosion has already been observed in many cities and places including Cebu and La Union. Changes in tides and salt water intrusion will affect the quality of water supply. Stress on water resources due to changing rain pattern and sea water intrusion can not be overruled. Mangrove belt will be lost due to sea level rise because the rise in the elevation of existing mudflats due to deposition of organic and inorganic deposition. The country will also be affected by shore line erosion and impact due to increased wave action. The country may face economic loss associated with increased flooding. The expected climate change can affect the health of country due to increase in infectious diseases and heat stress. Stress on mangroves and associates ecosystem will increase due to sea level rise. The changing climate also increases erosion of soils leading to depletion of fertility. The change in climate will result in threat to rich endemic biodiversity.

Saudi Arabia

Saudi Arabia, is the largest Arab country of the Middle East bordered by Jordan and Iraq on the north and northeast, Kuwait, Qatar, Bahrain and the United Arab Emirates on the east, Oman on the southeast, and Yemen on the south. It has an estimated population of 28 million, and its size is about 2,149,690 km^2. Native animals include the ibex, wildcats, baboons, wolves, and hyenas in the mountainous highlands. Small birds are found in the oases. The coastal area on the Red Sea, is cherished with its coral reefs, has a rich marine life. Chief agricultural products are dates, wheat, barely and fruits. Industrial activity includes petrochemicals, fertilizers, steel, gas, and plastics.

As could be seen in Table C.33 flood and storms are only two disasters country has witnessed till date from the beginning of last century. Due to proximity to red

Table C.33 Summarized table of natural disasters in Saudi Arabia from 1900 to 2010

		No. of events	Killed	Total affected	Damage (000 US$)
Epidemic	Bacterial infectious diseases	2	92	242	–
	Ave. per event		46	121	–
	Viral infectious diseases	1	76	329	–
	Ave. per event		76	329	–
Flood	Unspecified	1	32	5,000	450,000
	Ave. per event		32	5,000	450,000
	Flash flood	3	83	1,067	–
	Ave. per event		27.7	355.7	–
	General flood	5	197	23,480	900,000
	Ave. per event		39.4	4,696	180,000
Storm	Local storm	1	–	–	–
	Ave. per event		–	–	–

Created on: Oct-12-2010 – Data version: v12.07
Source: "EM-DAT: The OFDA/CRED International Disaster Database http://www.em-dat.net – Université Catholique de Louvain – Brussels – Belgium"

sea the costal are is likely to get affected by rise in sea level. The expected climate change can affect the health of country due to increase in infectious diseases and heat stress. There are no major forest and agricultural activity in the country. Native species can acclimatise themselves to harsh climates. The marine life will have severe consequences due to bleaching of coral reef and associated impact on food chain.

Singapore

Singapore is an island country off the southern tip of the Malay Peninsula. Singapore consists of 63 islands, including mainland Singapore. It has total land area of 647.5 km^2 and a population including foreigners working in Singapore, is estimated to be 3.9 million. Singapore has a tropical rainforest climate characterized by uniform temperature and pressure, high humidity, and abundant rainfall. Temperatures range from 22 to 34°C. The average daily temperature is 26.8°C and annual rainfall is 2,344 mm. On average, the relative humidity is around 90% in the morning and 60% in the afternoon. Average daily relative humidity is 84.3%. About 23% of Singapore's land area consists of forest and nature reserves. Singapore has a highly developed market-based economy, Singapore has one of the busiest ports in the world and popular travel destination. A great extent of the island is less than 15 m above sea level with considerable stretch of the 150 km coastline modified by reclamation work, embankments and redevelopment. More than 50% of its potable water is imported from Malaysia and nearly 90% of food consumed in Singapore is imported. Small agricultural sector in the country focuses mainly on produce such as eggs, fish and vegetables for local consumption, and on orchids and ornamental fish for export.

As can be seen in the Table C.34 the country is not affected by any climatic disasters. The country is likely to be affected by rise in sea level. According to Initial National Communication under the UNFCCC, Singapore is vulnerable to land loss and coastal erosion; loss of water resources; flooding; and public health impact from resurgence of diseases. Coastal erosion has already affected along the coasts at the East Coast Park. The expected climate change can affect the health of country due to increase in infectious diseases and heat stress.

Table C.34 Summarized table of natural disasters in Singapore from 1900 to 2010

		No. of events	Killed	Total affected	Damage (000 US$)
Epidemic	Viral infectious diseases	3	36	2,238	–
	Ave. per event		12	746	–

Created on: Oct-12-2010 – Data version: v12.07
Source: "EM-DAT: The OFDA/CRED International Disaster Database http://www.em-dat.net – Université Catholique de Louvain – Brussels – Belgium"

Sri Lanka

Sri Lanka is island country in the Indian Ocean, separated from the Indian subcontinent by the gulf of Mannar and the Palk Strait. The climate of Sri Lanka is tropical and warm. The mean temperature ranges from about 16 to 33°C (91.4°F) and average yearly temperature ranges from 28°C (82.4°F) to nearly 31°C (87.8°F). May, the hottest period, precedes the summer monsoon rains. The country has an area of about 65,610 km² with a population of about 21 million. Its terrain can be divided in to three zones based on elevation; the central highland, the plains, and the coastal belt. Sri Lanka's economy depends heavily on the export-oriented plantation crops like tea, rubber and coconut. Animal husbandry activities include rearing of live stocks which include cattle, buffaloes, sheep, goats, pigs and poultry. The country's fish production includes both marine and aqua culture fish production.

As can be seen by Table C.35 the country in the past was affected by drought, flood and mass movement and storm. Similar disasters will occur with changing

Table C.35 Summarized table of natural disasters in Sri Lanka from 1900 to 2010

		No. of events	Killed	Total affected	Damage (000 US$)
Drought	Drought	8	–	6,256,000	–
	Ave. per event		–	782,000	–
Earthquake (seismic activity)	Tsunami	1	35,399	1,019,306	1,316,500
	Ave. per event		35,399	1,019,306	1,316,500
Epidemic	Unspecified	1	1	5,936	–
	Ave. per event		1	5,936	–
	Bacterial infectious diseases	2	36	2,423	–
	Ave. per event		18	1,211.5	–
	Parasitic infectious diseases	1	2	200,000	–
	Ave. per event		2	200,000	–
	Viral infectious diseases	4	420	48,969	–
	Ave. per event		105	12,242.3	–
Flood	Unspecified	17	275	3,656,942	23,184
	Ave. per event		16.2	215,114.2	1,363.8
	Flash flood	8	279	2,014,720	29,050
	Ave. per event		34.9	251,840	3,631.3
	General flood	24	521	5,468,111	322,130
	Ave. per event		21.7	227,838	13,422.1
Mass movement wet	Landslide	3	119	130	–
	Ave. per event		39.7	43.3	–
Storm	Unspecified	1	–	8,000	–
	Ave. per event		–	8,000	–
	Tropical cyclone	5	1,160	2,060,000	137,300
	Ave. per event		232	412,000	27,460

Created on: Oct-12-2010 – Data version: v12.07
Source: "EM-DAT: The OFDA/CRED International Disaster Database http://www.em-dat.net – Université Catholique de Louvain – Brussels – Belgium"

climate. The proximity to sea in the periphery of country will result in impact due to rise in sea level. Sea level rise would result in inundation of low lying coastal areas, shoreline retreat, intrusion of salinity, coastal erosion, interference of long shore sediment transport rates and distribution. Increased mean sea level rise and increased wave heights would affect stability of coastal structures. As per initial national communication under UNFCCC, the country has been experiencing an erosion rate of 0.30–0.35 m/year for 45–55% of its coastline. Further acceleration of erosion due to sea level rise will result in increased loss of land, thereby affecting communities and economic activities. Sri Lanka is vulnerable to land loss and coastal erosion; loss of water resources; flooding; drop in fish catch, coral bleaching, imbalance of aquatic and terrestrial ecosystem. The expected climate change can affect the health of country due to increase in infectious diseases and heat stress. Both crops and livestock would be affected by means of increase in alien/invasive pests and diseases. An augment in temperature can reduce the ability of farmers to work.

Syria

Syria is a country in Western Asia, bordering Lebanon and the Mediterranean to the West, Turkey to the north, Iraq to the east, Jordan to the south, and Israel to the southwest. Syria consists mostly of arid plateau, although some part is fairly green. The Euphrates is Syria's most important river. The climate in Syria is dry and hot, and winters are mild with occasional snowfall. The country has an area of 185,180 m^2 with an estimated population of about 22 million as on 2009. Syria is a middle-income country, with an economy based on agriculture, oil, industry, and tourism.

As could be seen in Table C.36 historically the country has been affected by drought, flood, mass movement and storm. With changing climate country will face same disasters with change in frequency and magnitude. The presence of Meditarian Sea will affect the country through sea level rise. Both crops and livestock would be affected by means of increased pestilence of alien/invasive pests and diseases. An augment in temperature can reduce the ability of farmers to work.

Tajikistan

Tajikistan is a mountainous landlocked country in Central Asia bordered by Afghanistan to the south, Uzbekistan to the west, Kyrgyzstan to the north, and China to the east. The Amu Darya and Panj rivers mark the border with Afghanistan, and fed by the glaciers in Tajikistan's mountains. About 2% of the country's area is covered by lakes. Forests in Tajikistan occupy around 3% of country's area. The flora of Tajikistan comprise of 5,000 species of higher plants, over 3,000

Table C.36 Summarized table of natural disasters in Syria from 1900 to 2010

		No. of events	Killed	Total affected	Damage (000 US$)
Drought	Drought	2	–	1,629,000	–
	Ave. per event		–	814,500	–
Epidemic	Bacterial infectious diseases	1	88	2,865	–
	Ave. per event		88	2,865	–
	Parasitic infectious diseases	1	–	1,300	–
	Ave. per event		–	1,300	–
Flood	Unspecified	2	–	245,000	44,000
	Ave. per event		–	122,500	22,000
	General flood	1	6	–	–
	Ave. per event		6	–	–
Mass movement wet	Landslide	1	80	23	–
	Ave. per event		80	23	–
Storm	Local storm	2	32	352	–
	Ave. per event		16	176	–

Created on: Oct-12-2010 – Data version: v12.07

Source: "EM-DAT: The OFDA/CRED International Disaster Database http://www.em-dat.net – Université Catholique de Louvain – Brussels – Belgium"

species of lower plants. Endangered species of animals in Tajikistan include screw-horned goat, argali, urial, snow leopard, Central Asian cobra, desert monitor, peregrine and snow-cock. Tajikistan is a mountainous and landlocked country with mountains occupying nearly 93% of the terrain, while nearly half of the territory is situated at an altitude of 3,000 m above sea level. The annual mean temperature varies from 17 to −6°C. The annual precipitation varies from 70 mm to more than 1,800 mm a year. The major sources of income in Tajikistan are aluminium production, cotton growing and remittances from migrant workers.

The major impact due to climate change would be melting of ice and possible flood. As could be seen from Table C.37 the country in the past is affected by drought, land movement, extreme temperature and storms. The country will be affected due to change in temperatures, change in wind patterns; change in rainfalls, snowfalls and hailstones; floods and mudflows; strong winds and sandstorms; pests and diseases. These disasters could occur more frequently with higher magnitude due to change in climate. Both crops and livestock would be affected by means of increased pestilence of alien/invasive pests and diseases. An augment in temperature can reduce the ability of farmers to work. According to the first national communication of the Republic of Tajikistan under the UNFCCC, glaciers comprise 6% of the total country area and regulate river flow and climate of the country. Many small glaciers retreated due to climate warming. Snow stock is observed in the most of the foothills and low mountains where as the reduction of snow stock has been observed in many high altitude zones. Climate changes will have the impacts on the quantity and quality of water resources as well as flora and fauna. Mountain pastures and alpine meadows will be benefitted from increase in temperature where as winter pastures will degrade due to temperature rise and lack of precipitation. Flood plain ecosystem will degrade due to shortage of water

Table C.37 Summarized table of natural disasters in Tajikistan from 1900 to 2010

		No. of events	Killed	Total affected	Damage (000 US$)
Drought	Drought	2	–	3,800,000	57,000
	Ave. per event		–	1,900,000	28,500
Earthquake (seismic activity)	Earthquake (ground shaking)	7	17	38,000	23,500
	Ave. per event		2.4	5,428.6	3,357.1
Epidemic	Bacterial infectious diseases	4	171	23,590	–
	Ave. per event		42.8	5,897.5	–
Extreme temperature	Extreme winter conditions	1	–	2,000,000	840,000
	Ave. per event		–	2,000,000	840,000
Flood	Unspecified	5	1,367	67,139	300,000
	Ave. per event		273.4	13,427.8	60,000
	Flash flood	4	51	192,995	58,990
	Ave. per event		12.8	48,248.8	14,747.5
	General flood	12	170	497,918	99,000
	Ave. per event		14.2	41,493.2	8,250
Insect infestation	Locust	1	–	–	–
	Ave. per event		–	–	–
Mass movement dry	Avalanche	1	12	–	–
	Ave. per event		12	–	–
Mass movement wet	Avalanche	5	104	2,681	–
	Ave. per event		20.8	536.2	–
	Landslide	8	252	102,539	214,700
	Ave. per event		31.5	12,817.4	26,837.5
Storm	Unspecified	1	–	830	234
	Ave. per event		–	830	234
	Tropical cyclone	1	–	1,500	200
	Ave. per event		–	1,500	200

Created on: Oct-12-2010 – Data version: v12.07
Source: "EM-DAT: The OFDA/CRED International Disaster Database http://www.em-dat.net – Université Catholique de Louvain – Brussels – Belgium"

resources, rise in temperature. Climate warming will bring shift in phenological parameters of forest and affect agricultural productivity. Climate change will bring in deformation of road surfaces. Long-term drought and increased suspended solids will affect hydropower stations. The expected climate change can affect the health of country due to increase in infectious diseases and heat stress.

Thailand

Thailand is located in Southeast Asia. It is bordered by Myanmar, Laos, Cambodia, Malaysia, and Andaman Sea. Thailand has an area of approximately 513,000 km^2 with approximately 66 million people. Thailand has three distinct seasons: hot, wet and cool. The mean annual temperature varies between 22 and 32°C. Average annual precipitation is 1,692 mm. Forty percent of total land is used for agriculture,

Table C.38 Summarized table of natural disasters in Thailand from 1900 to 2010

		No. of events	Killed	Total affected	Damage (000 US$)
Drought	Drought	6	–	23,500,000	424,300
	Ave. per event		–	3,916,666.7	70,716.7
Earthquake (seismic activity)	Earthquake (ground shaking)	1	–	–	–
	Ave. per event		–	–	–
	Tsunami	2	8,845	67,007	1,000,000
	Ave. per event		4,422.5	33,503.5	500,000
Epidemic	Bacterial infectious diseases	2	189	4,746	–
	Ave. per event		94.5	2,373	–
	Viral infectious diseases	3	23	19	–
	Ave. per event		7.7	6.3	–
Flood	Unspecified	18	640	10,646,171	1,958,978
	Ave. per event		35.6	591,453.9	108,832.1
	Flash flood	17	645	6,138,849	1,473,571
	Ave. per event		37.9	361,108.8	86,680.6
	General flood	22	1,364	15,505,901	719,086
	Ave. per event		62	704,813.7	32,685.7
	Storm surge/coastal flood	3	41	170,200	267
	Ave. per event		13.7	56,733.3	89
Mass movement wet	Landslide	3	47	43,110	–
	Ave. per event		15.7	14,370	–
Storm	Unspecified	9	125	340,403	145,216
	Ave. per event		13.9	37,822.6	16,135.1
	Local storm	3	1	35,204	2,000
	Ave. per event		0.3	11,734.7	666.7
	Tropical cyclone	19	1,570	2,859,896	763,823
	Ave. per event		82.6	150,520.8	40,201.2
Wildfire	Forest fire	1	–	–	–
	Ave. per event		–	–	–

Created on: Oct-11-2010 – Data version: v12.07
Source: "EM-DAT: The OFDA/CRED International Disaster Database http://www.em-dat.net – Université Catholique de Louvain – Brussels – Belgium"

nearly 25% is forests, and around 16% is degraded land. Major crops included rice, corn, cassava, sugarcane, fruit, and vegetables. Thailand's forests consist of tropical evergreens, mixed deciduous, dipterocarps, pine and mangrove. Coastal aquaculture is the most important activity and Thailand is one of the world's leading exporters of shrimp. Major industry in the country includes textiles, garments, agricultural processing, cement, jewellery, integrated circuits, electronic goods, petrochemical, and auto assembly. Major natural resources include ores of tin, tungsten, tantalum and lead; gypsum; lignite; fluorite; natural gas; and timber.

As seen in Table C.38 the major threat to country due to climate change are storm, flood, drought and earth movements. Due to presence of coastal stretch, country is likely to affected impact due to sea level rise. An augment in temperature can reduce the ability of farmers to work. The country will be affected due to change in temperatures; change in wind patterns; change in rainfalls; stress on water resources; floods and mudflows; pests and diseases. According to initial national communication under UNFCCC rise in sea level will result in flooding and intrusion of seawater. Rise in temperature will result in coral bleaching.

Turkey

Turkey is a Eurasian country bordered by Bulgaria to the northwest; Greece to the west; Georgia to the northeast; Armenia, Azerbaijan (the exclave of Nakhchivan) and Iran to the east; and Iraq and Syria to the southeast. The Mediterranean and Cyprus are to the south; the Aegean Sea to the west; and the Black Sea is to the north. The Sea of Marmara, the Bosphorus and the Dardanelles demarcate the boundary between Eastern Thrace and Anatolia. Turkey has area of about 783,562 km^2 and population of 76.8 million (2010 estimates).

Turkey's varied landscapes and the Asian part of the country, Anatolia, consists of a high central plateau with narrow coastal plains, between the Köroğlu and Pontic mountain ranges to the north and the Taurus Mountains to the south. Eastern Turkey has a more mountainous landscape. The major rivers in the country are Euphrates, Tigris and Aras.

The coastal areas Aegean Sea and the Mediterranean Sea have a temperate Mediterranean climate, with hot, dry summers and mild to cool, wet winters. The coastal areas bordering the Black Sea have a temperate Oceanic climate with warm wet summers and cool to cold wet winters. The Black Sea coast receives the greatest amount of precipitation and is the only region of Turkey that receives high precipitation throughout the year. The eastern part of that coast averages 2,500 mm annually.

The coastal areas of Turkey bordering the Sea of Marmara have a transitional climate between a temperate Mediterranean climate and a temperate Oceanic climate. Snow does occur on the coastal areas of the Sea of Marmara and the Black Sea almost every winter, on the other hand snow is rare in the coastal areas of the Aegean Sea and very rare in the coastal areas of the Mediterranean Sea.

Winters on the plateau are especially severe. Temperatures of −30 to −40°C can occur in eastern Anatolia, and snow may lie on the ground at least 120 days of the year. In the west, winter temperatures average below 1°C. Summers are hot and dry, with temperatures generally above 30°C in the day. Annual precipitation averages to about 400 mm, with actual amounts determined by elevation. The driest regions are the Konya plain and the Malatya plain, where annual rainfall frequently is less than 300 mm. Turkey's economy is dependent on industry and less on agriculture. Major agricultural crops include tobacco, cotton, sugar beets, hazelnuts, wheat, barley and olives. Principal industries include automotive, electronic goods, food processing, textiles, basic metals, chemicals, and petrochemicals. Major natural resources in the country include coal; ores of chromium, mercury, copper, boron and gold; and oil.

As seen in Table C.39 the major threat to country due to climate change are extreme temperature, storms, and flood and earth movements. The country may witness more floods. The country due to its coastal stretch is likely to get affected by sea level rise. The country will be affected due to change in temperatures; change in wind patterns; change in rainfalls; stress on water resources; floods and mudflows; pests and diseases.

Table C.39 Summarized table of natural disasters in Turkey from 1900 to 2010

		No. of events	Killed	Total affected	Damage (000 US$)
Earthquake (seismic activity)	Earthquake (ground shaking)	72	88,589	6,879,696	22,941,400
	Ave. per event		1,230.4	95,551.3	318,630.6
Epidemic	Bacterial infectious diseases	1	11	150	–
	Ave. per event		11	150	–
	Parasitic infectious diseases	2	–	100,000	–
	Ave. per event		–	50,000	–
	Viral infectious diseases	5	602	104,705	–
	Ave. per event		120.4	20,941	–
Extreme temperature	Cold wave	3	69	–	–
	Ave. per event		23	–	–
	Extreme winter conditions	2	17	8,150	–
	Ave. per event		8.5	4,075	–
	Heat wave	2	14	300	1,000
	Ave. per event		7	150	500
Flood	Unspecified	11	897	372,617	65,000
	Ave. per event		81.5	33,874.3	5,909.1
	Flash flood	10	243	1,341,382	1,892,000
	Ave. per event		24.3	134,138.2	189,200
	General flood	16	181	64,518	238,500
	Ave. per event		11.3	4,032.4	14,906.3
Mass movement dry	Avalanche	1	261	1,069	–
	Ave. per event		261	1,069	–
Mass movement wet	Avalanche	2	146	6	–
	Ave. per event		73	3	–
	Landslide	8	273	13,275	26,000
	Ave. per event		34.1	1,659.4	3,250
Storm	Unspecified	4	49	3	–
	Ave. per event		12.3	0.8	–
	Local storm	5	51	13,636	2,200
	Ave. per event		10.2	2,727.2	440
Wildfire	Forest fire	5	15	1,150	–
	Ave. per event		3	230	–

Created on: Oct-11-2010 – Data version: v12.07

Source: "EM-DAT: The OFDA/CRED International Disaster Database http://www.em-dat.net – Université Catholique de Louvain – Brussels – Belgium"

Turkmenistan

The Turkmenistan is bordered by Afghanistan, Iran, Uzbekistan, Kazakhstan and Caspian Sea. Most of the country is covered by the Karakum (Black Sand) desert. The country has an area of 488,100 km² with a population of about five million. The Great Balkhan Range in the west of the country (Balkan Province) and the Köytendag Range on the south-eastern border with Uzbekistan (Lebap Province) are the significant hilly area. The rivers in country include the Amu Darya, the Murghab, and the Tejen. The climate is mostly arid subtropical desert, with little rainfall. Winters are mild and dry, with most precipitation falling between January and May. The desert occupies about 80% of Turkmenistan. Mountains and hills are mainly situated in the southern part of the country. Half of the country's irrigated land is planted with cotton. Other major crops include corn, wheat, barley, fruits and vegetables. Natural resources include ozocerite, oil, coal, sulphur, salt magnesium. Major industrial activity includes food, textile, chemical, cement, ferroconcrete, footwear and knitwear. Animal husbandry is also an important activity in the country.

As seen in Table C.40, the region was not affected by natural calamities in the past except flood. The impact of change in climate would affect cotton crop and economy which depends on it. Both crops and livestock would be affected by means of increased alien/invasive pests and diseases. An augment in temperature can reduce the ability of farmers to work. Water resources will be put into stress due to drop in precipitation, decline in glacier and anthropogenic activities. Critical ecological problems in Aral Sea will influence the economy of the Central Asian countries. According to initial national communication under UNFCCC loss of Pamir-Altai glaciers will have negative impact on the Amudarya water resources which contributes to about 90% of water resources of Turkmenistan. The droughts combined with high temperatures will affect agriculture and forestry. Sharp temperature fluctuations, will affect the human health.

Table C.40 Summarized table of natural disasters in Turkmenistan from 1900 to 2010

		No. of events	Killed	Total affected	Damage (000 US$)
Earthquake (seismic activity)	Earthquake (ground shaking)	1	11	–	–
	Ave. per event		11	–	–
Flood	General flood	1	–	420	99,870
	Ave. per event		–	420	99,870

Created on: Oct-11-2010 – Data version: v12.07

Source: "EM-DAT: The OFDA/CRED International Disaster Database http://www.em-dat.net – Université Catholique de Louvain – Brussels – Belgium"

United Arab Emirates

The United Arab Emirates (UAE) is a federation consisting of seven states: Abu Dhabi, Dubai, Sharjah, Ajman, Umm al-Quwain, Ras al-Khaimah and Fujairah. It is located in the southeast of the Arabian Peninsula in Southwest Asia on the Persian Gulf, bordering Oman and Saudi Arabia. It shares sea borders with Iraq, Kuwait, Bahrain, Qatar and Iran. The total area of the UAE is approximately 83,600 km^2 with population of about 4.5 million. South and west of Abu Dhabi, vast, rolling sand dunes merge into the Rub al-Khali of Saudi Arabia. Main crop in the area are date palms, acacia and eucalyptus trees. In the desert the flora is very sparse and consists of grasses and thorn bushes. The indigenous fauna had come close to extinction because of intensive hunting; some of the fauna surviving are Arabian oryx and leopards. Coastal fish/aquatic animals consist mainly of mackerel, perch, tuna, sharks and whales. Major resource in the country is oil and natural gas.

The climate of the UAE generally is hot and dry. The hottest months are July and August, when average maximum temperatures reach above 48°C on the coastal plain. In the Al Hajar Mountains, temperatures are considerably lower as a result of increased altitude. Average minimum temperatures in January and February are between 10 and 14°C. The average annual rainfall in the coastal area is less than 120 mm but in some mountainous areas annual rainfall often reaches 350 mm. The region is prone to occasional, violent dust storms, which can severely reduce visibility.

No disasters have been recorded in EM-DAT: The OFDA/CRED International Disaster Database making UAE safest country in the Asia. Country is not significant from biodiversity point of view and does not depend on Agriculture for economy and employment. The rise in coastal area may flood into the country. As the country depends hugely on imported of food grain, the shortage of food in other countries may disrupt the food availability in the country.

Uzbekistan

Uzbekistan has an area of 447,400 km^2 and has a population of 23.7 million in 1997. The country is located in the central part of the Eurasian continent between at the northern boundary of the subtropical and temperate climate zones. It, shares borders with Kazakhstan to the west and to the north; Kyrgyzstan and Tajikistan to the east; and Afghanistan and Turkmenistan to the south. It is dry, landlocked country. Irrigated agricultural land comprises of less than 10% of its territory. 78.8% of this country falls on the plains and the remaining part falls on the mountains and foothills. The Climate of Uzbekistan is arid continental. The average July temperature on the plains vary from 26 to 30°C, and the maximum temperature vary from 45 to 47°C. The average January temperature varies from 0 to −8°C in the North; the minimum temperature during some years will be as low as −38°C.

Precipitation mainly occurs during the winter-spring period with annual precipitation varying between 80 and 200 mm on the plains, 300 and 400 mm in the foothills area and 600 and 800 mm on the eastern and south-eastern slopes of the mountain ridges. The Aral Sea used to be the fourth-largest inland sea on Earth, has shrunk to less than 50% of its former area.

Uzbekistan is major cotton growing country. Other crops include potatoes, vegetables and fruit. Natural resources include oil, coal, copper and ozocerite. Industrial activities include agricultural machinery, cement, textiles and paper.

The country has many ecosystems which include desert ecosystems; semi-deserts and steppes; river and coastal ecosystems; wetland and delta zone ecosystems; and mountainous ecosystems. The country is habitat for 27,000 species of flora and fauna. Biological productivity in the desert is due to cattle breeding. The country has 5,000 species of wild animals, the vertebrates are represented which include birds; mammals; fish; reptiles; and amphibians. The country is home for about 53 endemic species.

The natural disasters witnessed by the country are depicted in the Table C.41. Irrigation from the flows of the Central Asian Rivers caused the drying-up of the Aral Sea level, and severe aggravation of the ecological situation in the Aral Sea Region. The climate change will cause additional adverse impacts due to rise in evaporation; decreasing ground water reserves; lessening of humid landscapes; salinity increase in the closed lakes; eutrophication in water bodies. As per the initial communication of the republic of Uzbekistan under UNFCCC, the country and Central Asia has shown warming trends in air temperature throughout the country both during the cold and the warm periods of the year. These changes in the climate will have considerable impact on the water balance and water resources of the region. Mountainous part of the region has shown degradation and reduction

Table C.41 Summarized table of natural disasters in Uzbekistan from 1900 to 2010

		No. of events	Killed	Total affected	Damage (000 US$)
Drought	Drought	1	–	600,000	50,000
	Ave. per event		–	600,000	50,000
Earthquake (seismic activity)	Earthquake (ground shaking)	1	9	50,000	–
	Ave. per event		9	50,000	–
Epidemic	Bacterial infectious diseases	1	40	148	–
	Ave. per event		40	148	–
Flood	Flash flood	1	–	1,500	–
	Ave. per event		–	1,500	–
Mass movement dry	Landslide	1	1	400	–
	Ave. per event		1	400	–
Mass movement wet	Avalanche	1	24	–	–
	Ave. per event		24	–	–

Created on: Oct-11-2010 – Data version: v12.07
Source: "EM-DAT: The OFDA/CRED International Disaster Database http://www.em-dat.net – Université Catholique de Louvain – Brussels – Belgium"

of the glaciers. There is increase in the annual volume of precipitation in some parts of the country. Temperature increase can result in decrease in productivity of crops due to rise in evaporation and decline in water supply. There could be increase in vegetation period due to increase in temperature.

Vietnam

Vietnam is the easternmost country on the Indochina Peninsula in Southeast Asia. It is bordered by China in the north, Laos in the northwest, Cambodia in the southwest, and the South China Sea, in the east. Vietnam has population of over 86 million and Area of about 331,114 km^2. The country has hills and densely forested mountains along with plain land. The south is divided into coastal lowlands and Annamite Chain peaks characterized by extensive forests. The country has delta of the Red River (also known as the Sông Hông), and Mekong River Delta. Due to huge variation in topographical relief, the climate tends to vary considerably throughout the country. During the winter, the monsoon winds usually blow from the northeast. The average annual temperature is higher in the plains than in the mountains and plateaus. Temperatures in plains vary between 21 and 28°C but temperatures may vary from 5 to 37°C in highlands. Vietnam has six world biosphere reserves and belongs to Indomalaya ecozone. It is 1 of the 25 countries that possess a high level of biodiversity, and is ranked 16th in biological diversity having 16% world's species. Major agricultural products in the country include rice, coffee, cashews, maize, pepper, sweet potato, pork, groundnut and cotton. Major industries include mining and quarrying, manufacturing, electricity, gas, water supply, cement, phosphate, and steel. Natural resources include coal; crude oil; ores of zinc, copper, silver, gold, manganese and iron.

The natural disasters witnessed by the country are depicted in the Table C.42. Storms and floods have caused wide and repeated damages to buildings and infrastructure; and significant losses to the agriculture and fisheries. Country is important from biodiversity point of view and much of the population depends on agriculture with rice as main crop. The change in climate will affect biodiversity and agriculture. The country will also be affected by rise in sea level especially in delta regions. Augmented inundation of crop land in the rainy season and rise in salinity intrusion in the dry season will both impact crop productions. The possible strategy for adaptation with respect to agriculture include changes sowing dates, switching to drought-tolerant crops, adaptation of salinity-tolerant varieties of rice, adaptation of new varieties for other crops. Adaptation to climate change in Mekong delta will require investments in coastal and flood defence to minimize impact due to saline intrusion and flooding. Both crops and livestock would be affected by means of increased pestilence of alien/invasive pests and diseases. An augment in temperature can reduce the ability of farmers to work.

Table C.42 Summarized table of natural disasters in Vietnam from 1900 to 2010

		No. of events	Killed	Total affected	Damage (000 US$)
Drought	Drought	5	–	6,110,000	649,120
	Ave. per event		–	1,222,000	129,824
Epidemic	Unspecified	1	16	83	–
	Ave. per event		16	83	–
	Bacterial infectious diseases	1	598	10,848	–
	Ave. per event		598	10,848	–
	Parasitic infectious diseases	1	200	–	–
	Ave. per event		200	–	–
	Viral infectious diseases	7	368	17,823	–
	Ave. per event		52.6	2,546.1	–
Flood	Unspecified	7	836	1,150,175	13,400
	Ave. per event		119.4	164,310.7	1,914.3
	Flash flood	9	293	213,603	59,200
	Ave. per event		32.6	23,733.7	6,577.8
	General flood	38	3,210	20,627,410	1,946,925
	Ave. per event		84.5	542,826.6	51,234.9
	Storm surge/coastal flood	6	804	4,353,316	749,000
	Ave. per event		134	725,552.7	124,833.3
Insect infestation	Unspecified	1	–	–	–
	Ave. per event		–	–	–
Mass movement wet	Avalanche	1	200	38,000	–
	Ave. per event		200	38,000	–
	Landslide	5	130	1,074	2,300
	Ave. per event		26	214.8	460
Storm	Unspecified	9	298	36,780	1,035
	Ave. per event		33.1	4,086.7	115
	Local storm	7	144	4,450	10,100
	Ave. per event		20.6	635.7	1,442.9
	Tropical cyclone	66	18,425	44,885,509	4,334,470
	Ave. per event		279.2	680,083.5	65,673.8
Wildfire	Forest fire	1	–	–	–
	Ave. per event		–	–	–

Created on: Oct-11-2010 – Data version: v12.07
Source: "EM-DAT: The OFDA/CRED International Disaster Database http://www.em-dat.net –
Université Catholique de Louvain – Brussels – Belgium"

Yemen

Yemen is located on the Arabian Peninsula in Southwest Asia and is bordered by
Saudi Arabia to the north, the Red Sea to the west, the Arabian Sea and Gulf of
Aden to the south, and Oman to the east. Yemen's land area is about 530,000 km^2
with an estimated population of more than 23 million people. The country has
coastal plains as well as high lands. The western highlands receive the highest
rainfall in Arabia, varying from 100 to 1,000 mm per anum. Temperatures are hot in
the day but fall dramatically at night. There are perennial streams in the highlands
but due to high evaporation they do not reach sea. The central highlands are over

Table C.43 Summarized table of natural disasters in Yemen from 1900 to 2010

		No. of events	Killed	Total affected	Damage (000 US$)
Earthquake (seismic activity)	Earthquake (ground shaking)	1	10	40,039	–
	Ave. per event		10	40,039	–
Epidemic	Viral infectious diseases	2	32	468	–
	Ave. per event		16	234	–
Flood	Unspecified	3	2	23,450	–
	Ave. per event		0.7	7,816.7	–
	Flash flood	6	246	27,928	400,000
	Ave. per event		41	4,654.7	66,666.7
	General flood	13	541	298,747	1,211,500
	Ave. per event		41.6	22,980.5	93,192.3
Mass movement wet	Landslide	2	76	11	–
	Ave. per event		38	5.5	–
Storm	Unspecified	2	30	–	–
	Ave. per event		15	–	–
Volcano	Volcanic eruption	1	6	15	–
	Ave. per event		6	15	–

Created on: Oct-11-2010 – Data version: v12.07
Source: "EM-DAT: The OFDA/CRED International Disaster Database http://www.em-dat.net – Université Catholique de Louvain – Brussels – Belgium"

2,000 m in elevation. This area is drier than the western highlands because of rain-shadow influences. Diurnal temperature ranges from 30°C in the day to 0°C at night. The Rub al Khali in the east is generally below 1,000 m with almost no rain. Yemen's economy depends heavily on the oil it produces and agriculture. Agriculture in Yemen is diverse, with crops such as sorghum, Cotton, wheat, barley and fruits.

The natural disasters witnessed by the country are depicted in the Table C.43. The country has been affected by earthquake, flood, mass movement, storm, and volcano. Drop in yield of crop and increase in disease burden with changing climate can not be over ruled. The rise in sea level will have impact as well.

Index